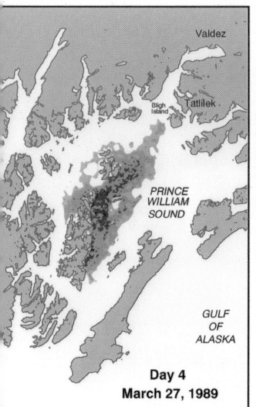

Valdez

Bligh Island

Tatlilek

PRINCE
WILLIAM
SOUND

GULF
OF
ALASKA

Day 4
March 27, 1989

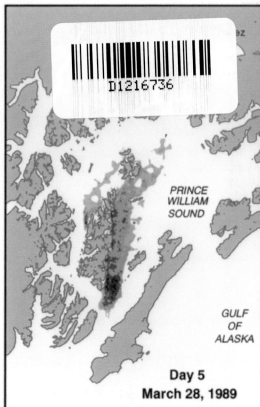

PRINCE
WILLIAM
SOUND

GULF
OF
ALASKA

Day 5
March 28, 1989

Valdez

Bligh Island

Tatlilek

PRINCE
WILLIAM
SOUND

GULF
OF
ALASKA

Day 19
April 12, 1989

NOAA HAZMAT HINDCAST
TRAJECTORY MODEL

Gallons of Oil per Square Mile

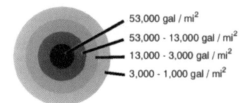

53,000 gal / mi^2

53,000 - 13,000 gal / mi^2

13,000 - 3,000 gal / mi^2

3,000 - 1,000 gal / mi^2

The NOAA HAZMAT trajectory model, called
the On-Scene Spill Model (OSSM) was developed
by NOAA Hazardous Material Response Branch for
estimating the quantitative distribution of
surface oiling over time. Tabular results of the
OSSM model have been processed by Technical
Services 3 (TS3) of the Natural Resource Damage
Assessment Studies for displaying relative
oil densities.

Alaska Department of Natural Resources
Land Records Information Section
Technical Services Number 3
September 24, 1993

Marine Mammals
and
the
Exxon Valdez

JR. Gilbert
1/10/93

The T/V *Exxon Valdez* hard aground on Bligh Reef, Prince William Sound, Alaska. The ship hit the reef at 2404 on Good Friday, 24 March 1989, resulting in the largest oil spill in U.S. history. The photograph was taken by Dr. S. Zimmerman, National Marine Fisheries Service, Juneau, Alaska, on Easter Sunday, 26 March when the weather was calm. Later that day a storm hit causing the spilled oil to be driven southwest away from the ship and onto the shores of western Prince William Sound and into the Gulf of Alaska. The *Exxon Baton Rouge* is tied alongside to transfer the remaining oil.

Marine Mammals and the *Exxon Valdez*

Edited by

Thomas R. Loughlin

National Marine Mammal Laboratory
Alaska Fisheries Science Center
National Marine Fisheries Service
Seattle, Washington

Sponsored by

The *Exxon Valdez* Oil Spill Trustee Council:
Alaska Department of Environmental Conservation
Alaska Department of Fish and Game
Alaska Department of Law
National Oceanic and Atmospheric Administration
U.S. Department of Agriculture
U.S. Department of the Interior

and the National Marine Mammal Laboratory, AFSC, NMFS

Academic Press

San Diego   New York   Boston   London   Sydney   Tokyo   Toronto

Funding support for the publication of *Marine Mammals and the Exxon Valdez* was provided, in part, by the *Exxon Valdez* Oil Spill Trustee Council. The findings and conclusions presented in the chapters included within this book are those of the individual investigators or authors and do not necessarily reflect the views of the Trustee Council.

Front cover photographs: (Top) A harbor seal and its pup that were classified as "oiled" at Applegate Rocks, Prince William Sound, 8 June 1989. Note the dark coloration of the pelage behind the neck of the adult and on the flank of the pup (see Figure 12-2, by Lowry *et al). (Bottom)* The T/V *Exxon Valdez* at anchor near Naked Island awaiting repairs after running aground on Bligh Reef on 24 March 1989. Photograph by John Hyde, Alaska Department of Fish and Game, Juneau, Alaska.

This book is printed on acid-free paper.

Academic Press, Inc.
A Division of Harcourt Brace & Company
525 B Street, Suite 1900, San Diego, California 92101-4495

United Kingdom Edition published by
Academic Press Limited
24-28 Oval Road, London NW1 7DX

Library of Congress Cataloging-in-Publication Data

Marine mammals and the Exxon Valdez / edited by Thomas R. Loughlin.
     p.   cm.
  Includes index.
  ISBN 0-12-456160-8
    1. Marine mammals--Effect of oil spills on-Alaska--Prince William
Sound. 2. Oil spills and wildlife--Alaska--Prince William Sound.
3. Exxon Valdez (ship) I. Loughlin, Thomas R.
QL713.2.M353 1994
599.5'09798'3--dc20   94-20833
                                            CIP

PRINTED IN THE UNITED STATES OF AMERICA
94 95 96 97 98 99 EB 9 8 7 6 5 4 3 2 1

Contents

Contributors

Numbers in parentheses indicate the pages on which the authors' contributions begin.

George A. Antonelis (227), National Marine Mammal Laboratory, Alaska Fisheries Science Center, National Marine Fisheries Service, Seattle, Washington 98115

David E. Bain (243), Marine World Foundation, Vallejo, California 94589

Jennifer Balke (227), Denman Island, British Columbia, Canada V0R 1T0

Brenda E. Ballachey (47, 265, 313), Alaska Fish and Wildlife Research Center, National Biological Survey, Anchorage, Alaska 99503

Earl Becker (119), Alaska Department of Fish and Game, Anchorage, Alaska 99502

James L. Bodkin (47, 81, 193), Alaska Fish and Wildlife Research Center, National Biological Survey, Anchorage, Alaska 99503

Douglas M. Burn (81), Marine Mammals Management, U.S. Fish and Wildlife Service, Anchorage, Alaska 99508

Donald G. Calkins (119), Alaska Department of Fish and Game, Anchorage, Alaska 99502

Marilyn E. Dahlheim (141, 163, 173, 243, 257), National Marine Mammal Laboratory, Alaska Fisheries Science Center, National Marine Fisheries Service, Seattle, Washington 98115

Anthony R. DeGange (47, 61), Marine Mammals Management, U.S. Fish and Wildlife Service, Anchorage, Alaska 99508

Angela M. Doroff (193), Marine Mammals Management, U.S. Fish and Wildlife Service, Anchorage, Alaska 99508

Graeme M. Ellis (141), Pacific Biological Station, Nanaimo, British Columbia, Canada V9R 5K6

Kathryn J. Frost (97, 209, 281, 331), Alaska Department of Fish and Game, Fairbanks, Alaska 99701

Joseph R. Geraci (371), Department of Pathology, Ontario Veterinary College, University of Guelph, Guelph, Ontario, Canada N1G 2W1

Carol S. Gorbics (23), Marine Mammals Management, U.S. Fish and Wildlife Service, Anchorage, Alaska 99508

Romona J. Haebler (265), Environmental Protection Agency, Narragansett, Rhode Island 02882

Richard K. Harris (265), Department of Veterinary Pathology, Armed Forces Institute of Pathology, Washington, D.C. 20306

James T. Harvey (257), Moss Landing Marine Laboratories, Moss Landing, California 95039

Thomas P. Lipscomb (265), Department of Veterinary Pathology, Armed Forces Institute of Pathology, Washington, D.C. 20306

Thomas R. Loughlin (1, 119, 359, 377), National Marine Mammal Laboratory, Alaska Fisheries Science Center, National Marine Fisheries Service, Seattle, Washington 98115

Lloyd F. Lowry (23, 97, 209, 281), Alaska Department of Fish and Game, Fairbanks, Alaska 99701

Carol-Ann Manen (331), National Ocean Service, Ocean Assessments Division, Rockville, Maryland 20852

Craig O. Matkin (141, 163), North Gulf Oceanic Society, Homer, Alaska 99603

Dennis C. McAllister (97), Alaska Department of Fish and Game, Anchorage, Alaska 99502

Elizabeth Miller (173), Wildlife and Visual Enterprises, Seattle, Washington 98116

Byron F. Morris (1), National Marine Fisheries Service, Alaska Region, Juneau, Alaska 99802

Daniel M. Mulcahy (313), BioVet Services, Anchorage, Alaska 99524

Kenneth W. Pitcher (209), Alaska Department of Fish and Game, Anchorage, Alaska 99502

Alan H. Rebar (265), School of Veterinary Medicine, Purdue University, West Lafayette, Indiana 47909

Elizabeth H. Sinclair (97, 377), National Marine Mammal Laboratory, Alaska Fisheries Science Center, National Marine Fisheries Service, Seattle, Washington 98115

Terry R. Spraker (119, 281), Department of Pathology, College of Veterinary Medicine, Colorado State University, Fort Collins, Colorado 80523

David J. St. Aubin (371), Mystic Life Aquarium, Mystic, Connecticut 06355

Mark S. Udevitz (81), Alaska Fish and Wildlife Research Center, National Biological Survey, Anchorage, Alaska 99503

Jay Ver Hoef (97), Alaska Department of Fish and Game, Fairbanks, Alaska 99701

Terry L. Wade (331), Geochemical and Environmental Research Group, Texas A & M University, College Station, Texas 77845

Terrie M. Williams (227), Office of Naval Research, Arlington, Virginia 22271

Judy Zeh (141), Statistics Department, University of Washington, Seattle, Washington 98195

Steven T. Zimmerman (23), National Marine Fisheries Service, Alaska Region, Juneau, Alaska 99802

Olga von Ziegesar (173), North Gulf Oceanic Society, Homer, Alaska 99603

Foreword

The grounding of the *Exxon Valdez* resulted in the largest oil spill in United States history. Approximately 11 million gallons of crude oil spilled from the ship and was then transported by winds and currents throughout much of western Prince William Sound and into the Gulf of Alaska and lower Cook Inlet. More than 1100 km of coastline was oiled, including parts of Chugach National Forest; Kodiak, Alaska Maritime, and Alaska Peninsula/Becharof National Wildlife Refuges; Kenai Fjords National Park; Katmai National Park and Preserve; and Aniakchak National Monument and Preserve. Detectable amounts of oil were transported to shorelines nearly 900 km from the spill site.

The contributors to this book played major roles in the design and execution of programs to assess and mitigate the impacts of the spill on marine mammals. The nature and results of those programs are described herein, including data and analyses that indicate clear and sometimes surprising impacts on some species, yet no or uncertain impacts on others. While these studies advanced our knowledge of the impacts of oil on marine mammals, much remains unknown.

For example, little was learned about the indirect effects of the spill (including containment and clean-up operations) through perturbation of the food chain. The effects of attraction to, or repulsion from, noise generated by containment and clean-up operations on species or specific age/sex classes were equivocal. The effects of contact with oil on cetaceans (particularly killer whales) could not be determined.

Except in the case of sea otters, the possible effects of the spill on long-term survival and productivity remain unknown. Even the overall biological significance of the damage caused by the oil spill in terms of the numbers of marine mammals directly or indirectly impacted was not well documented.

Even though all the related questions were not answered, the studies provided much new and important information. In fact, probably more was learned about the possible effects and ways to minimize and mitigate the effects of oil on marine mammals from the studies described in this book than from all previous studies of the effects of oil on marine mammals.

We learned that sea otters can be killed by breathing fumes from evaporating oil, as well as by heat loss when oil compromises the insulating capacity of the animal's fur. It was shown that a single large oil spill could affect an area greater than the sea otter range in California, thus confirming that the small sea otter population in California could be jeopardized by a tanker accident. It was documented that humpback whales, gray whales, killer whales, Dall's porpoise, and harbor porpoise will not necessarily avoid swimming in spilled oil.

That so much was learned was due largely to three factors: (1) the expertise and dedication of those who planned and carried out the work; (2) the studies were initiated soon after the spill occurred and were not severely limited by funding or logistic constraints; and (3) the early establishment of an independent peer review process to help identify the most critical research needs and how those needs could best be met.

That more was not learned may be due principally to the lack of an adequate contingency plan developed and implemented prior to the event. This problem was compounded by: (1) limited information on the seasonal distribution and movement patterns, abundance, and vital rates of the affected species and populations prior to the spill; (2) constraints on communications imposed by lawyers seeking to build and to defend against claims for damages caused by the spill; and (3) reluctance or difficulty getting authority to capture or sacrifice live animals to look for evidence of exposure to and impacts from the spilled oil.

Sea otters were the most abundant marine mammal in the spill area and the species for which both immediate and long-term effects were best documented. Efforts were initiated within days of the spill to capture and rehabilitate oiled otters. Two rehabilitation centers were established and a total of 343 otters were taken to and held at those centers for cleaning and treatment. Eighteen pups were born while their mothers were being held. Of the 361 otters handled, 123 died in captivity, 196 were cleaned, treated, and eventually released back to the wild, and 37 were judged unlikely to survive if released and sent to public display facilities. Radio transmitters were surgically implanted in 45 of the adult sea otters returned to the wild. At least 12, and possibly as many as 21, of these animals died within 8 months following release, suggesting that the rescue and rehabilitation program was not very effective.

An estimated 3500–5500 otters from a total population of about 30,000 in Prince William Sound and the Gulf of Alaska may have died as a direct result of the oil spill. Oiling and ingestion of oil-contaminated shellfish may have affected reproduction and had a variety of long-term sublethal effects as well. While some population and habitat assessment studies are continuing, the duration and adequacy of these studies is uncertain.

Effects on harbor seals and Steller sea lions were even more difficult to determine because, unlike sea otters, they usually sink when dead, making accurate direct mortality estimates impossible. Furthermore, harbor seal and Steller sea lion populations in Prince William Sound and adjacent areas had been declining prior to the spill, confounding proximal versus ongoing effects.

Many harbor seals were heavily oiled, and at least 302 in Prince William Sound died after they were oiled. Heavily oiled seals behaved abnormally. They did not, for example, try to escape into the water when approached. Four types of lesions characteristic of hydrocarbon toxicity were found in the brains of heavily oiled seals. These lesions occurred principally in the thalamus and may explain the abnormal behavior.

Oil did not appear to persist on sea lions, as it did on harbor seals. Likewise, there were no indications of significant oil-caused sea lion mortality, although it could have been masked by the continuing population decline.

Five cetacean species — humpback, gray, and killer whales, and Dall's and harbor porpoises — were observed swimming in and near areas affected by the spill. However, fewer individuals were sighted than expected from prespill observation data, suggesting that many cetaceans may have left or avoided spill areas. When present in the spill area, cetaceans did not appear to behave abnormally or to obviously avoid contact with oil.

The cetacean studies provided circumstantial evidence that at least 14 killer whales may have died, directly or indirectly, as a result of the spill. One previously studied resident killer whale pod, which had 36 members in September 1988, was missing seven members when photographed on 31 March 1989. An additional six members were lost between September 1989 and June 1990; one more was lost in 1991. However, this same group of animals has been observed taking sablefish from longlines in Prince William Sound and, in 1985 and 1986, six whales were lost from the pod, possibly as a result of shooting by fishermen. Since no carcasses have been recovered there is no way to ascertain whether either shooting or the oil spill were responsible.

Aerial surveys were conducted in June 1989 to search for cetaceans that might have died due to contact with oil, and washed up on beaches. Over six thousand miles of coastline were surveyed and 37 carcasses were located. Of these, 26 were gray whales, 5 were harbor porpoises, 2 were minke whales, 1 was a fin whale, and 3 were unidentified. Only 7 animals — 3 gray whales, 3 harbor porpoises, and 1 minke whale — had not decomposed beyond use for detection of hydrocarbon contamination. Two of these animals had hydrocarbon residues in their blubber, but the levels were low and provided no evidence of impact.

This book clearly provides much new and useful information concerning the actual and potential effects of oil spills on marine mammals. However, its greatest value may be that it illustrates that damage caused by events like the *Exxon Valdez* oil spill cannot be assessed and mitigated effectively and economically without prior planning and preparation. The scientists who designed and carried out the studies described in this book no doubt learned much from the experience. Those who are responsible for planning and preparing for future oil spills should take advantage of this hard-earned expertise by asking these scientists: (1) what they view as the remaining critical uncertainties concerning the possible direct and indirect effects of spilled oil and related containment and clean-up activities on marine mammals; and (2) what could be done to resolve these uncertainties and to be better prepared to assess, minimize, and mitigate damages to marine mammals and their habitat when oil spills occur in the future.

Robert J. Hofman
Marine Mammal Commission
Washington, D.C. 20009

Preface

When the 300-m supertanker *Exxon Valdez* grounded on Bligh Reef in Prince William Sound, Alaska, on 24 March 1989, approximately 258,000 barrels (11 million gallons) of crude oil were spilled onto the water. Federal, state, and non-government scientists were on the scene within 24 hours to determine the trajectory of the spilled oil and its possible impact on wildlife. The chapters included here represent the combined efforts of these scientists to assess the impacts of the spill on marine mammals within and adjacent to Prince William Sound.

Activity during the aftermath of the spill was divided into response and natural resource damage assessment (NRDA). The response activities included monitoring the spill, cleanup, mitigation measures (such as identifying priority areas for deployment of protection booms), and rehabilitation of oiled wildlife. The NRDA program assessed injury to wildlife from 1989 to 1992 and included studies on marine and terrestrial mammals, birds, fish and shellfish, and coastal habitat. Both response activities and NRDA studies on marine mammals are included in this book and represent multidisciplinary topics covering population biology, behavior, pathology, and toxicology.

The work presented here is unique because it was initiated as a result of an accident. Thus, many activities were guided by the needs of legal council, which did not always involve the rigors of scientific methodology [e.g., Baffin Island Oil Spill project, summarized in *Arctic* **40** (Suppl 1), 1987]. All of these studies were conducted by dedicated individuals who frequently labored under difficult circumstances but never compromised the quality of their work during this time of environmental crisis.

In some cases the marine mammal studies show a clear cause-and-effect relationship between exposure to the spilled oil and the death of some marine mammals. However, in most situations the results were equivocal and a cause-and-effect relationship could not be demonstrated.

I proposed to Academic Press that these studies be consolidated into a single volume to facilitate access of results to the scientific community and to state and federal agencies charged with the responsibility of conserving and protecting marine mammals. It is my hope that this volume will be useful to marine scientists when (not if) the next spill occurs and that our efforts to understand the effects of the *Exxon Valdez* oil spill on marine mammals will help those responding to similar events.

T. R. Loughlin

Acknowledgments

I am indebted to Dr. C. Crumly, editor at Academic Press, and his team to whom I extend my appreciation for support, encouragement, and advice. The staff of the Alaska Fisheries Science Center (AFSC), National Marine Fisheries Service (NMFS), Seattle, provided a significant contribution to the book; I thank G. Duker and J. Lee for many hours spent reviewing each chapter and suggesting many helpful improvements, C. Leap and W. Carlson for preparation of most of the graphics, M. Muto for checking the final copy, and K. Cunningham for preparation of the tables. Dr. H. Braham, Director, National Marine Mammal Laboratory, and the Oil Spill Trustee Council provided funds to defray costs of the volume. E. Sinclair provided insight during all phases of the process. D. Matson and R. McMahon, Alaska Department of Natural Resources, prepared the endplates. The frontispiece is a photograph by Dr. Steven Zimmerman, NMFS, Juneau, Alaska. Reviewers of individual chapters include G. Antonelis, W. Au, B. Ballachey, J. Barlow, H. Braham, R. Brown, P. Boveng, J. Calambokidis, L. Dierauf, D. DeMaster, B. Fenwick, R. Ferrero, L. Fritz, D. Garshelis, M. Gosho, J. Hall, R. Hobbs, S. Insley, J. Laake, C. Manen, S. Melin, R. Merrick, R. V. Miller, S. Mizroch, D. Potter, M. Riedman, J. Rice, D. Rugh, J. Sease, J. Short, B. Stewart, S. Swartz, G. VanBlaricom, W. Walker, J. Ward, T. Williams, T. Work, B. Wright, B. Würsig, P. Yochem, and A. York.

I thank the contributors for their willingness to participate in this project and for their patience and understanding in its production.

Finally, I am grateful to S. Calderón (AFSC) for her considerable time, diligence, and expertise in the design, production, and management of the in-house production of the book.

Chapter 1

Overview of the *Exxon Valdez* Oil Spill
1989–1992

Byron F. Morris and Thomas R. Loughlin

INTRODUCTION

Alaska is the leading producer of crude oil in the United States, supplying roughly 20% of the Nation's output. Most of this oil comes from two major oil fields on Alaska's North Slope, the Prudhoe Bay and Kuparuk fields, the two largest fields in North America. North Slope crude is delivered to tidewater facilities by the Trans-Alaska Pipeline System, a 1287-km-long pipeline from Prudhoe Bay to the Port of Valdez in Prince William Sound (PWS). Valdez is an ice-free port year-round and is at the head of the narrow Valdez Arm in the northeast corner of PWS. Currently, 2.1 million barrels of crude oil are piped to the Port of Valdez every day.

For almost 12 years, tankers had safely transported crude oil through PWS more than 8700 times until the supertanker *Exxon Valdez* left the Valdez oil terminal at 9:12 P.M. on 23 March 1989. The *Exxon Valdez* was a supertanker of single hull, high-strength steel construction, built in 1986. The ship was 300 m long and 51 m wide, about the size of an aircraft carrier (Trustee Council 1989). At 24:04 A.M. on 24 March 1989, the *Exxon Valdez*, loaded to a draft of 17 m, ran hard aground on Bligh Reef, 40 km from the Port of Valdez.

Bligh Reef is well charted, well marked, and a known navigational hazard for tankers. Icebergs calved from the Columbia Glacier, about 24 km to the northwest of Bligh Reef, are another hazard to navigation. The glacier has been actively retreating in recent years, creating many icebergs with the potential to drift into tanker lanes. It was reported that in order to avoid icebergs the *Exxon Valdez* changed course and ended up on Bligh Reef.

Heavy, black crude oil began to spew out of the ship when eight of the eleven cargo tanks and three of the five ballast tanks were punctured on the rock reef. Of the 1,480,000 barrels (approximately 62 million gallons) the tanker carried, some

1

258,000 barrels (approximately 11 million gallons; Table 1-1) spilled into the surrounding water in less than 5 hours, resulting in the largest oil spill in United States history.

GEOGRAPHIC SETTING OF THE OIL SPILL AREA

Prince William Sound lies near the top of the 1370-km arc of the Gulf of Alaska, which extends from the Aleutian Islands on the west to the islands of southeastern Alaska (Fig. 1-1). The Gulf of Alaska is remote, rugged country of great natural beauty. Much of the region was pristine before the spill.

Prince William Sound is an enclosed sea with many mountainous islands, glacial fiords, and protected embayments (Fig. 1-2). The east-central portion contains the main open body of water. There are two main entrances from the surrounding Gulf of Alaska: Hinchinbrook Entrance on the east, which most of the tankers use, and Montague Strait on the west.

The Sound is one of the largest undeveloped marine ecosystems in the United States. It has one of the continent's largest tidal estuary systems, a productive environment with abundant marine life created by the mingling of rivers, tides, and ocean currents. The water surface area is about the size of Chesapeake Bay. Prince William Sound's many islands, bays, and fiords give it a shoreline more than 3200 km long, nearly equal to that of the entire coasts of California, Oregon, and Washington combined.

Prince William Sound lies within the boundaries of the Chugach National Forest. To the southwest is the Kenai Peninsula, location of the Kenai Fiords National Park. The western portion of the Sound is within the Nellie Juan–College Fiord Wilderness Study Area. Both the National Forest and National Park are accessible by air and boat from Anchorage, Alaska's major population center, making the area popular with recreationists. State ferries that run between the larger communities of the region make it easy for people to visit PWS and adjacent areas. In the late 1980s, there was a steady increase in the number of cruise ship and other tourist visits. Bears, whales, bald eagles, puffins, seals, sea lions, and sea otters are

Table 1-1. Amount of Prudhoe Bay crude oil estimated to be on the *Exxon Valdez* before and after the spill on 24 March 1989.

Before the spill	After the spill	Estimated spilled
1,480,000 barrels	1,222,000 barrels	258,000 barrels
62,160,000 gallons	51,324,000 gallons	10,836,000 gallons
235,275,600 liters	194,261,340 liters	41,014,260 liters

⋊⋉   ⋊⋉   ⋊⋉

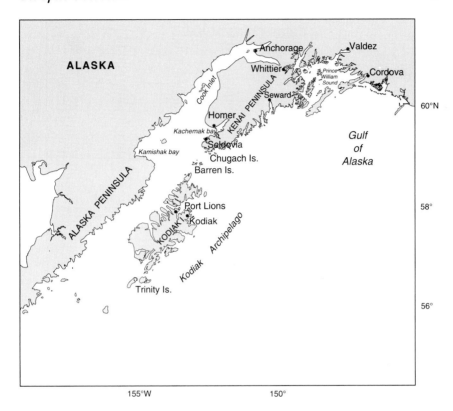

Figure 1-1. Map of the north central Gulf of Alaska showing the location of major Alaskan cities and islands impacted by the oil spill.

among the wildlife most viewed by visitors. The climate of the region is maritime, nourishing a lush and green boreal rain forest of spruce and hemlock.

The Gulf of Alaska's weather is primarily caused by low-pressure storm systems passing along the Aleutian storm track. Throughout the year, offshore winds blow predominantly from the south in the eastern Gulf, from the east in the north–central region, and from the west (but are highly variable) near the Aleutian Islands. Winds are strongest from October through April. Storms that cross the Gulf drop as much as 7.6 m of rain and snow annually in the high coastal mountains.

The weather of the Gulf affects the regional oceanography. There are both wind-induced currents and coastal currents driven by differences in water density from the large runoff of fresh water in this region.

Major currents in the northern Gulf of Alaska flow from east to west (Fig. 1-3). They enter PWS through Hinchinbrook Entrance and lesser openings to the east,

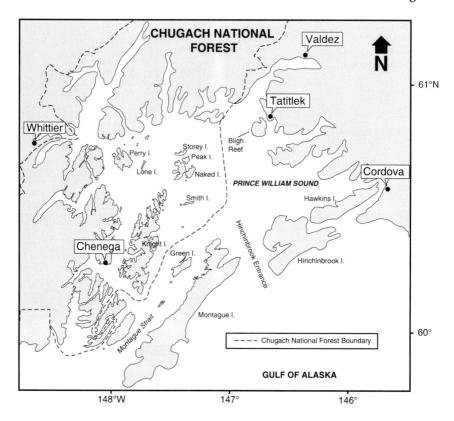

Figure 1-2. Map of Prince William Sound showing the location of Valdez and Bligh Reef, the site of the grounding of the *Exxon Valdez*.

then sweep counterclockwise through the Sound, exiting mainly through Montague Strait. Surface water temperatures in PWS range from about 0.5° to 12° C. Surface salinity fluctuates, especially in the north where glacial runoff (fresh water) pours into salt water.

The Alaska Coastal Current moves southwest along the Kenai Peninsula to the Chugach and Barren Islands. There the current splits, with some water entering Cook Inlet and the rest moving into a system of gyres along the east and south shores of Kodiak Island. Currents in lower Cook Inlet move generally northward along the east side of the Inlet, rotating counterclockwise into Kachemak Bay, and across the inlet to Kamishak Bay. The currents then flow southwest through Shelikof Strait, along the coast of the Alaska Peninsula toward the Aleutian Islands chain. This was the path followed by the oil spill.

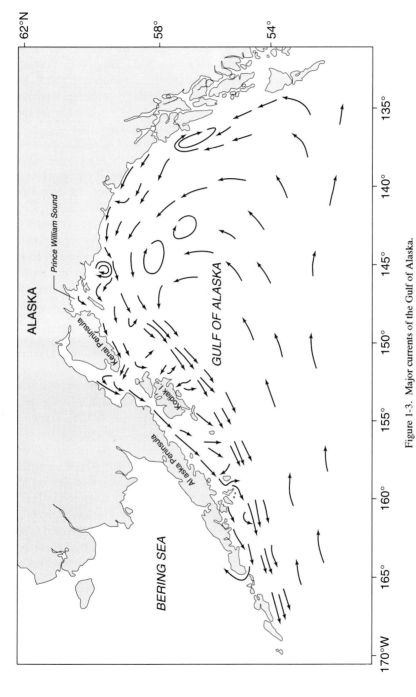

Figure 1-3. Major currents of the Gulf of Alaska.

INITIAL RESPONSE EFFORT

The serious damage to the *Exxon Valdez* sent crude oil pouring into the water. Within 5 hours, approximately 11 million gallons had been spilled. Fortunately, about 80% of the ship's cargo (51 million gallons) remained on board (Alaska Department of Environmental Conservation 1992). There were three top priorities for dealing with the spill: (1) to contain the initial spill; (2) to prevent the spilling of more crude oil; and (3) to remove the oil remaining on the ship.

Under federal law, the U.S. Coast Guard is in charge of oil spill response in the marine waters of the United States. If a spill occurs and the responsible party (the spiller) decides it can pay for the response and wants to direct the response, it has the right to do so. Exxon Corporation and their contractor, Alyeska Pipeline Service Company (the consortium of oil companies using the North Slope pipeline), assumed this responsibility and were given the leading response and subsequent clean–up role. Alyeska was responsible only for the immediate response to the spill. The U.S. Coast Guard and the State of Alaska's Department of Environmental Conservation (ADEC) were responsible for overseeing the industry's response and cleanup of the spill. For the most complete report written to date on the response effort, the reader should refer to Piper (1993).

The actions that unfolded in the first few days of the spill can be summarized as follows (Trustee Council 1989):

1. The Port of Valdez was closed 24 minutes after the grounding and remained closed to all tanker traffic for 4 days.

2. Soon after the accident, the Alyeska Pipeline Service Company was notified of the spill and was instructed to activate its oil spill response plan. (It was not until 7:30 A.M. that the company had a helicopter aloft with a U.S. Coast Guard investigator aboard. Videotape recorded during that flight showed an oil slick about 300 m wide and 6 km long.)

3. A barge transporting the containment equipment left Alyeska's docks 10 hours after the oil spill and arrived on the scene 2 hours after that. The barge was supposed to provide the containment equipment within 5 hours, but it had been stripped of its gear prior to the spill. Reloading the barge consumed valuable time and was further delayed when cranes loading the barge were redirected to load a tug bound for the stricken ship. The number and size of containment booms arriving at the spill site were insufficient to contain the slick. Few skimmers were put to work during the first 24 hours, and Alyeska did not have a tank barge into which the skimmers could discharge any recovered oil.

4. Environmentally sensitive areas needed protection. The National Oceanic and Atmospheric Administration (NOAA) had identified those areas before the spill (NOAA 1988). In addition, Alyeska's contingency plan had identified those areas and Alyeska knew that those areas required protective booming. When

⋈   ⋈   ⋈

Alyeska failed to protect these areas, a group composed of state agency personnel and local fishing groups chose four key fish hatchery sites and deployed equipment to protect them. (This action was successful, but there was not enough equipment left to contain the oil or to protect other sensitive areas.)

5. Exxon Shipping Company began cleanup on the second day after the spill. The *Exxon Baton Rouge* was redirected to the scene to off-load the remaining oil. This critical work was begun within 3 days of the grounding. (The last of the oil remaining on the *Exxon Valdez* was removed on 4 April. The following day, the vessel was refloated and towed to a sheltered harbor on Naked Island for temporary repairs.)

6. Wildlife rescue operations began as early as 25 March. The International Bird Rescue Research Center was contacted and their personnel arrived on the scene on 25 March and began to set up a bird cleaning and rehabilitation center in Valdez. A specialist from Hubbs Sea World Marine Research Institute in San Diego arrived in Valdez to set up a sea otter facility. Bird collection and sea otter collection, both supervised by the U.S. Fish and Wildlife Service, began on 29 March. Later, as the oil moved southwest, more centers were opened in Seward and Homer.

During the first 60 hours of the spill, the weather conditions were nearly perfect for oil spill response—the wind was no more than 5 knots, visibility was excellent, and seas were calm. Under these conditions, there was little wind or wave activity to affect the movement of the spilled oil. During this time, the oil spread into a large, more or less contiguous pool that moved slowly to the southwest (Galt and Payton 1990). Conditions were perfect for conducting mechanical cleanup with booms and skimmers. The 18-hour delay in this equipment reaching the spill represented a lost opportunity to contain the spill and limit its damage.

Winds were not, however, suitable for the use of oil dispersants, which require 10- to 20-knot winds. Dispersants were tested on the first day, with a trial application by helicopter at 3:10 P.M. on 24 March. However, the dispersants did not work due to the lack of mixing energy in the calm seas. Two subsequent tests on 25 and 26 March also did not work for similar reasons. A fourth test on the afternoon of 26 March did work because the winds and seas were rising. This brief window did not last, for by late afternoon the winds increased to 25 knots and eventually reached 50 knots by the end of the day. The State of Alaska has argued that Exxon did not have the proper aircraft or enough chemical dispersants on hand to disperse the massive leading edge of the slick even if the storm had not arrived (State of Alaska 1989).

On 26 March, attempts were also made to burn the oil on the water's surface. This resulted in the removal of an estimated 350 barrels. Experts later estimated that about 30% of the oil (77,000 barrels) evaporated in the first few days after the spill.

⋊⋉ ⋊⋉ ⋊⋉

In the first 3 days, the spilled oil remained in a relatively small area near the site of the grounding. By the time the response effort was fully mobilized with booms and skimmers deployed and personnel on the scene in effective numbers, the winter windstorm hit with winds of 50 knots or more. The 6-km-long slick spread into a 64-km-long slick within hours. By the time the winds and seas subsided 2 days later, the spill had spread over hundreds of square kilometers to the west and south of Bligh Reef, toward Naked Island, Smith and Lone Islands, and along both sides of Knight Island, which were in the direct path of the spill. Within three more days, the spill contacted many kilometers of shoreline along the islands of western PWS. In these few, short days following the spill, the containment option was lost and only the clean-up option was left.

The remaining response effort consisted of deploying booms and skimmers along local shorelines and embayments to prevent oil from entering strategic areas such as salmon hatcheries and boat harbors. Some of the oil that could be contained locally was removed by skimmers and absorbents where possible. U.S. Coast Guard On-Scene Commander, Vice-Admiral Robbins, later described this effort as "trying to empty the ocean with a teaspoon" in its effect on decreasing the amounts of oil remaining in the environment.

Alyeska has received blame for the lack of preparedness for the initial spill response (State of Alaska 1989). After the *Exxon Valdez* ran aground, the Alyeska Pipeline Service Company did not respond according to the contingency plan. The mobilization of response equipment was delayed. The barge that was supposed to be loaded and ready was not, and loading and deployment took four times longer than called for in the plan. Skimmers and booms did not arrive at the spill site until nearly 18 hours after the grounding—at least three times longer than specified in the plan. At 70 hours, the point at which the plan stated a spill of more than 200,000 barrels would be picked up, no more than 3000 barrels had been recovered. However, whether a properly deployed response could have prevented the massive spill from spreading remains a subject of debate among oil spill experts.

The Massive Clean-up Phase, 1989

Once oil contacts water, it is extremely difficult to contain and collect, even under ideal conditions (Alaska Oil Spill Commission 1989). The U.S. General Accounting Office suggests that no more than 10–15% of oil lost in a major spill is ever recovered. The Alaska Office of Technology Assessment estimated that only 3–4% of the oil spilled from the *Exxon Valdez* was recovered despite the massive beach clean-up and skimming efforts that were mounted. The oil moved across nearly 26,000 km^2 of water in PWS and the Gulf of Alaska and along over 1100 km of coastline. Up to 20% (581 km) of the shorelines of PWS were heavily oiled (ADEC 1992). Eventually, an additional 208 km of shoreline outside the

Sound (130 km along the Kenai Peninsula and 78 km on Kodiak Island) were also coated with oil.

Exxon initiated the first shoreline clean-up attempts on 2 April, placing work crews on Naked, Peak, and Smith Islands. The rocky shoreline with cobble beaches made cleanup difficult and labor intensive. Exxon temporarily suspended these initial clean-up efforts on 13 April while it developed a comprehensive shoreline clean-up plan for approval by the federal On-Scene Commander. U.S. Coast Guard approval was granted on 17 April after Exxon agreed to hire more workers and speed up the work. Interagency Shoreline Cleanup committees (whose membership included the Trustee Council agencies and other federal, state, and local experts) were formed to advise on areas requiring priority attention and on clean-up techniques that would minimize environmental harm. [The Trustee Council included NOAA, the Department of Interior, the U.S. Forest Service, the Alaska Department of Fish and Game, and the U.S. Environmental Protection Agency (in advisory capacity). After settlement the Alaska Department of Law and the ADEC were added.] Despite these efforts, progress remained slow, handicapped by the nature of the shoreline and by reoiling whenever the tides brought in new oil.

By mid-May, most of the oil slick had left PWS and was spread across the western Gulf of Alaska toward Kodiak Island and the Alaska Peninsula. By then, local communities had built their own defenses with commercial booms and absorbent pads, and had built their own booms from local lumber as well. Eventually, 22,000 barrels of oil were recovered by clean-up activities in 1989 (ADEC 1992). More than 147,000–158,000 barrels remained unrecovered in the environment of PWS and the Gulf of Alaska after the 1989 cleanup ended.

Over 800 km of oiled shoreline in PWS was treated in 1989 using a variety of hydraulic wash and bioremediation (fertilization) techniques (Lees et al. 1991). Additional mechanical cleanup and bioremediation occurred during the summers of 1990 and 1991.

Cold-, warm-, and hot-water (to 60°C) washing from a variety of devices were the standard methods used in most of the heavily oiled beaches within the Sound and along parts of the Kenai Peninsula coast in 1989. On the most heavily oiled beaches, workers used high-pressure (50–100 psi) hoses to get as much oil as possible washed off the beaches and into the water where it was contained by booms and picked up by skimmers and suckers. This method worked best for getting thick oil off the shoreline surface, but could not remove all the oil from rocks and gravel, and often left a tarry residue behind. The ADEC estimated that the washing removed 20–25% of the oil actually in and on the beaches (State of Alaska 1989). High-pressure hosing may actually have driven oil deeper into the beach deposits. Mousse (water-in-oil emulsion), tar mats, oil-contaminated drift seaweed, and other debris that had come ashore were manually removed.

><>   ><>   ><>

Some shorelines were washed followed by major nonmechanical means of cleanup. The chemical dispersant COREXIT 7664 and the chemical cleaning agent COREXIT 9580 were applied to some oiled beaches. These methods were designed to remobilize oil that coated the substrate and to facilitate its removal from beaches. However, their use was not considered effective because it required large amounts of the kerosene-based chemical to remove small amounts of oil, and because clean-up crews had difficulties containing the chemical–oil mixture that washed off the beaches into the water. Their use was discontinued in favor of washing with water (State of Alaska 1989).

Application of bioremediation fertilizer was designed to encourage natural bacterial breakdown of the oil. Bioremediation appears to have been partially successful, and helped to remove oil from the surface of rocks (U.S. EPA 1990).

Outside PWS, workers conducted primarily mechanical cleanup, which consisted of picking up what oil could be raked, wiped, shoveled, or removed by hand. Oiled debris, along with carcasses of seabirds, was picked up, bagged, and transferred to four sites around the state. At these sites, oil was either burned or packaged for shipment to an approved hazardous waste dump in Arlington, Oregon.

It has been estimated that 65,000 barrels of oil/water emulsions were recovered in 1989, which represented between 18,000 and 22,000 barrels of the original 258,000 barrels spilled.

During the 1989 clean-up effort, around 23,000 metric tons of oily sediments and solid wastes were removed from beaches in the spill area by clean-up crews. These oily wastes were also transported by barge and truck to the industrial landfill in Oregon. No estimates were given for the volume of oil contained in this material.

By the end of 1989, 322 km of heavily oiled shoreline in PWS still had substantial amounts of oil buried deep in the sand and gravel. This oil continued to leach to the surface and form sheens.

Later Cleanup, 1990–1992

The oil clean-up effort was put on hold over the winter of 1989/1990. Oil was left on the beaches, and oily debris continued to wash up at new sites. Additional clean-up planning continued over the winter for the 1990 season.

Shoreline treatment for 1990 included the manual pickup and removal of oiled debris and trash, and continued application of bioremediation fertilizer. Residents of local communities in the spill area did their own clean-up work during the spring and fall of 1990 in PWS and on Kodiak Island (ADEC 1992). These efforts removed approximately 214 metric tons of oily sediment (199 metric tons in PWS and 15 metric tons of oiled sediments and tar balls near Kodiak villages). In 1990, a total of about 4500 metric tons of oiled waste was collected, and in 1991 an additional 544 metric tons of oily waste were recovered. After a brief final field

⋈ ⋈ ⋈

assessment and minor oil removal in 1992, the U.S. Coast Guard officially declared that the clean-up efforts were over.

The winter storms of 1989/1990 appear to have removed about 90% of the oil that remained in surface beach sediments (<25 cm), but removed only about 40% of the more deeply buried subsurface oil (Wolfe et al. 1993). By 1992, the combination of natural processes and continued cleanup had eliminated nearly all the remaining surface oil, with isolated exceptions.

Today, most beaches are or appear clean. Oil remains only in isolated patches, under rocks, buried in beach sediment, under mussel bed mats, or as asphalt coatings on rock surfaces. The massive oil slicks are gone, as are the sheens that washed off oily beaches for at least 2 years after the spill. Although largely invisible to the eye, trace amounts of oil can still be found in shallow subtidal sediments, and in the tissues of some shallow-water species such as clams and mussels. Subsurface oil persists at some sites, buried beneath beach sediments. An estimated 36.8 km of shoreline in PWS still contained subsurface oil through 1991 (ADEC 1992).

FATE AND EFFECT OF SPILLED OIL

Effects of oil spills on the environment depend on location and conditions at the time a spill occurs. Geographic location, weather, and season affect both the kinds and numbers of marine life in the vicinity, and also the movements, weathering, and subsequent toxicity of the oil itself. Figure 1-4 is a schematic representation of the behavior of the oil spill in the Alaska marine environment.

Immediately after the *Exxon Valdez* oil spill (EVOS), wind conditions were calm, and the floating slick spread as a large, more or less contiguous pool that slowly moved to the west and southwest with the currents (Fig. 1-5). During this time, the oil showed no tendency to form mousse. The more volatile or soluble components evaporated or dissolved, but this was probably limited by surface diffusion processes because there was no stirring or rupturing of the oil surface by wave action (Galt and Payton 1990). In general, because of the cold water temperature in PWS, the oil was more stable than in warmer climates. Rates of physical weathering (evaporation, dissolution), chemical weathering (oxidation), and biodegradation were also slower.

On the afternoon of the third day, 26 March, the winter windstorm struck the Sound and the oil moved between Naked Island and Smith Island toward Montague Strait (Figs. 1-6 and 1-7). The first effect of the storm was to transform the slick into bands and streaks spread over a significantly larger area (Galt and Payton 1990). Evaporation of the lighter and more toxic fractions of the oil was enhanced. An estimated 15–20% of the total oil evaporated by the end of the storm. High-wave energy increased dissolution and, especially near shore, created droplets of

⧓ ⧓ ⧓

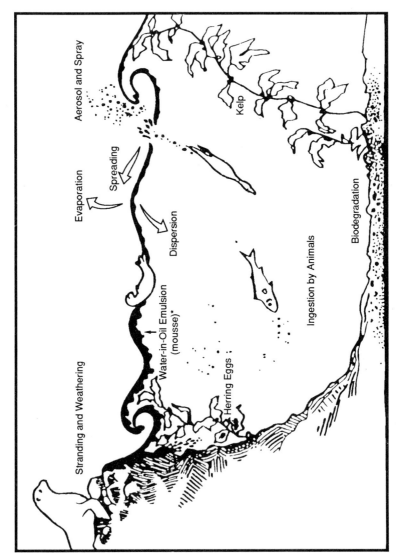

Figure 1-4. Schematic representation of the behavior of oil in the Alaska marine environment.

Aerosol and Spray

Kelp

Evaporation

Spreading

Dispersion

Stranding and Weathering

Water-in-Oil Emulsion (mousse)*

Herring Eggs

Ingestion by Animals

Biodegradation

12

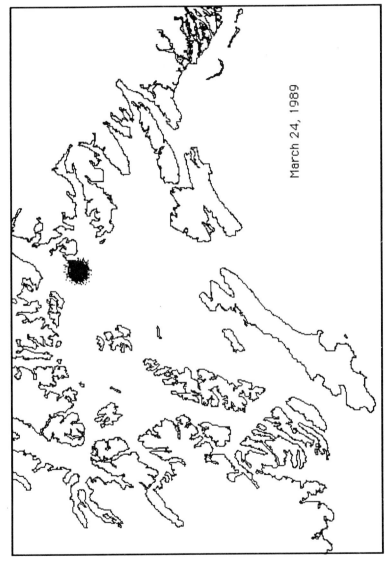

March 24, 1989

Figure. 1-5. Approximate distribution of the floating oil on 24 March 1989, day 1 of the spill. (From Galt and Payton 1990.)

13

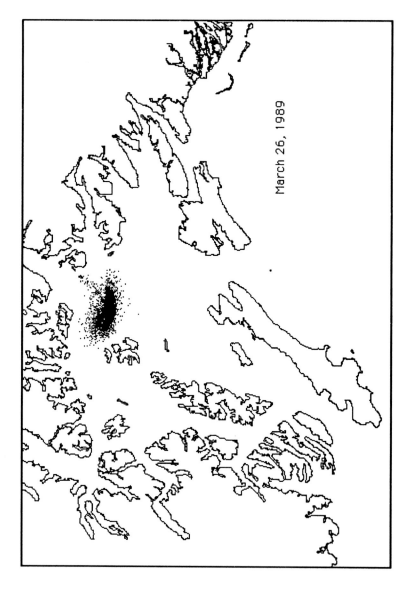

March 26, 1989

Figure 1-6. Approximate distribution of the floating oil on the afternoon of 26 March 1989, showing the oil as it began moving to the southwest from the storm on day 3. (From Galt and Payton 1990.)

March 30, 1989

Figure 1-7. Approximate distribution of the floating oil on 30 March 1989, beginning to exit the Sound on day 6. (From Galt and Payton 1990.)

15

oil which were beaten by waves into the surface waters, increasing both the speed at which oil entered the water column and also its potential toxicity to plankton and fish. Dissolution by the storm may have accounted for another 15–20% of the spilled oil.

The floating slick spread as it was carried along the surface, mixing by emulsi-fication with the seawater into mousse as it moved. After several days, this oily mousse may have been as much as 75% water, more than tripling the volume of the oil slick remaining.

The oil then moved quickly through the island passages to the south and west, particularly Knight Island Passage and Montague Strait. By 30 March, the slick's center of mass was near the northern end of Montague Strait where the main current exits the Sound into the Gulf of Alaska (Fig. 1-7). The first floating oil exited that day, and a more or less continuous passage of oil flowed into the Gulf of Alaska for the next two or more weeks (Galt and Payton 1990). By late April, the majority of the remaining floating oil had exited the Sound and was spreading westward along the coastal currents of the Gulf of Alaska.

The oil–water emulsifications lost their cohesiveness and broke up into floating patches of mousse. By the time the oil reached the Barren Islands (south end of Kenai Peninsula) in mid-April, it was in the form of widely scattered patches and lines of sheen. Affected further by wave energy, the patches eventually became tarry masses or tar balls varying in size from 2.5 cm to nearly 1 m in diameter. Eventually, as the oil passed through Shelikof Strait, the patches of mousse and tar balls may have accumulated sufficient amounts of dense material to make them neutrally buoyant. The farthest visible extent of floating oil was north of Chignik on the Alaska Peninsula. Figure 1-8 depicts the furthest extent of the oil as it spread throughout the oil spill area.

When the floating oil or mousse contacted land, it often stranded in the intertidal zone. In high-energy environments the stranded oil coated the rocks and hardened into a tough, tarry skin. Much of this oil coating weathered and washed away in the following months, but pools of oil collected in hollows among the rocks, where it remained for years. On cobble or coarse sand beaches, the oil sank deeply into the sediments. Wave erosion is less active in these environments, and degradation of the oil under biological, chemical, and physical processes was slow. Some of the buried oil later became mobilized by storms or extreme tides and reentered the water as new slicks or sheens, or as oil-laden sediments that sank to the nearshore bottom. In protected muddy bays, such as a few estuarine wetlands, oil penetration was minimal, but the stranded surface oil might have persisted for decades if it had not been removed.

Prince William Sound is generally a fiord/estuary system not subject to the high energy of the open coastal environment which characterizes much of Kenai Peninsula's outer coast. Since the oil was relatively fresh when it hit the coastal

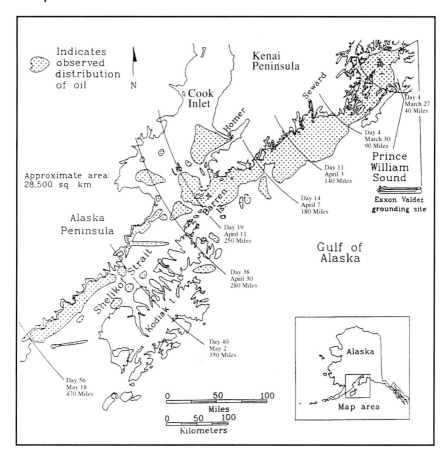

Figure 1-8. The overall cumulative extent of the aerial oil slick observations.

areas of western PWS, it penetrated readily into porous sediments. This oil entered many low-energy bays containing sensitive, higher risk environments. In contrast, the rocky headlands along the Gulf of Alaska's outer coast underwent self-cleansing by wave action fairly rapidly. As the oil moved outside PWS, it was transformed into mousse and tar balls. This weathered oil was less able to penetrate into sediments, decreasing the effects in relation to the distance and time from the origin of the spill.

Oil usually is most toxic at the early stages of the spill. The aromatic constituents (e.g., benzene, toluene, xylene, and naphthalene) are the first compounds to dissipate. As they do, the oil becomes less toxic. The greatest toxicity was in the upper few meters of the water surface. The oil was also most toxic when it beached in the intertidal zone soon after the spill. Naked and Knight Islands were two of the many nearby islands and mainlands that were oiled only a few days after the spill.

Modeling techniques, combined with observational data, were used (Galt and Payton 1990) to reconstruct the surface movement of the spilled oil. It was estimated that by the second week of the spill, about 30% of the spilled oil was lost to weathering processes, 40% beached within PWS, 25% exited into the Gulf of Alaska, and about 5% remained floating in the Sound. Of the oil leaving PWS, it was estimated that about 10% passed the Kenai Peninsula, and only about 2% reached Shelikof Strait. The amount of oil recovered by skimming operations in 1989 accounted for about 8.5% (22,000 barrels or 918,000 gallons) of the original volume, and clean-up operations recovered about 5–8% (13,000–21,000 barrels) of the oil from the beaches.

RESULTS AND PREVENTION ACTIVITIES

Shoreline clean-up activities are over. The summer of 1992 was the last year that oil removal activities were conducted by Exxon under the supervision of the U.S. Coast Guard and the State of Alaska. Of the original 788 km of oiled shoreline, approximately 124 km of shoreline still remained oiled at the end of 1991 (ADEC 1992). However, the extent of the shoreline considered heavily or moderately oiled decreased significantly (from 86 km to 1.6 km for heavy oiling and from 79 km to 12.4 km for moderate oiling).

The State of Alaska, the federal government, and the oil industry have all implemented new prevention and response strategies since the grounding of the *Exxon Valdez*. Together these new programs decrease the risks of another oil spill in PWS or elsewhere in the state, and increase the ability of state and federal agencies and the industry to respond should another spill occur. Chief among these preventative measures are SERVS (Ship Escort and Response Vessel System) emergency response and tug vessels which now escort all outbound tankers through PWS past Hinchinbrook Entrance; the U.S. Coast Guard improved radar capabilities to track tankers in PWS and has installed a permanently lighted marker on Bligh Reef; the oil industry increased Alyeska's response capability to include the ability to deploy high-capacity skimmers to any site in PWS within 6 hours, and to maintain caches of boom equipment in easily accessible locations around the Sound.

The Oil Pollution Act of 1990 (OPA 90), signed into law on 18 August 1990, overhauled previous federal oil spill liability and compensation laws. The bill established new standards for the oil industry to lower the incidence of catastrophic oil spills (Steiner and Byers 1990). In general, the law increased the liability of companies that ship and transport oil, established a federal trust fund for financing

✄ ✄ ✄

clean-up operations, and mandated new preventive measures. Among other prevention strategies, the Act mandates drug and alcohol testing procedures for
persons holding licenses to operate tanker vessels, and provides a schedule that
requires virtually every vessel that carries oil in United States waters to have a
double hull by the year 2015.

Marine Mammal Assessments

The EVOS set into motion damage assessment studies designed to assess injury
to the natural resources of PWS (Fig. 1-9). Although the most visible impact of
the EVOS was the large number of dead sea otters, other marine mammal species
were potentially injured by the spill, including Steller sea lions, harbor seals, killer
whales, and humpback whales. In 1989, seven studies were approved to gather
information on injury to marine mammals. Aerial surveys for stranded cetaceans
were also conducted.

In 1990, most of these studies were continued to refine the information documenting injury resulting from the spill. Three of these studies were continued into
1991, including studies on killer whales, harbor seals, and sea otters. In addition,
the study of sea lions conducted in 1989 and 1990 was completed in 1991 with final
data analysis and report preparation.

In many cases, the original 1989 studies were expanded and modified in
response to knowledge gained in subsequent years and comments from reviewers
and the public. The killer whale study was continued to provide information on
changes in killer whale use of the spill zone, to assess long-term impacts, and to
corroborate information on injury to killer whales gathered during the 1989 and
1990 studies. Harbor seal studies were continued to provide information on the
toxicological effects of the spill. Sea otter studies continued to look at population
effects and possible physiological and toxicological impacts that could result in
long-term, sublethal injuries.

OIL SPILL SETTLEMENT

On 8 October 1991, a settlement agreement was approved in U.S. District Court
that required Exxon Corporation to pay $1 billion in criminal restitution and civil
damages to both the state and federal governments. This settlement provided an
extraordinary opportunity to address the restoration of injuries resulting from the
spill.

The settlement specified that Exxon pay the governments $900 million over the
next 10 years in the settlement of civil claims under the following terms: $90
million on 9 December 1991; $150 million (minus the cost of cleanup in 1991 and
1992) to be paid on 1 December 1992; $100 million in September 1993; and $70

⋖⋗ ⋖⋗ ⋖⋗

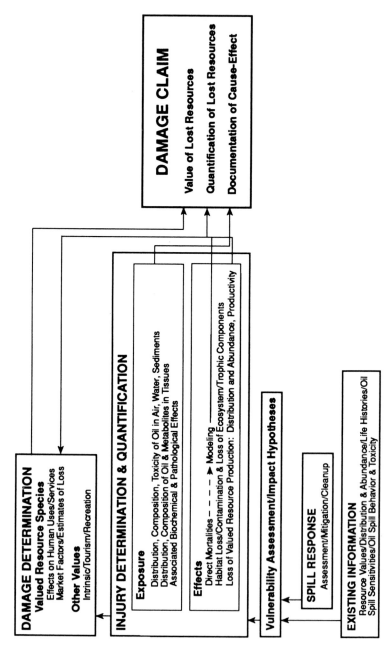

DAMAGE DETERMINATION
Valued Resource Species
Effects on Human Uses/Services
Market Factors/Estimates of Loss

Other Values
Intrinsic/Tourism/Recreation

INJURY DETERMINATION & QUANTIFICATION

Exposure
Distribution, Composition, Toxicity of Oil in Air, Water, Sediments
Distribution, Composition of Oil & Metabolites in Tissues
Associated Biochemical & Pathological Effects

Effects
Direct Mortalities – – – – ▶ Modeling
Habitat Loss/Contamination & Loss of Ecosystem/Trophic Components
Loss of Valued Resource Production: Distribution and Abundance, Productivity

Vulnerability Assessment/Impact Hypotheses

SPILL RESPONSE
Assessment/Mitigation/Cleanup

EXISTING INFORMATION
Resource Values/Distribution & Abundance/Life Histories/Oil
Spill Sensitivities/Oil Spill Behavior & Toxicity

DAMAGE CLAIM

Value of Lost Resources

Quantification of Lost Resources

Documentation of Cause-Effect

Figure 1-9. Schematic representation of the Trustee Council's approach to the design and implementation of the natural resource damage assessment studies.

million to be paid each September through the year 2001. An additional $100 million in criminal restitution penalties was also split equally between the federal and state governments to be used for restoration purposes.

RESTORATION AND OTHER POSTSETTLEMENT ACTIVITIES

When the quantification of injury resulted in the settlement of the damage claim, the emphasis shifted to developing restoration plans for the affected natural resources. Restoration includes actions undertaken to restore an injured resource to its baseline condition, as measured in terms of the injured resources' physical, chemical, or biological properties or the services that they previously provided. The Trustees recognized from the beginning that restoration of the ecological health of areas affected by the oil spill is the fundamental purpose of conducting the Natural Resource Damage Assessment (NRDA) studies. Beginning in late 1989, considerable effort has gone into the specific restoration planning activities.

The Trustee Council appointed a Restoration Team to complete the damage assessment studies that remained, develop a Restoration Plan, and make recommendations for restoration projects to be conducted. The Trustees and the EPA began preliminary restoration planning through the Restoration Planning Work Group from late 1989 until December 1991. This group carried out several scoping activities, including a series of public meetings and consultations with technical experts. The restoration group also developed the draft criteria for evaluating restoration options, and began analyzing the many restoration ideas suggested by the public, resource managers, and scientists.

A two-volume document consisting of a Draft Restoration Framework (EVOS Trustees 1992a) and the 1992 Work Study Plan (EVOS Trustees 1992b) was released in April 1992 for public review and comment. The Draft Restoration Framework outlined the process that would be followed to develop a restoration plan for the oil spill area. A 1993 Work Plan was developed and released to the public in October 1992. The Draft Restoration Plan was released in late 1993.

ACKNOWLEDGMENTS

We are grateful to G. Duker, M. Gosho, and J. Lee for valuable comments on an early draft of this chapter.

>◇ >◇ >◇

REFERENCES

Alaska Department of Environmental Conservation (ADEC). 1992. *Exxon Valdez* Oil Spill Update. Fact Sheet, 24 March 1992. 8 p

Alaska Oil Spill Commission. 1989. Spill: The Wreck of the *Exxon Valdez*. Implications for Safe Transportation of Oil. Final Report, 1989. Anchorage: Alaska Oil Spill Commission, 1989. 187 p.

Exxon Valdez Oil Spill Trustees. 1992a. *Exxon Valdez* Oil Spill Restoration Framework. Anchorage, Alaska. 52 p.

Exxon Valdez Oil Spill Trustees. 1992b. 1992 Draft Work Plan. Anchorage, Alaska. 302 p.

Galt, J. A., and D. L. Payton. 1990. Movement of oil spilled from the T/V *Exxon Valdez*. Pages 4–17, *in* K. Bayha and J. Kormendy (eds.), Sea otter symposium: Proceedings of a symposium to evaluate the response effort on behalf of sea otters after the T/V *Exxon Valdez* oil spill into Prince William Sound. Anchorage, Alaska, 17-19 April 1990. U.S. Fish and Wildlife Service Biological Report 90(12).

Lees, D. C., J. P. Houghton, H. Teas, Jr., H. Cumberland,, S. Landino, W. Driskell, and T. A. Ebert. 1991. Evaluation of the condition of intertidal and shallow subtidal biota in Prince William Sound following the *Exxon Valdez* oil spill and subsequent shoreline treatment; based on the NOAA *Exxon Valdez* Shoreline Monitoring Program, Summer 1990. Unpublished Final Report, NOAA, WASC Contract No. 50ABNC-0-00121 and 50ABNC-0-00122. January 1991.

NOAA. 1988. Prince William Sound environmentally sensitive areas by season. RPI, Inc. for Hazard Materials Branch, NOAA, Seattle, Washington. 4 maps.

Piper, E. 1993. The *Exxon Valdez* oil spill, final report, state of Alaska response. Alaska Department of Environmental Conservation, Juneau, Alaska. 184 p.

State of Alaska. 1989. *Exxon Valdez* Oil Spill Information Packet. State of Alaska, Office of the Governor. 1 September 1989. 36 p.

Steiner, R., and K. Byers. 1990. Lessons of the *Exxon Valdez*. University of Alaska Sea Grant College Program, Sea Grant Education Publication No. 8. 32 p.

Trustee Council. 1989. State/Federal Natural Resource Damage Assessment Plan for the *Exxon Valdez* Oil Spill. August 1989. Public Review Draft. 258 p.

United States Environmental Protection Agency. 1990. Alaskan Oil Spill Bioremediation Project 1990 Update. Office of Research and Development, Washington D.C. 20 p.

Wolfe, D. A., M. J. Hameedi, J. A. Galt, G. Watabayashi, J. Short, C. O'Clair, S. Rice, J. Michel, J. R. Payne, J. Braddock, S. Hanna, and D. Sale. 1993. Fate of the oil spilled from the T/V *Exxon Valdez* in Prince William Sound, Alaska. Pages 7–9, *in Exxon Valdez* oil spill symposium, 2-5 February, Anchorage, Alaska. Available Oil Spill Public Information Center, 645 G Street, Anchorage, Alaska

Chapter 2

Response Activities

Steven T. Zimmerman, Carol S. Gorbics, and Lloyd F. Lowry

INTRODUCTION

At approximately 12:04 A.M. on 24 March 1989, the oil tanker *Exxon Valdez* ran aground on Bligh Reef in Prince William Sound (PWS), Alaska, resulting in the largest oil spill to date in United States waters (Fig. 2-1). Approximately 11 million gallons of the approximately 62 million gallons of oil on board were spilled from the tanker during the days following the spill (Townsend and Heneman 1989; Morris and Loughlin, Chapter 1). Federal and state agencies and the oil industry immediately began sending response personnel to Valdez. Initial objectives were to determine the extent of the spill and to identify resources, including marine mammals, that were at risk. When the oil spread to waters outside PWS, these agencies expanded their efforts to include the Gulf of Alaska, lower Cook Inlet, the Kodiak Archipelago, and the Alaska Peninsula. In this chapter we summarize response activities by the Trustee agencies resulting from the spill.

The Regulatory Framework for Response

The Federal Clean Water Act (CWA—Section 311) and the Comprehensive Environmental Response, Compensation and Liability Act (CERCLA—Section 105) establish a framework under which federal agencies are responsible for public or trust resources that can be affected by an oil spill. These acts require establishment of a National Oil and Hazardous Substances Pollution Contingency Plan to provide the organizational structure and procedures for responding to discharges of oil. In addition, Alaska state regulations require an Oil and Hazardous Substances Pollution Contingency Plan which is parallel to and compatible with contingency plans developed under federal regulation.

At the time of the *Exxon Valdez* oil spill (EVOS), the Environmental Protection Agency (EPA) had prepared a National Contingency Plan. Title 40 of The Code

23

Figure. 2-1. Aerial view of the *Exxon Valdez*, still leaking oil at midday on 26 March 1989. The smaller vessel is the *Exxon Baton Rouge*, which was off-loading the remaining oil from the *Exxon Valdez*.

of Federal Regulations[1] details the organizational structure under which this plan is to be carried out. Response operations at the scene of a discharge will be directed by a Federal On-Scene Coordinator (FOSC) with advice from National and Regional Response Teams (NRT and RRT) and from Scientific Support Coordinators (SSCs).

Soon after the ship grounded, the Exxon Corporation notified the U.S. Coast Guard (USCG) that it intended to accept responsibility for the spill and its cleanup. In accordance with the CERCLA, the federal government determined that Exxon was financially sound and competent to respond. The CERCLA also authorizes the federal government to assume certain responsibilities in support of Exxon's

---

[1] 40 CFR 300 Subpart B.

∽ ∽ ∽

response. Two weeks after the spill, the federal government assumed responsibility for directing the spill response. Thereafter, the FOSC was responsible for determining what actions were necessary to respond to and clean up the spill; Exxon was responsible for carrying out these actions.

The FOSC relied heavily on input from the Alaska RRT and the SSCs. The Alaska RRT included representatives from several federal and state agencies including the U.S. Department of Commerce (National Oceanic and Atmospheric Administration—NOAA), the U.S. Department of the Interior (DOI), and the Alaska Department of Fish and Game (ADF&G).[2] During the early days of the spill the Alaska RRT convened several times by teleconference and provided advice to the FOSC on such topics as the use of dispersants, areas to be boomed or otherwise protected, and coordination of the response effort. The Alaska RRT did not mobilize operations on-scene.

The SSCs, on the other hand, were headquartered throughout the spill area and provided on-scene advice to the FOSC. As required by the Code of Federal Regulations, SSCs are provided by NOAA during spills which may affect coastal and marine areas. The SSCs use a variety of scientific tools, including oil spill trajectory estimates and assessments of the sensitivity of natural resources, to help the FOSC make timely operational decisions regarding how best to respond.

Other laws and regulations dictate which agencies are responsible for certain aspects of spill response. In the case of marine mammals, the Federal Marine Mammal Protection Act (MMPA) designates the DOI responsible for sea otters, walrus, polar bears, dugongs, and manatees, and the U.S. Department of Commerce responsible for seals, sea lions, whales, dolphins, and porpoises. These responsibilities are carried out by the U.S. Fish and Wildlife Service (USFWS) and the National Marine Fisheries Service (NMFS), respectively. By Alaska statute, the ADF&G has responsibility for protecting and managing all wildlife resources of the state. Under provisions of the CERCLA, state and federal agencies serve as cotrustees and advise the FOSC, through the SSC and RRT, on protection of marine mammals, on the rescue and rehabilitation of marine mammals, and on documenting the damage to marine mammals caused by the spill (Parker 1991).

Start-up Problems

Valdez is a small town with a population of approximately 4000 people. As such, it was unprepared to handle the huge influx of people and demands that converged on it following the EVOS. Of the principal agencies responding to the

---

[2] Responsibilities of and membership on the RRT are described at 40 CFR 300.105 (b) (2) and 40 CFR 300.115

⚬⚬ ⚬⚬ ⚬⚬

spill, only the Alaska Department of Environmental Conservation (ADEC) maintained a regular office in Valdez. State and federal staff with expertise in marine mammals had to mobilize from throughout Alaska and the nation.

Although at least eight contingency plans were in place for responding to oil spills in PWS, no agency or organization was prepared to respond to a spill the magnitude of the EVOS. Many of the agencies and people who responded had not been involved in the formulation of these plans or in the drills that had been carried out in preparation for such an event. Thus, there was some initial confusion regarding how the response would be funded, who would respond, and under what authority. Wildlife agencies had few oil spill experts, pathologists, veterinarians, or animal rescue and rehabilitation experts available. Accordingly, the agency staff initially sent to Valdez were required to make a myriad of decisions regarding how to proceed in unfamiliar situations.

Another problem was that virtually all federal and state agency personnel still had their regular jobs to do, and devoting large blocks of time to oil-spill tasks was often difficult or impossible. As a result, there was a substantial turnover of people. In some cases personnel from as far away as the New England area were detailed to Valdez to cover for Alaska staff that had other commitments. This turnover of personnel occasionally caused confusion. Also, the relationships between the various agencies and Exxon were often awkward, with roles and responsibilities sometimes confused or controversial. In spite of this, there was generally a strong feeling of teamwork and close cooperation between wildlife agency personnel.

Logistics in Valdez were particularly limited. Because the area affected by the spill was far removed from any roads, and few suitable vessels were immediately available, initial surveys were mostly done from aircraft. During the first days of the spill, it was difficult to charter aircraft, especially helicopters. Initially, scientists attempting to survey the distribution of marine mammals had to compete for the use of the limited number of Alaska Air National Guard, state-leased, or USCG helicopters that had been mobilized. The emergency nature of the event caused continual and sudden changes in the availability of aircraft. By the second day of the spill, daily coordination meetings were initiated to summarize logistic needs and to develop plans for obtaining and sharing aircraft. Through this process, USCG helicopters, Civil Air Patrol aircraft, and fixed-wing aircraft leased by the state, along with two NOAA helicopters, were made available and within a few days the logistic crisis was somewhat alleviated.

Finding work space was also a problem. The USFWS marine mammal staff were provided limited space in the local USCG office. The NMFS personnel rented facilities or used space leased by NOAA in support of its SSC-related responsibilities. The ADF&G marine mammal scientists used ADEC facilities in Valdez for offices, laboratories, and storage of equipment and specimens.

✗✗✗ ✗✗✗ ✗✗✗

Federal agencies also had to deal with the lack of any previously identified funding for response activities. Federal contracting procedures require the identification of existing funds for most expenditures. As a result, procurement of logistics and supplies moved slowly until Exxon promised to make response funds available. Unlike the federal agencies, Alaska agencies were not hampered by funding problems. State of Alaska account codes were issued immediately, allowing for reasonable expenditures related to spill response activities.

When the USCG, local communities, and resource agencies realized that on-scene response management would be needed outside PWS, the USCG established four "zones"—Valdez, Seward, Homer, and Kodiak—with USCG on-scene coordinators in each location reporting to the FOSC in Valdez. Federal and state resource agencies assigned staff to each of these locations to participate in oil-spill response activities, oversee wildlife rescue and rehabilitation efforts, and participate in planning shoreline cleanup.

INITIAL SURVEYS

Federal Regulations[3] require trustee agencies to conduct preliminary surveys of the area affected by a discharge of oil to determine if trust resources, including marine mammals, are or may be affected. Accordingly, during the first 3 weeks following the spill, the NMFS and ADF&G cooperated in a series of surveys to determine the distribution and presence of pinnipeds and cetaceans in PWS. Initial baseline surveys began 2 days after the spill and were conducted from helicopters and fixed-wing aircraft. Vessel surveys were begun 4 days after the spill.

The USFWS began fixed-wing surveys within hours of the spill to determine the number and distribution of sea otters in areas that could be affected. Within 3 days, the USFWS began vessel surveys of sea otters in the path of the spill.

Steller Sea Lion Surveys

Steller sea lions (*Eumetopias jubatus*) occur throughout the Gulf of Alaska. Two major rookeries (Outer and Sugarloaf Islands) and several haulout sites (The Needle, Point Elrington, and Point Eleanor) were in the area directly affected by the spill. Several thousand sea lions breed or haul out in the area that the oil eventually impacted (Loughlin et al. 1992).

Wildlife surveys flown on 25 March by USFWS personnel found Steller sea lions in the area of the spill. Three or four animals were seen on a buoy in the center

---

[3] 40 CFR 300.615 (c) (1).

✖◇ ✖◇ ✖◇

of the oil slick (C. Monnett, Enhydra Research, Homer, Alaska, personal communication) and approximately 90 others were seen in or near the slick, some in very thick oil (L. Rotterman, Enhydra Research, Homer, Alaska, personal communication).

Aerial surveys of pinnipeds by NMFS personnel began in the central and western areas of PWS on 26 March using a helicopter provided by the Alaska Air National Guard. Pinniped surveys were also flown in the eastern part of PWS by NMFS personnel on 26 March in a fixed-wing aircraft that was chartered by the sea otter survey program. These surveys, and the ones that followed over the next few days, covered large parts of PWS in an attempt to document the distribution of animals. When the oil began moving to the southwest, more and more attention was focused on rookeries and haulout areas that lay in the path of the spreading oil.

Short-term response efforts focused on identifying areas most heavily used by sea lions and determining if their distribution changed when those areas were oiled. Observations of sea lions in oil, research to determine the degree of oiling on animals and the substrate, and scat collection were also undertaken. During this time, NMFS and ADF&G scientists advised the SSC in Valdez on protective measures that should be taken to lessen impacts of the spill on sea lions.

Photographic surveys were conducted five times at major rookery and haulout sites: The Needle, Point Elrington, Wooded Islands, and Seal Rocks. Two of these sites (Point Elrington and The Needle) were eventually oiled. However, the number of animals there did not appear to decrease relative to unoiled sites (Wooded Islands, Seal Rocks; Table 2-1). Only 5–10% of the animals present at oiled sites appeared to be oiled and none appeared to be debilitated (National Marine Mammal Laboratory, unpublished data). Based on these observations, the preliminary conclusion was that Steller sea lions were not being acutely affected by the oil spill.

Harbor Seal Surveys

Harbor seals (*Phoca vitulina richardsi*) occur year-round in PWS and the Gulf of Alaska. Unlike Steller sea lions, which have strong affinities to a limited number of haulout and rookery sites, harbor seals use dozens of haulout sites in PWS (Pitcher and Calkins 1979).

The NMFS and ADF&G worked cooperatively to assess whether harbor seals in PWS were affected by the spill. Initial aerial surveys began on 26 March using NOAA helicopters. Emphasis was on determining the distribution and numbers of seals on haulout sites in areas that could be affected. Boats and a helicopter were used for access to haulout areas that were oiled or were in the path of the oil, and for observation of the impact of oil on seals and their habitats. Of particular interest was the impact on seal pups born in areas heavily contaminated with oil. Detailed observations from shore, boats, and aircraft continued through August.

⤙⤚ ⤙⤚ ⤙⤚

Table 2-1. Counts of Steller sea lions at haulout sites in the Prince William Sound area following the *Exxon Valdez* oil spill (29 March 1989 to 19 April 1989).[a]

Location	Date	Count
The Needle	29 March	1064
The Needle	1 April	867
The Needle	4 April	661
The Needle	13 April	876
The Needle	19 April	1281
Point Elrington	29 March	425
Point Elrington	1 April	963
Point Elrington	4 April	551
Point Elrington	13 April	437
Point Elrington	19 April	728
Wooded Is.	1 April	962
Wooded Is.	5 April	907
Wooded Is.	13 April	844
Wooded Is.	19 April	814
Seal Rocks	1 April	1477
Seal Rocks	5 April	1710
Seal Rocks	13 April	1380
Seal Rocks	19 April	1709
Cape St. Elias	19 April	1521

[a] Data provided courtesy of T. R. Loughlin, National Marine Mammal Laboratory, 7600 Sand Point Way, N.E., Seattle, WA 98115.

ADF&G personnel had considerable expertise on the distribution of harbor seals in PWS and adjacent areas. They were able to provide detailed information to the FOSC and RRT on the historical distribution of animals and the likely significance of impacts on their habitats. Information was also provided on the location of pupping areas and the normal timing of seal births. The ADF&G and NMFS additionally provided recommendations to the FOSC on priority areas for cleanup, and evaluations of the effectiveness of clean-up treatments on seal haulout sites.

Within a few days of the spill, hundreds of harbor seals were oiled in PWS. Animals were seen swimming through oil and several important haulout sites (Bay of Isles, Herring Bay, Seal Rocks, Green Island, Smith Island, Little Smith Island, Applegate Rocks) were heavily oiled. At some sites (Smith Island, Green Island, Applegate Rocks), virtually all animals were oiled, many of them heavily (Townsend and Heneman 1989). Scientists observed that some oiled animals could be

><> ><> ><>

approached easily, and some did not react when approached (Lowry et al., Chapter 12).

Sea Otter Surveys

Sea otters (*Enhydra lutris*) occur year-round in the shallow coastal waters of PWS and the Gulf of Alaska. Within 24 hours of the spill, the USFWS began fixed-wing aircraft flights along the coastline of PWS to record the distribution and number of sea otters. Four planes were used for 5 days during the first week of the spill. The pilots and biologists surveyed the entire shoreline of PWS. Their reports provided key information which was used in making response decisions.

On 27 March, the USFWS began vessel-based shoreline surveys of birds and sea otters in the immediate path of the spill (e.g., Naked, Peak, and Storey Islands), counting all birds and mammals within 200 m of the shoreline. By 6 April, USFWS biologists were conducting vessel-based shoreline surveys for live sea otters throughout PWS (DeGange et al. 1990). The initial surveys were completed on 18 April. In the first week of April, USFWS biologists also flew along PWS shorelines that were oiled after the March survey to assess changes in wildlife numbers. These newly oiled shorelines were surveyed again by air on 20 April. Another complete survey of PWS was flown between 1 and 10 May.

Spilled oil moved out of PWS about 1 April. Three USFWS biologists arrived in Seward on 1 April, ahead of the oil, and flew sea otter surveys and conducted shoreline boat surveys along the Kenai coastline in cooperation with National Park Service personnel. On 16 April, before the oil hit, the entire Kenai coastline was systematically surveyed. The area was flown again a month later after the oil came ashore. On 11 April, the USFWS initiated aerial and boat surveys of the Kodiak Island shoreline to identify sea otter concentrations. Survey activities there continued through the spring of 1989.

The oil spill surveys and previous surveys (Irons et al. 1988) provided information on the distribution and concentration of sea otters in the path of the spill. The post–oil spill surveys counted a minimum of 4500 sea otters inhabiting the nearshore waters of PWS. An additional 1330 sea otters were counted from helicopters along the coast of the Kenai Peninsula; 3500 were counted in coastal surveys from helicopters at Kodiak Island; and about 6500 sea otters were counted along the Alaska Peninsula between Kamishak Bay and Unimak Pass.

Cetacean Surveys

Cetacean species that occur commonly in PWS and adjacent waters include Dall's porpoise (*Phocoenoides dalli*), harbor porpoise (*Phocoena phocoena*), killer whale (*Orcinus orca*), minke whale (*Balaenoptera acutorostrata*), humpback whale (*Megaptera novaeangliae*) and gray whale (*Eschrichtius robustus*). The USFWS aerial surveys on 25 March indicated that both killer and humpback whales were in the area of the spill. Between 26 March and 9 April, 22.4 hours of aerial searching and 950 nautical miles of dedicated shipboard surveys were completed by NMFS personnel. Cetaceans observed during these vessel and aerial surveys are listed in Table 2-2.

Many cetaceans observed during the vessel surveys were swimming through areas of oil, making no obvious attempt to avoid contaminated areas. One Dall's porpoise calf was observed with light oil adhering to its back and upper portion of its dorsal fin (Harvey and Dahlheim, Chapter 15). This animal's behavior was unusual—it was lying close to the surface for extended periods of time. All other cetaceans seen, both in and out of the oiled area, did not appear to be externally oiled and no unusual behavior was observed. During vessel surveys within oil-covered areas of PWS, approximately 80% of all cetaceans observed were in areas of light sheen, 10% were in moderate sheen, and 10% were in areas of heavy sheen (National Marine Mammal Laboratory, unpublished data).

RESPONSE ACTIVITIES SUPPORTING THE FOSC

Scientists from NOAA were immediately on the scene to assess the spill's trajectory and potential impact.[4] Beginning on 24 March, tide, current, and wind data were entered into a mathematical oil trajectory model to predict the path the oil would take. These data were supplemented with daily overflights by NOAA, Exxon, and ADEC to track the oil and empirically determine what areas were being impacted. Maps portraying this information were prepared and disseminated daily. By the end of the summer, approximately 260 overflight maps of the location and general concentration of floating oil had been prepared and distributed (Fig. 2-2).

Areas to Be Protected

The data generated by NOAA, Exxon, and ADEC were used to determine what kind of actions should be undertaken to prevent or minimize threats to the environment. With regard to marine mammals, the USFWS, NMFS, and ADF&G personnel had to determine: (1) which areas should be protected from the oil by booming or other means; and (2) which areas should be protected from human impacts related to the cleanup.

---

[4] NOAA's responsibilities in responding to an oil spill are outlined in 40 CFR 300.175 (b) (97).

⬯⬯⬯ ⬯⬯⬯ ⬯⬯⬯

Table 2-2. Cetacean sightings in Prince William Sound (PWS) and adjacent waters following the *Exxon Valdez* oil spill (25 March 1989 to 9 April 1989).[a]

	Number		Adjacent water		
Species	Vessel	Aerial	PWS	(Outside PWS)	Total
Dall's Porpoise	80	0	80	0	80
Harbor Porpoise	2	0	2	0	2
Killer Whale	20	0	20	0	20
Humpback Whale	3[b]	12	5	10	15
Gray Whale	2[b]	54	2	54	56

[a] Data provided courtesy of T. R. Loughlin, National Marine Mammal Laboratory, 7600 Sand Point Way N.E., Seattle, WA 98115.
[b] Nondedicated vessel.

Areas to Be Protected from Oil

Sufficient quantities of oil booms were not available in Valdez to deal with a major spill (Townsend and Heneman 1989). Exxon eventually assembled 190,500 m of boom, and, during the early days of the spill, state, federal, and local agencies, with participation by the USCG and Exxon, formed the Interagency Shoreline Cleanup Committee to recommend where it should be deployed (Fig. 2-3). The committee also recommended other protective measures based on wildlife information gathered during the response effort or provided by the resource agencies.

Areas to Be Protected from Human Disturbance

Decisions also had to be made regarding which marine mammal use areas to clean. With the exception of Point Eleanor, Steller sea lion haulout areas had not been heavily oiled. Thus, the risk of disturbance seemed greater than the threat posed by oil and cleanup of sea lion haulout areas was given a low priority (Townsend and Heneman 1989).

The situation at harbor seal haulout sites and pupping areas was different; several important haulout sites were heavily oiled (Fig. 2-4). Cleanup of these sites prior to the onset of pupping was given a very high priority. In a 21 April memorandum to the FOSC from the Interagency Shoreline Cleanup Committee, it was requested that Exxon clean areas used by harbor seals on Applegate Rocks, Seal Island, Green Island, Smith Island, and Little Smith Island prior to 10 May in order to avoid disturbing seals that would begin pupping soon thereafter. Pools of oil and heavily oiled algae were removed from these islands (Figs. 2-5 and 2-6). However, some of these sites were reoiled as incoming tides floated the interstitial oil that had lain

≫ ≫ ≫

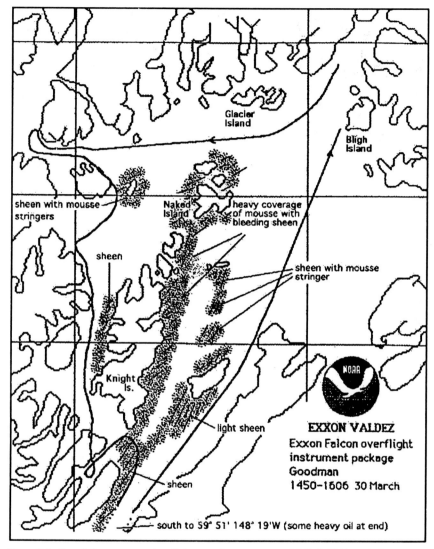

Figure 2-2. One of the many overview flight maps that were prepared daily in Valdez to indicate where the oil was moving.

below the surface, which recoated the rocks as the tides ebbed. Heavy films of oil were found on some sites just prior to the pupping period.

On 10 May, in conjunction with the NMFS and ADF&G, the FOSC issued an advisory stating that cleanup of harbor seal haulout areas would stop by 15 May. After that date, unauthorized persons could not approach within 805 m of any seal

Figure 2-3. Cabin Bay on Naked Island was one of the many areas in Prince William Sound that was recommended for, and received, some measure of protection from oil.

or sea lion haulout site or rookery area. Aircraft were cautioned not to fly lower than 305 m over such areas. At the request of the USCG, the NMFS sent Federal Agents to help enforce this advisory. The advisory was to remain in effect through 31 July 1989; however, the NMFS and ADF&G found that pupping and weaning was essentially over by early July. On 6 July, the NMFS issued an advisory indicating that activities directly related to the cleanup of seal haulout sites and sea lion rookeries would be authorized during the period of 6 July to 10 August. The 10 August closure date marked the estimated beginning of the annual harbor seal molting period.

Advisories were also issued to reduce vessel traffic in areas frequented by humpback and killer whales. On 27 June, the NMFS recommended to Exxon that vessels in the area from the Pleiades Islands southeast through lower Knight Island Passage and into Montague Strait should exercise caution to avoid unnecessary harassment of whales. Approaches closer than 137 m were discouraged and vessels were advised to operate at speeds that would allow sufficient time for personnel to sight whales and make necessary course adjustments to avoid the animals.

As the summer progressed, the NMFS and ADF&G became involved in determining whether haulout sites had been adequately cleaned, and, if not, what

⋈ ⋈ ⋈

Figure 2-4. View of a heavily oiled beach near a harbor seal haulout site on Smith Island.

types of clean-up methods should be used. For example, on 6 July, the NMFS sent a memorandum to the FOSC indicating that several haulout areas identified by Exxon and the USCG as being "environmentally clean" still contained considerable amounts of oil. Potential adverse effects of chemicals such as "Inipol" or "Corexit" on marine mammals had to be considered by the NMFS, USFWS, and ADF&G even though the effects of such chemicals on marine mammals were largely unknown. Such topics were often handled at meetings of the Interagency Shoreline Cleanup Committee. Additional USFWS and ADF&G personnel were present on clean-up boats to monitor activities near marine mammal haulout sites.

WILDLIFE RESCUE AND REHABILITATION

Federal regulations[5] require the DOI, the U.S. Department of Commerce, and state representatives to the RRT to arrange for the collection, cleaning, rehabilita-

---

[5] 40 CFR 300.330.

Figure 2-5. Pressurized hot water was used as part of the clean-up process on Seal Rock in April 1989.

Figure 2-6. Oil washed from the beach was collected behind booms and then was collected by skimmer vessels, April 1989.

≫ ≫ ≫

tion, and recovery of wildlife affected by oil spills. Accordingly, both the USFWS and NMFS established rehabilitation centers and prepared permits necessary to allow for rehabilitation of live marine mammals and collection and disposal of dead animals.

Sea Otters

Because the spill was not "federalized," the USFWS had no authority or funding to initiate rescue and rehabilitation of sea otters. The Alaska RRT's Wildlife Protection Guidelines and Alyeska Oil Spill Contingency Plan named Sea World as technical advisor for sea otter rehabilitation. Accordingly, Exxon contracted with Sea World to implement the sea otter rehabilitation program (Fig. 2-7). Under provisions of the Marine Mammal Protection Act, the USFWS authorized Dr. R. Davis and his designees from Sea World, on behalf of Exxon, to implement the rescue effort and perform necropsies on animals that died during rehabilitation (Bayha 1990).

Sea World personnel initiated the sea otter rescue effort on 29 March using two boats supervised by a Cordova veterinarian with experience in handling sea otters. On 30 March, Sea World requested USFWS assistance in the rescue effort. On 1 April, the USFWS contracted two boats and a helicopter, staffed them with USFWS and California Department of Fish and Game biologists, and helped coordinate the rescue effort.

During the first days following the spill, approximately 30 fishing vessels were chartered by Exxon to pick up sea otters. The USFWS received numerous reports of mishandled sea otters, sea otters chased to exhaustion, and human injury as a result of picking up live sea otters. In addition, by 5 April the Valdez otter center was inundated. For these reasons, under the authority of the MMPA, the USFWS limited the otter rescue effort to personnel on USFWS-approved capture boats after 6 April. From that date on, the otter rescue effort was jointly coordinated by the USFWS in Anchorage and Sea World in Valdez. As the rescue effort expanded into the Gulf of Alaska, it required coordination of 14 capture crews with air support out of Valdez, Seward, Homer, and Kodiak (Parker 1991).

The first sea otter rehabilitation center opened in Valdez on 30 March. In total, the Valdez center received 156 sea otters of which 63 survived, 85 died in the center, and 8 died after being transported to sea aquaria (Davis and Williams 1990).

On 5 April, before oil reached sea otters on the Kenai Peninsula, the USFWS requested that Exxon build a fully equipped otter rehabilitation center in Seward. The center was completed on 10 May by which time the rescue effort had shifted from PWS to the Kenai coast (Parker 1991). In total, the Seward center received 187 otters, of which 151 survived and 36 died; 21 of the 151 survivors completed their rehabilitation at the Valdez center (Davis and Williams 1990).

><> ><> ><>

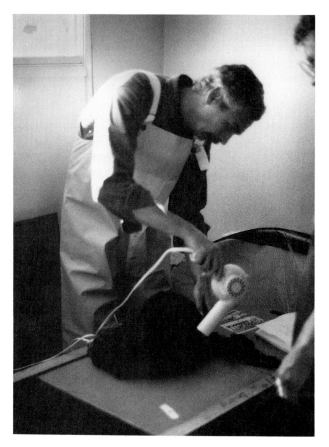

Figure 2-7. A sea otter being dried after washing had removed external oil, March 1989.

A local resident in Homer started a temporary care facility with a capacity for 10 sea otters. Another temporary care center was built at the NMFS office on Kodiak Island in late April. This center held otters for 1 to 2 days until they were flown to the Seward center (Davis and Williams 1990). Later, an Exxon-financed holding center capable of housing 100 sea otters was built at Jakolof Bay outside of Homer. By late May, rehabilitated sea otters from Valdez and Seward were shipped to Jakolof pending their release in late August. The Jakolof center eventually received 125 otters (Davis and Williams 1990).

The sea otter rescue effort peaked in May 1989, tapered off in June, and ended in late July. By the end of the summer, approximately 360 live sea otters had been received and 18 pups were born in the rehabilitation centers. A total of 123 otters died at the rehabilitation centers.

Figure 2-8. A formerly oiled harbor seal pup being fed at a rehabilitation center in Homer, Alaska, May 1989.

The USFWS developed a Sea Otter Program Strategy to return rehabilitated sea otters to their natural environment. On 23 June, the ADF&G recommended that no animals be released based on the possibility that released animals would introduce diseases or parasites to wild sea otter populations, or other species such as land otters. On 12 July the USFWS responded, indicating that it felt that the procedures used in the capture, handling, and maintenance of rehabilitated sea otters should have minimized their exposure to diseases. The USFWS indicated that careful disease and parasite screening would be conducted to determine which animals were suitable for release.

To test the plan, one unoiled otter and six rehabilitated sea otters were fitted with flipper-tag radio transmitters, released into the eastern, oil-free portion of PWS, and followed for 30 days. An additional 28 sea otters were released in the clean waters of eastern PWS in late July. Twenty-one of these sea otters carried implanted radios to monitor their movements for at least 20 days. Most remained in clean water habitats, and the few that did return to oiled water subsequently returned to the vicinity of the release sites. Consistent with the release strategy, the remaining rehabilitated sea otters were released in clean water habitats near their capture locations. The last of a total of 196 sea otters were released on 30 August

✖ ✖ ✖

1989. Thirteen pups, the last of 37 sea otters not fit for release, were sent to aquaria on 11 September (Parker 1991).

In 1990, the USFWS again coordinated with Exxon to prepare for the possibility of sea otters requiring rehabilitation as a result of impacts from *Exxon Valdez* oil. The Seward rehabilitation center reopened in April 1990, staffed by a veterinary technician and a local rehabilitator. A veterinarian remained on standby in Anchorage. Rehabilitation contacts were named in all the major towns in the spill zone. A list of contacts and a protocol for handling oiled animals were passed out to all Exxon contractors, government employees, and volunteer groups working on the spill. In August 1990 the Seward center was closed permanently after 5 months of no activity (Parker 1991).

Harbor Seals

Harbor seals were the only other marine mammal species that appeared to have been affected by oil and that were small enough that rehabilitation was feasible. The initial policy of the NMFS was that any animals affected by oil should be captured and rehabilitated.[6] Accordingly, in early May the NMFS issued permits to the sea otter rehabilitation facilities in Valdez and Seward to receive and treat harbor seals and hold them until they were ready for release. As pupping began in early May, people began bringing harbor seal pups into rehabilitation centers in Valdez and Seward. Several of the pups were oiled; however, in the frenzy to collect potentially affected wildlife, unoiled seals were also collected. Therefore, on 10 May 1989, NMFS issued an advisory stating that only heavily oiled pups or those that had been permanently abandoned by their mothers should be captured. Guidelines were provided to assist people in determining which animals were candidates for capture.

As harbor seal pupping progressed, ADF&G biologists informed the NMFS that the condition of pups in oiled areas appeared similar to that of pups in unoiled areas. Based on this information, NMFS and ADF&G biologists concluded that the collection of oiled seals and pups would probably do more harm than good. Accordingly, on 10 June, the NMFS issued a second advisory stating that live seal pups were to be left undisturbed, whether or not they had oil on them. This advisory stated that even heavily oiled pups would fare better if left with their mothers and that the rehabilitation process could pose a serious risk to the wild seal population through the possible introduction of diseases.

A total of 19 harbor seal pups were brought into rehabilitation centers for treatment. In addition to the sea otter facilities in Seward and Valdez, two

---

[6] This initial policy was undertaken in conformance with 40 CFR 300.330.

⋙   ⋙   ⋙

additional facilities, a veterinary clinic in Anchorage and a private facility in Homer, were used in treating these 19 seals (Fig. 2-8). One seal had been mortally injured by an outboard motor during capture and two others died while in treatment. The remaining 16 seals were eventually released (Williams et al., Chapter 13).

In a 6 July 1989 letter to the NMFS, the ADF&G expressed concern regarding whether rehabilitated harbor seals should be returned to the wild. This concern was founded on a disease-related die-off of harbor seals in the North Atlantic. The NMFS responded that it had developed a protocol for the release of harbor seals, and, in accordance with the MMPA, intended to release the harbor seals back to the wild, provided that blood tests confirmed that the seals had not been exposed to parvo or distemper viruses. All pups were tested for these diseases, and two veterinarians inspected each pup for visible lesions or behavioral abnormalities before the animals were released.

Permits and Protocols for the Collection of Animals

The MMPA prohibits the "taking" of any species of marine mammal. The definition of "take" includes to harass, hunt, capture, or kill. There are provisions in the MMPA, however, that allow the taking of marine mammals for their protection and welfare [Section 109(h)] or for scientific research [Section 104(c)]. Accordingly, the NMFS and USFWS authorized the taking of marine mammals for rehabilitation, the collection and study of dead animals, the disposal of dead animals, and the taking of marine mammals for research. Necessary authorizations were also issued for the rehabilitation facilities for washing, rehabilitating, and holding animals. Authorizations to destroy carcasses once full necropsies had been performed were also issued to Exxon and to the rehabilitation facilities.

Because the process whereby permits are issued to scientists to conduct research on marine mammals takes several months, it was impossible to issue new research permits quickly enough to allow any meaningful studies of the effects of the spill. To get around this problem, the NMFS National Marine Mammal Laboratory, which already held appropriate research permits, designated several biologists to act as agents under its permits. Thus, marine mammal scientists from the ADF&G were authorized to collect sea lions and harbor seals for research under a NMFS permit.

COLLECTION AND DISPOSAL OF DEAD ANIMALS

On 28 March 1989, ADF&G biologists began boat surveys looking for live and dead animals in oiled areas of PWS. By 6 April, USFWS biologists were conduct-

⋉ ⋉ ⋉

ing beach surveys on foot throughout PWS to look for dead animals. Randomly selected beaches and shorelines were surveyed in the spill area to provide an estimate of the number of animals killed. Boat and ground searches for dead and dying animals in the Kodiak Island area began on 17 April. These activities continued throughout the spring of 1989.

Agencies were also expected to respond immediately to anecdotal reports of dead or distressed animals. Although such responses occasionally led to the collection of carcasses, in many cases the reports were naive or incorrectly portrayed. Reports of "several dead, floating sea lions" inevitably turned out to be normal basking animals. In one case, a report of several dead seals turned out to be a misreported observation of several healthy seals.

On 5 April the USFWS established a wildlife morgue for the processing and storage of dead animals that were brought to Valdez. The morgue, consisting of a freezer van and work bench, was staffed full time by two USFWS biologists for 2 months, after which it was staffed as needed. Similar receiving stations were later established in Homer, Seward, and Kodiak.

During the second week of April the EPA sent a veterinary pathologist to Valdez to assist in the wildlife pathology work. The pathologist conducted necropsies and collected samples at the Exxon-operated rehabilitation center and laboratory in Valdez. As a temporary USFWS employee, the pathologist served as the official government representative for oil spill response pathology work in Valdez. Working in conjunction with the pathologist, ADF&G staff, assisted by NMFS staff, took a lead role in receiving, examining, and collecting samples from harbor seal, northern fur seal, and Steller sea lion carcasses.

The USFWS also opened a laboratory at the Exxon facility in Seward to perform necropsies and other pathology work. The laboratory was staffed by pathologists from the USFWS's National Wildlife Health Laboratory, the USFWS's Patuxent Wildlife Research Center, the Armed Forces Institute of Pathology, and the EPA. A protocol was devised in 1990 by the USFWS and Exxon to handle dead wildlife collected the year after the spill. Freezers were set up in all major coastal towns in the impact zone and lists of contacts were distributed to all private and government employees and interest groups involved in the 1990 cleanup.

By the end of the summer of 1989, the USFWS had processed over 1000 dead sea otters. Seventy-four sea otter carcasses were collected during the 1990 season. Tissue samples were taken from all dead sea otters and shipped to an analytical laboratory in Texas to determine the extent and type of oil contamination (Parker 1991). Necropsies were performed on all sea otter carcasses.

Only small numbers of dead sea lions and harbor seals were found during the weeks following the oil spill. Approximately six aborted sea lion fetuses were found at The Needle (D. Calkins, ADF&G, Anchorage, Alaska, personal communication). Without baseline data there was no way to know if this was abnormally

≫ ≫ ≫

high. During the first 4 months after the spill, only 14 other dead sea lions were found. Several of these were judged to have died before the spill. During the same period approximately 16 dead harbor seals were found, mostly from the PWS area.

Early reports of dead killer whales led to rapid response from biologists, but no dead animals were ever found. On 1 April a dead gray whale was reported in the Gulf of Alaska east of Hinchinbrook Entrance. The animal was found and determined to have been dead for several days, making it unlikely that it had been affected by oil. This same animal was reported a few days later as it floated westward along the outer coast of PWS. The NMFS asked that it be marked and released by the vessel that was towing it toward shore. When found a third time, it was reported to have a dead deer tied to it.

Although there was no evidence during the first weeks of the spill that whales had been killed by oil, the NMFS eventually began receiving reports of stranded animals. By mid-June, it was apparent that dead gray whales were stranding on Kodiak, Tugidak, and Sitkinak Islands. Between 13 May and 31 July, 14 dead gray whales were found on these three islands, and by the end of October a total of 26 dead gray whales had been found in the Gulf of Alaska and southern Bering Sea (Loughlin, Chapter 20). The number of dead whales (20) found in the Gulf of Alaska during 1989 exceeded the total number of stranded gray whales reported during the entire 1975–1987 period (Zimmerman 1991). However, the lack of systematic prespill surveys, especially in the Kodiak Island area, precluded any determination as to whether this was an abnormal number of strandings.

Marine mammal carcasses were kept frozen in Anchorage, Alaska, until 1992 when many were finally incinerated. The salvage and incineration process began on 29 October 1992 and took 9 days. A crew of about 15 scientists initially sorted through the carcasses and salvaged many (primarily birds and approximately 220 sea otters) for research or education.[7] Many of these will be prepared as skeletons and used to augment museum and university reference collections.

INITIAL DECISIONS REGARDING ASSESSMENT STUDIES

In addition to responding to oil spills, trustee agencies are required by the Clean Water Act to assess the damage that has occurred to natural resources, as well as what restoration might be undertaken.[8] On 31 March 1989 the heads of all of the

---

[7] Only a few of the Steller sea lions and none of the whales were brought in for storage.

[8] The guidelines for Natural Resource Damage Assessment (NRDA) studies are described in 43 CFR Subtitle A.

✽ ✽ ✽

trustee agencies met in Cordova to develop an initial assessment plan. At that meeting, it was stated that Exxon would pay "several million dollars up front" for the initial assessment studies. During the next 2 days, preliminary study plans for assessing the effects of the oil spill on sea otters, harbor seals, Steller sea lions, killer whales, and humpback whales were developed by the response teams in Valdez.

On 28 April the Federal Trustees signed a memorandum of agreement providing a framework to begin assessing damages for injuries to marine mammals and other natural resources caused by the oil spill. Signatories were Manual Lujan, Secretary of Interior; Clayton Yeutter, Secretary of Agriculture; and William Evans, Under Secretary for Oceans and Atmosphere in the U.S. Department of Commerce. The State of Alaska declined to sign the Memorandum of Agreement because the state felt it should be the lead trustee. With the signing of the agreement by the Federal Trustees, the damage assessment activities began in earnest. Much of the rest of this book is devoted to describing the Natural Resource Damage Assessment (NRDA) marine mammal studies that were undertaken as a result of the *Exxon Valdez* oil spill.

ACKNOWLEDGMENTS

In many cases, the facts presented in this chapter were derived from on-site discussions, from personal observation, from meeting notes, or from other interactions which cannot be easily referenced or acknowledged. The authors therefore wish to thank the many people that we worked with during the response activities described in this chapter. We also wish to thank J. Parker, A. DeGange, C. Berg, and D. Kennedy for their help in accurately recalling the chaotic events of the oil spill response and in contributing to or reviewing early drafts. R. Miller and J. Sease provided comments on early drafts.

REFERENCES

Bayha, K. 1990. Role of the U.S. Fish and Wildlife Service in the sea otter rescue. Pages 26–28, *in* K. Bayha and J. Kormendy (eds.) Sea otter symposium: Proceedings of a symposium to evaluate the response effort on behalf of sea otters after the T/V *Exxon Valdez* oil spill into Prince William Sound, Anchorage, Alaska, 17–19 April 1990. U.S. Fish and Wildlife Service Biological Report 90(12).

Davis, R. W., and T. M. Williams. 1990. Valdez otter rehabilitation program. Pages 158–166, *in* K. Bayha and J. Kormendy (eds.) Sea otter symposium: Proceedings of a symposium to evaluate the response effort on behalf of sea otters after the T/V *Exxon Valdez* oil spill into Prince William Sound, Anchorage, Alaska, 17–19 April 1990. U.S. Fish and Wildlife Service Biological Report 90(12).

⤜⤛   ⤜⤛   ⤜⤛

DeGange, A. R., D. H. Monson, D. B. Irons, C. M. Robbins, and D. C. Douglas. 1990. Distribution and relative abundance of sea otters in south-central and southwestern Alaska before or at the time of the T/V *Exxon Valdez* oil spill. Pages 18–25, *in* K. Bayha and J. Kormendy (eds.) Sea otter symposium: Proceedings of a symposium to evaluate the response effort on behalf of sea otters after the T/V *Exxon Valdez* oil spill into Prince William Sound, Anchorage, Alaska, 17–19 April 1990. U.S. Fish and Wildlife Service Biological Report 90(12).

Irons, D. B., D. R. Nysewander, and J. L. Trapp. 1988. Prince William Sound sea otter distribution in relation to population growth and habitat type. Unpublished Report, U.S. Fish and Wildlife Service, Alaska Investigations, Anchorage, Alaska, 31 p. (unpublished report).

Loughlin, T. R., A. S. Perlov, and V. A. Vladimirov. 1992. Range-wide survey and estimation of total number of Steller sea lions in 1989. Marine Mammal Science 8:220–239.

Parker, J. 1991. U.S. Fish and Wildlife service response activities following the *Exxon Valdez* oil spill. Pages 243–245, *in* Proceedings of the 1991 International Oil Spill Conference (Prevention, Behavior, Control, Cleanup) San Diego, California, 4–7 March 1991.

Pitcher, K. W., and D. G. Calkins. 1979. Biology of the harbor seal, *Phoca vitulina richardsi*, in the Gulf of Alaska. U.S. Department of Commerce, NOAA, Environmental Assessment of the Alaskan Continental Shelf, Final Reports of Principal Investigators 19:231–310.

Townsend, R. and B. Heneman. 1989. The *Exxon Valdez* oil spill: A management analysis. Prepared for: Center for Marine Conservation, 1725 DeSales Street NW, Washington, D.C. 20036. 239 p. plus appendices.

Zimmerman, S. T. 1991. A history of marine mammal stranding networks in Alaska, with notes on the distribution of the most commonly stranded cetacean species, 1975–1987. Pages 43–53, *in* J. Reynolds and D. Odell (eds.), Marine mammal strandings in the United States: Proceedings of the Second Marine Mammal Workshop, 3–5 December 1987, Miami, Florida. NOAA Technical Report NMFS 98.

Chapter 3

An Overview of Sea Otter Studies

Brenda E. Ballachey, James L. Bodkin, and Anthony R. DeGange

INTRODUCTION

The *Exxon Valdez* oil spill (EVOS) on 24 March 1989 threatened extensive areas of prime sea otter (*Enhydra lutris*) habitat along the coasts of south-central Alaska. The spill occurred in northeastern Prince William Sound (PWS), and oil moved rapidly south and west through PWS into the Gulf of Alaska. Much of the coastline of western PWS was heavily oiled, and the slick eventually spread as far southwest as Kodiak Island and the Alaska Peninsula (Galt and Payton 1990; Morris and Loughlin, Chapter 1). All coastal waters affected by the spill were inhabited by sea otters.

Concern for the survival of sea otters following the oil spill was immediate and well founded. Sea otters are particularly vulnerable to oil contamination because they rely on pelage rather than blubber for insulation, and oiling drastically reduces the insulative value of the fur (Costa and Kooyman 1982; Siniff et al. 1982; Geraci and Williams 1990). Within days of the spill, recovery of oiled live otters and carcasses began. During the several months following the spill, sea otters became symbolic of the mortality associated with the spilled oil, and of the hope for rescue and recovery of injured wildlife (Batten 1990).

An extensive sea otter rescue and rehabilitation effort was mounted in the weeks and months following the spill. Handling and treatment of the captive sea otters posed an enormous and difficult challenge, given the large number of otters held at the facilities and minimal prior experience in caring for oiled sea otters. Rehabilitation of sea otters was a separate effort from the postspill studies designed to evaluate injury to the otter populations and is not addressed in this chapter only as it relates to evaluation of damage assessment studies. Detailed information on the rehabilitation effort is presented in Bayha and Kormendy (1990) and Williams and Davis (1990).

Sea otters retained a high profile in the Natural Resource Damage Assessment (NRDA) studies largely because the initial injury to the sea otter population was readily demonstrable, but also because of concerns about long-term damages. The scope of the postspill studies to assess oil-related damages to sea otters was extensive: From 1989 through 1993, more than $3,000,000 was spent, and more than 20 scientists were involved in a comprehensive research program. The studies were predominantly directed at sea otter populations in PWS.

Damages to sea otters generally can be classified as either acute, defined as spill-related deaths occurring during the spill, or chronic, defined as longer term lethal or sublethal oil-related injuries. Studies of acute damages focused on estimating the total initial loss of sea otters. Characterization of the pathologies associated with exposure to oil was a secondary goal of studies of acute effects. Chronic or longer term damages may have resulted from sublethal initial exposure or continued exposure to hydrocarbons persisting in the environment. Studies of chronic effects included evaluating abundance and distribution, survival and reproduction rates, foraging behavior, and pathological, physiological, and toxicological changes in the years following the spill.

The objective of this chapter is to review the studies conducted on sea otters in response to the EVOS and to synthesize the major findings of those studies relative to injury to the sea otter population associated with exposure to oil. We also provide recommendations for research to improve our understanding of the effects of future oil spills on sea otter populations.

STUDIES OF ACUTE INJURY TO SEA OTTERS

Acute losses to the sea otter population were reflected in the carcasses collected in the spring and summer following the spill. Four hundred ninety-three sea otter carcasses were collected from PWS, 181 from the Kenai Peninsula, and 197 from the Kodiak Island/Alaska Peninsula area, with a total recovery of 871 carcasses. In addition, 123 sea otters died at the rehabilitation centers, bringing the total number of carcasses accumulated in the 6 months following the spill to 994 (Doroff et al. 1993; DeGange and Lensink 1990). Undoubtedly, some additional number of sea otters were exposed to oil and subsequently died but were not recovered during the clean-up efforts. Studies of carcass recovery efforts have shown relatively low recovery rates (Wendell et al. 1986; Piatt et al. 1990; Ford et al. 1991).

Quantifying the total number of sea otter deaths related to the oil spill was of extreme importance for the U.S. Fish and Wildlife Service (USFWS) and U.S. Department of Justice in light of pending litigation against Exxon Corporation and the potential monetary value associated with each sea otter mortality. Consequently, several approaches were taken to enumerate the total acute loss of sea otters; these are described below.

Surveys of Sea Otter Abundance and Distribution

One approach to enumeration of the total sea otter loss was comparison of pre- and postspill estimates of otter abundance. At the time of the spill, knowledge of sea otter abundance along the Kenai and Alaska Peninsulas or in the area of Kodiak Island was limited. Helicopter surveys were implemented in April 1989, concurrent with or prior to the arrival of oil in these areas (DeGange et al. 1994), to obtain "prespill" estimates of sea otter abundance and distribution. These surveys were repeated in the fall of 1989, providing "post-spill" counts for comparison with the prespill data.

In PWS, a boat-based survey of sea otters had been conducted in 1984-1985 (Irons et al. 1988), providing a relatively recent prespill count for comparison. A boat-based survey was also conducted in PWS in the summer of 1989 to estimate postspill sea otter abundance and distribution (Burn, Chapter 4); the survey was repeated in the spring and summer of 1990, 1991, and 1993. Pre- and postspill survey methods were similar, except that the entire shoreline of PWS was surveyed prespill, whereas only a portion (approximately 25% or less) of the coastline in PWS was surveyed postspill (Burn, Chapter 4).

The helicopter surveys did not detect acute losses for the Kenai Peninsula, Kodiak Island, and Alaska Peninsula areas (DeGange et al. 1994). Although declines in estimated abundance of sea otters were seen between the pre- and postspill helicopter surveys in all three regions, none of the decreases was statistically significant. Additionally, no changes were observed in the distribution of sea otters among heavily, lightly, and nonoiled areas within each region (DeGange et al. 1994).

Burn (Chapter 4) analyzed the data collected in the boat surveys and found a 35% decline (793 fewer otters) in abundance of sea otters in oiled areas of PWS in 1989, and a 13% increase (234 more otters) in abundance in nonoiled areas. Garrott et al. (1993), who used the analysis of these data outlined by Burn (Chapter 4) and adjusted the counts for population growth and detectability of sea otters (Udevitz et al. 1994), estimated that 2800 sea otters died acutely in PWS following the oil spill. However, differences in the areas where the pre- and postspill surveys were conducted complicated analysis and interpretation of the data. In a separate analysis of the boat-survey data set, D. Garshelis (Minnesota Department of Natural Resources, personal communication) concluded that sea otters were at least as abundant in western PWS after the spill as in the mid-1980s. He recognized, based on recovery of carcasses, that significant mortality of otters had occurred after the spill. To explain these observations, Garshelis speculated that the otter population in western PWS may have increased in abundance between 1984-1985 and 1989.

⋊⋉ ⋊⋉ ⋊⋉

Evaluation of the Probability of Carcass Recovery

A second approach to quantifying acute losses involved estimating the probability of recovery of sea otter carcasses. This was accomplished by a trial release of 25 marked sea otter carcasses that had been recovered by clean-up crews, and then releasing them back into the water in the vicinity of northern Kodiak Island in May and June 1989 (DeGange et al. in press; Doroff and DeGange 1994). Clean-up efforts were ongoing in the area. Five of these carcasses were relocated and returned to the collection center, for a recovery rate of 20%.

Due to constraints existing in 1989 associated with the overall effort to collect carcasses, the sample size used in this study was small and several factors limit the validity of extrapolating to sea otter carcasses collected throughout the spill zone. Nevertheless, this effort provided the only quantitative estimate of the probability of recovery of sea otter carcasses following the oil spill (DeGange et al. in press; Doroff and DeGange 1994). Applying the recovery probability of 20% to the overall spill area, the number of sea otters estimated to have died acutely following the oil spill is 4028 (3905 plus 123 dead at rehabilitation centers).

Intersection Model for Estimation of Total Acute Mortality

Development of an analytical model to estimate sea otter mortality based on oil exposure levels was the third approach to quantifying total acute losses (Bodkin and Udevitz, Chapter 5). The Kenai Peninsula sea otter population was modeled because data were most recent and complete for that area. The model integrated oil abundance and movements, sea otter distribution and abundance, and estimates of mortality associated with exposure to varying amounts of oil (including data from rehabilitated sea otters). Simulations indicated potential exposure to some degree of oiling of approximately half (1200 of 2330) of the sea otters along the Kenai Peninsula. By assigning exposure-specific mortality rates to those sea otters exposed to oil, acute mortality can be estimated. However, accurate estimates of total Kenai Peninsula mortality were not possible because of an apparent nonlinear relation between the amount of oil otters encountered and the survival of those otters following the rehabilitation process. Given current limitations of the data, the intersection model may prove to be of greater application in risk assessment than for estimation of the total spill-related mortality.

Necropsies of Sea Otter Carcasses

Necropsies of sea otter carcasses were performed to maximize our understanding of pathological mechanisms contributing to death following oil exposure and to estimate the proportion of deaths that could be classified as oil related (Lipscomb et al., Chapter 16). During the necropsy procedures, tissues were collected for hydrocarbon assays (Ballachey and Mulcahy 1994a; Mulcahy and

✄   ✄   ✄

Ballachey, Chapter 18) and reproductive studies (Bodkin et al. 1993), and morphometric data were collected to complement other studies.

Pathologic and histopathologic changes of the lung, liver, and kidney associated with oil exposure were observed in many of the sea otter carcasses (Lipscomb et al. 1993, Chapter 16). For 71% of the carcasses, death of the otter was judged to be oil related. In only 1% of the cases was the cause of death clearly due to causes other than the oil spill. These findings substantiate the conclusion that most deaths of sea otters recovered as carcasses in the spring and summer of 1989 were oil related rather than incidental mortalities.

Hydrocarbon Assays

Elevated but variable levels of hydrocarbons were found in tissues of heavily oiled sea otters from PWS that died soon after the spill, as compared to control samples from southeastern Alaska (Mulcahy and Ballachey, Chapter 18). As time and distance from the spill increased, hydrocarbon concentrations in the tissues tended to decrease. Hydrocarbon levels in sea otter tissues had seemingly returned to normal levels by the summer of 1990 (USFWS, unpublished data).

Summary of Acute Injury to Sea Otters

Considerable effort was put into the determination of acute losses of sea otters related to the oil spill. However, because of prespill data limitations and a lack of or limitations on studies implemented immediately after the spill, a reliable and defensible estimate of the total number of sea otters that died because of the oil spill cannot be generated. A synthesis of available information leads us to conclude that the total acute loss was in the range of several thousand sea otters.

STUDIES OF CHRONIC INJURY TO SEA OTTERS

Chronic or long-term damages may have limited recovery of sea otters in oiled areas. Several mechanisms for long-term injury can be postulated: (1) sublethal initial exposure to oil causing pathological damage to the otters; (2) continued exposure to hydrocarbons persisting in the environment, either directly or through ingestion of contaminated prey; and (3) altered availability of sea otter prey as a result of the spill.

A variety of approaches, as described below, were taken to evaluate chronic injury to sea otter populations. The design for many of these studies was based on comparisons of sea otters in eastern PWS, a region considered a "control" area because it was not directly affected by the spilled oil, with sea otters in western PWS, much of which was directly contaminated with oil. Differences in population demography and habitat between eastern and western PWS, however, provide

⋙   ⋙   ⋙

difficulties when making comparisons between areas. As in studies of acute losses, efforts to quantify chronic damage suffered from a lack of baseline data and from unavoidable delays in implementation of postspill studies.

Patterns of Mortality

Beach-cast sea otter carcasses were routinely collected in western PWS for approximately 10 years prior to the spill and ages of the otters at the time of death estimated (Johnson 1987). Mortality of sea otters during this period occurred mostly in very young (2 years of age) and older (8 years of age) animals. Only 15% of the carcasses recovered during the prespill period were prime-aged otters (2 to 8 years of age). Similar natural mortality patterns have been described in other sea otter populations (Bodkin and Jameson 1991; Kenyon 1969). From 1989 to 1993, beach-cast sea otter carcasses were collected in western PWS providing comparable postspill data on ages at death.

As expected, given the catastrophic nature of the spill, the sea otter carcasses recovered in 1989 included a much higher proportion of prime-aged sea otters. However, this pattern persisted in 1990 and 1991 (Monson 1994). This shift from a low prespill incidence to a relatively high postspill incidence of prime-aged animals suggests prolonged spill-related mortality. The proportion of prime-aged sea otters recovered from beaches in 1992 decreased relative to the previous years, and by 1993 was similar to prespill values, indicating that perhaps the population is returning to a prespill pattern of mortality. Sample sizes from 1992 and 1993 were relatively small, however, and compared to collections in 1990 and 1991, included fewer carcasses collected from the most heavily oiled areas of PWS.

Survival and Reproduction Rates

Telemetry studies were used by C. Monnett and L. Rotterman (1984-1991) and by the USFWS (1992-1993) to evaluate survival and reproductive rates of sea otters in PWS. Three segments of the population were studied: (1) juvenile sea otters in their first year of life; (2) adult female sea otters; and (3) sea otters released from rehabilitation centers.

Survival of juvenile sea otters (normally born in the spring) through their first winter was measured as an index of population status. Differences were observed in the overwinter survival rate of juvenile sea otters in eastern and western PWS in 1990-1991 (Rotterman and Monnett 1991) and again in 1992-1993 (USFWS, unpublished data). Both studies demonstrated lower survival for pups inhabiting western areas of PWS, relative to the eastern areas. However, the survival rates for both eastern and western areas were lower in 1990-1991 relative to 1992-1993. The influence of the oil spill on survival in subsequent years, relative to factors

⋈ ⋈ ⋈

such as habitat quality, overall population status, and severity of weather conditions in the different years is difficult to ascertain.

Adult female sea otters were monitored from late 1989 to the summer of 1991 (Monnett and Rotterman 1992a). The survival rate of adult female sea otters was not observably diminished by oil exposure (Monnett and Rotterman 1992a). Reproduction of the adult females did not differ: similar pupping rates were observed in eastern and western PWS in 1991 and 1992 (Monnett and Rotterman 1992a). This observation was supported by similar ratios of independent to dependent sea otters counted in surveys in eastern and western PWS in 1991 (Bodkin and Udevitz 1991).

Sea otters released from the rehabilitation centers in the summer of 1989 and monitored until the summer of 1991 exhibited relatively low survival and reproduction rates (Monnett et al. 1990; Monnett and Rotterman 1992b). Interpretation of these results is complicated by the inability to differentiate effects of oil exposure, treatment, prolonged captivity, and translocation (animals were not released in the areas where they had originally been captured) on the sea otters. Although rehabilitated sea otters may not be representative of the larger sea otter population, the relatively low reproduction and survival rates indicate potential long-term damages may result from oil exposure, rehabilitation, or translocation.

Boat-based Surveys of Sea Otters in Prince William Sound

The boat-based surveys conducted in 1989 were repeated in 1990, 1991, and 1993. In addition, a pilot study to evaluate detectability of sea otters in boat surveys was conducted (Udevitz et al. 1994) in conjunction with the summer 1990 boat survey. Counts of sea otters in PWS in 1990 and 1991 were slightly lower but not significantly different from the 1989 counts (Burn, Chapter 4), suggesting that recovery of the population was not occurring. Preliminary analysis of the 1993 data, however, suggests a slight (but, again, not significant) increase in numbers of sea otters inhabiting oiled areas relative to numbers obtained in 1990 and 1991.

Foraging Behavior and Hydrocarbon Levels of Prey

Prey selection and foraging success of sea otters at two oiled sites and one nonoiled site in western PWS were examined in the summer of 1991 (Doroff and Bodkin, Chapter 11). Clams, a common prey item, were collected from subtidal areas at each site for hydrocarbon analyses. Prey selection and foraging success were similar among areas, and hydrocarbon levels in clams were generally low and similar among areas.

Intertidal mussels were reported to retain relatively high levels of hydrocarbon following oiling (Babcock et al. 1993). Mussels contributed about 20% of the total

× × ×

prey items recovered at each site (Doroff and Bodkin, Chapter 11), providing an avenue of continued oil exposure for otters. Sea otters consuming relatively large quantities of mussels may be subjected to higher exposure rates of hydrocarbons.

Physiological and Toxicological Measures of Oil Exposure

Blood analyses

Blood sampled from live sea otters in PWS in 1989 to 1992 was submitted for hematologic and serum chemistry analyses (Rotterman and Monnett, unpublished data; Rebar et al. 1994; Lipscomb et al., Chapter 16; USFWS, unpublished data). Several differences in blood parameters were noted between eastern and western PWS, but in most cases the biological significance of the differences was difficult to ascertain. Elevated levels of serum transaminases measured in the samples collected in western PWS in 1992 (USFWS, unpublished data) were consistent with liver damage but cannot be linked to oil exposure without histopathological examination of tissue samples. Unfortunately, little information is available on pathologies observed in sea otters in the years subsequent to 1989 because fresh carcasses suitable for necropsy are rarely recovered.

Bioindicators of genotoxicity

Methods for detection of genotoxic damage at the cellular or DNA level (McBee and Bickham 1990; Ballachey et al. 1986; Evenson 1986) were applied to evaluate chronic oil-related injury to sea otters. Cells used in this component of the studies included blood lymphocytes, sperm, and testicular cells from adult male sea otters in eastern and western PWS. The cell samples were analyzed by flow cytometry, a technique that allows rapid measurement of large numbers of cells per sample (Melamed at al. 1979). For all three cell types, DNA was the object of the flow cytometry measurements (Ballachey 1994).

No differences in DNA characteristics were noted in sperm or testicular samples from sea otters living in eastern PWS versus western PWS, suggesting that oil exposure did not adversely affect male germ cells. For blood lymphocytes, mean values did not differ but variance of the sample values was higher for otters from oiled versus nonoiled areas, suggesting chemical exposure and genotoxic damage to a proportion of the western otters (Ballachey 1994). Because exposure of individual otters to oil was not known, linking this observation to oil exposure is difficult.

Hydrocarbon residues in tissue samples

Prior to studies conducted in response to the *Exxon Valdez* spill, little was known about hydrocarbon residues in mammalian tissues. Since the oil spill, approximately 500 sea otter tissue samples (primarily liver, blood, and fat) from both

≫ ≫ ≫

oil-exposed and nonexposed animals have been analyzed for levels of aliphatic and aromatic hydrocarbons (Ballachey and Mulcahy 1994a,b; Mulcahy and Ballachey, Chapter 18). This effort has generated a large and unique data base and provides a basis for comparison in future studies on contamination of tissues by hydrocarbons. Analyses of the data suggest that sea otters sampled after 1989 did not have elevated levels of hydrocarbons (USFWS, unpublished data).

Summary of Chronic Injuries to Sea Otters

By late 1991, three findings indicated that chronic damages were limiting recovery of the sea otter population in PWS: patterns of mortality were abnormal when compared to prespill data (Monson 1994), surveys showed no increase in abundance (Burn, Chapter 4), and juvenile survival was low in oiled areas of western PWS (Rotterman and Monnett 1991). These results generated concern about the status of the population and were the impetus for continued research. However, continuing studies of mortality patterns, abundance, and survival rates indicate that by 1993 chronic damages may be subsiding and recovery of the sea otter population may be under way.

Possible mechanisms for prolonged oil-related injury to sea otters were substantiated by several studies. Pathologies of the liver and kidney were observed in oiled sea otters recovered dead (Lipscomb et al., Chapter 16), and it seems reasonable to assume that similar pathological changes may have occurred in sea otters that were exposed to oil and survived. Residual oil persisted in intertidal (Roberts et al. 1993) and subtidal (O'Clair et al. 1993) areas of PWS, providing the potential for continued direct exposure of sea otters to oil and possible ingestion through grooming. In addition, particularly high concentrations of oil were found in mussel beds 2 and 3 years after the spill (Babcock et al. 1993). Because mussels are common prey for sea otters, particularly juveniles, the contaminated mussel beds may have been a source of continued indirect exposure to oil.

CONTINUING POSTSPILL STUDIES OF SEA OTTERS

Studies after 1992 focused on restoration of injured populations rather than on further damage assessment. The primary direction of restoration efforts for sea otters has been to monitor the PWS population for evidence of recovery. As with many other species injured by the oil spill, little if anything can be done to speed the recovery. Habitat protection through establishment of marine sanctuaries or parks may have merit, depending on developing patterns of use in affected areas (e.g., increased recreation, logging). Subsistence take by native hunters, as allowed under the Marine Mammal Protection Act, is apparently increasing throughout Alaska. With the cooperation of native hunters, harvests could be managed to afford additional protection and an opportunity for increasing the rate of recovery for sea otters.

Studies in 1993-1994 include further development of aerial survey methods that should result in more accurate and precise estimates of sea otter abundance and distribution, continued evaluation of patterns of mortality through recovery of beach-cast carcasses in PWS, and modeling of the PWS sea otter population as a tool to predict recovery. We anticipate some level of population monitoring, including abundance estimates, mortality monitoring, and possibly evaluation of physiological parameters, will continue for the next several years.

CONCLUSIONS

Oil spilled from the T/V *Exxon Valdez* caused a significant proportion of the exposed sea otter population to die within several months of the spill. The total acute loss probably numbered several thousand otters; however, due to a lack of prespill data on population abundance and distribution and inability to implement appropriate studies immediately following the oil spill, a precise estimate cannot be made.

Studies of long-term effects of the spill indicate that the sea otter population in PWS suffered from chronic effects of oil exposure, at least through 1991. Data collected in 1992 and preliminary results from 1993 suggest a decline in chronic effects, which may lead to recovery. Monitoring continues to provide additional information regarding the status of recovery of the sea otter population.

In retrospect, the studies on sea otters (and most of the postspill studies) were driven largely by the impending litigation against the Exxon Corporation. Political and scientific considerations influenced the direction of the research. Strong emphasis was placed on estimating acute mortality and documenting continuing damages. These tasks were rendered from difficult to impossible by a lack of prespill data and a lack of contingency plans for research, along with the failure to implement appropriate studies in a timely manner.

Our experience at describing the effects of the EVOS on the PWS sea otter population has provided an opportunity to consider how the damage assessment process may provide improved results in future similar events. It is critical to have studies designed and equipment ready in advance of an oil spill. Standardized protocols for the collection of carcasses and biological specimens should be established. Accurate and precise estimates of abundance should be obtained at regular intervals in areas of petroleum recovery, storage, or transportation, and repeated following a spill. A rigorous study to estimate the probability of recovering a carcass, using telemetry and ratio estimators, should be implemented at the time of the spill. Survival and reproduction of exposed animals should be estimated by marking animals prior to and during the spill and following those individuals

⋈ ⋈ ⋈

over time by telemetry.   Marked individuals should provide evaluation of exposure-dependent effects on survival, reproduction, and behavior.
Although there are many uncertainties in describing the effects of this oil spill on sea otter populations, there is incontrovertible evidence that the population suffered a major perturbation. We will likely never know the full effect of this spill on PWS sea otters. A long-term commitment to monitoring change in the affected sea otter population through the recovery process may provide our final insight into the effects of the EVOS on sea otters.

REFERENCES

Babcock, M., G. Irvine, S. Rice, P. Rounds, J. Cusick, and C. Brodersen. 1993. Oiled mussel beds two and three years after the Exxon Valdez oil spill. Pages 184–185, *in Exxon Valdez* oil spill symposium, 2–5 February 1993, Anchorage, Alaska. (Available, Oil Spill Public Information Center, 645 G Street, Anchorage, Alaska 99501.)

Ballachey, B. E., H. L. Miller, L. K. Jost, and D. P. Evenson. 1986. Flow cytometry evaluation of testicular and sperm cells obtained from bulls implanted with zeranol. Journal of Animal Science 63:995–1004.

Ballachey, B. E. 1994. Biomarkers of damage to sea otters in Prince William Sound, Alaska, following potential exposure to oil spilled from the T/V *Exxon Valdez*. NRDA Report, Marine Mammal Study No. 6. U.S. Fish and Wildlife Service, Anchorage, Alaska.

Ballachey, B. E., and D. M. Mulcahy. 1994a. Hydrocarbons in hair, liver and intestines of sea otters (*Enhydra lutris*) found dead along the path of the *Exxon Valdez* oil spill. NRDA Report, Marine Mammal Study No. 6. U.S. Fish and Wildlife Service, Anchorage, Alaska.

Ballachey, B. E., and D. M. Mulcahy. 1994b. Hydrocarbon residues in tissues of sea otters (*Enhydra lutris*) collected from southeast Alaska. NRDA Report, Marine Mammal Study No. 6. U.S. Fish and Wildlife Service, Anchorage, AK.

Batten, B. T. 1990. Press interest in sea otters affected by the T/V *Exxon Valdez* oil spill: A star is born. Pages 32–40, *in* K. Bayha and J. Kormendy (eds.), Sea otter symposium: Proceedings of a symposium to evaluate the response effort on behalf of sea otters after the T/V *Exxon Valdez* oil spill into Prince William Sound, Anchorage, Alaska, 17–19 April 1990. U.S. Fish and Wildlife Service, Biological Report 90(12).

Bayha, K., and J. Kormendy (eds.). 1990. Sea otter symposium: Proceedings of a symposium to evaluate the response effort on behalf of sea otters after the T/V *Exxon Valdez* oil spill into Prince William Sound, Anchorage, Alaska, 17–19 April 1990. U.S. Fish and Wildlife Service, Biological Report 90(12).

Bodkin, J. L., D. M. Mulcahy, and C. J. Lensink. 1993. Age-specific reproduction in female sea otters (*Enhydra lutris*) from southcentral Alaska: analysis of reproductive tracts. Canadian Journal of Zoology 71:1811–1815.

Bodkin, J. L., and R. J. Jameson. 1991. Patterns of seabird and marine mammal carcass deposition along the central California coast, 1980-1986. Canadian Journal of Zoology 69:1149–1155.

Bodkin, J. L., and M. S. Udevitz. 1991. Development of sea otter survey techniques. Section 2 of NRDA Draft Preliminary Status Report, Marine Mammals Study Number 6 (November 1991, Revised May 1992). U.S. Fish and Wildlife Service, Anchorage, Alaska.

Costa, D. P., and G. L. Kooyman. 1982. Oxygen consumption, thermoregulation, and the effect of fur oiling and washing on the sea otter, *Enhydra lutris*. Canadian Journal of Zoology 60:2761–2767.

◇ ◇ ◇

DeGange, A. R., A. M. Doroff, and D. H. Monson. In press. Experimental recovery of sea otter carcasses at Kodiak Island, Alaska, following the *Exxon Valdez* oil spill. Marine Mammal Science.

DeGange, A. R., D. C. Douglas, D. H. Monson, and C. Robbins. 1994. Surveys of sea otters in the Gulf of Alaska in response to the *Exxon Valdez* oil spill. NRDA Report, Marine Mammal Study No. 6. U.S. Fish and Wildlife Service, Anchorage, Alaska.

DeGange, A. R., and C. J. Lensink. 1990. Distribution, age and sex composition of sea otter carcasses recovered during the response to the T/V *Exxon Valdez* oil spill. Pages 124–129, *in* K. Bayha and J. Kormendy (eds.), Sea otter symposium: Proceedings of a symposium to evaluate the response effort on behalf of sea otters after the T/V *Exxon Valdez* oil spill into Prince William Sound, Anchorage, Alaska, 17–19 April 1990. U.S. Fish and Wildlife Service, Biological Report 90(12).

Doroff, A., and A. R. DeGange. 1994. Experiments to determine drift patterns and rates of recovery of sea otter carcasses following the *Exxon Valdez* oil spill. NRDA Report, Marine Mammal Study No. 6. U.S. Fish and Wildlife Service, Anchorage, Alaska.

Doroff, A., A. R. DeGange, C. Lensink, B. E. Ballachey, and J. L. Bodkin. 1993. Recovery of sea otter carcasses following the Exxon Valdez oil spill. Pages 285–287, *in Exxon Valdez* oil spill symposium, 2–5 February 1993, Anchorage, Alaska. (Available, Oil Spill Public Information Center, 645 G Street, Anchorage, Alaska 99501.)

Estes, J. A. 1991. Catastrophes and conservation: Lessons from sea otters and the *Exxon Valdez*. Science 254:1596.

Evenson, D. P. 1986. Male germ cell analysis by flow cytometry: Effects of cancer, chemotherapy and other factors on testicular function and sperm chromatin structure. Pages 350–367, *in* M. A. Andreeff (ed.), Clinical Cytometry. New York Academy of Sciences.

Ford, G. F., M. L. Bonnell, D. H. Varoujean, G. W. Page, B. E. Sharp, D. Heineman, and J. L. Casey. 1991. Assessment of direct seabird mortality in Prince William Sound and the Western Gulf of Alaska resulting from the *Exxon Valdez* oil spill. Final Report to the U.S. Fish and Wildlife Service, Anchorage, Alaska. Ecological Consulting, Inc., Portland, Oregon. 153 p.

Galt, J. A., and D. L. Payton. 1990. Movement of oil spilled from the T/V *Exxon Valdez*. Pages 4–17, *in* K. Bayha and J. Kormendy (eds.), Sea otter symposium: Proceedings of a symposium to evaluate the response effort on behalf of sea otters after the T/V *Exxon Valdez* oil spill into Prince William Sound, Anchorage, Alaska, 17–19 April 1990. U.S. Fish and Wildlife Service, Biological Report 90(12).

Garrott, R. A., L. L. Eberhardt, and D. M. Burn. 1993. Mortality of sea otters in Prince William Sound following the *Exxon Valdez* oil spill. Marine Mammal Science 9:343–359.

Geraci, J. R., and T. D. Williams. 1990. Physiologic and toxic effects on sea otters. Pages 211–221, *in* J.R. Geraci and D.J. St. Aubin (eds.), Sea mammals and oil: Confronting the risks. Academic Press, Inc., San Diego, California.

Irons, D. B., D. R. Nysewander, and J. L. Trapp. 1988. Prince William Sound sea otter distribution in respect to population growth and habitat type. Unpublished Report, U.S. Fish and Wildlife Service, Anchorage, Alaska.

Johnson, A. M. 1987. Sea otters of Prince William Sound, Alaska. Unpublished Report, U.S. Fish and Wildlife Service, Anchorage, Alaska.

Kenyon, K. W. 1969. The sea otter in the eastern Pacific Ocean. North American Fauna 68. 352 p.

Lipscomb, T. P., R. K. Harris, R. B. Moeller, J. M. Pletcher, R. J. Haebler, and B. E. Ballachey. 1993. Histopathologic lesions in sea otters exposed to crude oil. Veterinary Pathology 30:1–11.

McBee, K., and J. W. Bickham. 1990. Mammals as bioindicators of environmental toxicity. Pages 37–88, *in* Hugh H. Genoways, (ed.), Current Mammalogy Vol. 2. Plenum Publishing Corporation, New York.

Melamed, M. R., P. F. Mullaney, and M. L. Mendelsohn (eds.). 1979. Flow cytometry and sorting. J. Wiley and Sons, New York.

Monnett, C., and L. M. Rotterman. 1992a. Mortality and reproduction of female sea otters in Prince William Sound, Alaska. NRDA Report, Marine Mammal Study No. 6. U.S. Fish and Wildlife Service, Anchorage, Alaska.

Monnett, C., and L. M. Rotterman. 1992b. Mortality and reproduction of sea otters oiled and treated as a result of the *Exxon Valdez* oil spill. NRDA Report, Marine Mammal Study No. 7. U.S. Fish and Wildlife Service, Anchorage, Alaska.

Monnett, C., L. M. Rotterman, C. Stack, and D. Monson, D. 1990. Postrelease monitoring of radio-instrumented sea otters in Prince William Sound. Pages 400–420, *in* K. Bayha and J. Kormendy (eds.), Sea otter symposium: Proceedings of a symposium to evaluate the response effort on behalf of sea otters after the T/V *Exxon Valdez* oil spill into Prince William Sound, Anchorage, Alaska, 17–19 April 1990. U.S. Fish and Wildlife Service, Biological Report 90(12).

Monson, D. H. 1994. Age distributions and sex ratios of sea otters found dead in Prince William Sound, Alaska, Following the *Exxon Valdez* Oil Spill. NRDA Report, Marine Mammal Study No. 6. U.S. Fish and Wildlife Service, Anchorage, Alaska.

O'Clair, C. E., J. W. Short, and S. D. Rice. 1993. Contamination of subtidal sediments by oil from the *Exxon Valdez* in Prince William Sound, Alaska. Pages 55–56, *in Exxon Valdez* oil spill symposium, 2–5 February 1993, Anchorage, Alaska. (Available, Oil Spill Public Information Center, 645 G Street, Anchorage, Alaska 99501.)

Piatt, J. F., C. J. Lensink, M. Butler, M. Kendziorek, and D. Nysewander. 1990. Immediate impact of the *Exxon Valdez* oil spill on marine birds. Auk 92:387–397.

Rebar, A. H., B. E. Ballachey, D. L. Bruden, and K. Kloecker. 1994. Hematology and clinical chemistry of sea otters captured in Prince William Sound, Alaska, following the *Exxon Valdez* oil spill. NRDA Report, Marine Mammal Study No. 6. U.S. Fish and Wildlife Service, Anchorage, Alaska.

Roberts, P. O., C. B. Henry, Jr., E. B. Overton, and J. Michel. 1993. Characterization of residual oil in Prince William Sound, Alaska--3.5 years later. Pages 46–47, *in Exxon Valdez* oil spill symposium, 2–5 February 1993, Anchorage, Alaska. (Available, Oil Spill Public Information Center, 645 G Street, Anchorage, Alaska 99501.)

Rotterman, L. M., and C. Monnett. 1991. Mortality of sea otter weanlings in eastern and western Prince William Sound, Alaska, during the winter of 1990-91. NRDA Report, Marine Mammal Study No. 6. U.S. Fish and Wildlife Service, Anchorage, Alaska.

Siniff, D. B., T. D. Williams, A. M. Johnson, and D. L. Garshelis. 1982. Experiments on the response of sea otters, *Enhydra lutris*, to oil. Biological Conservation 23:261–272.

Udevitz, M. S., J. L. Bodkin, and D. P. Costa. 1994. Sea otter detectability in boat-based surveys of Prince William Sound, Alaska. NRDA Report, Marine Mammal Study No. 6. U.S. Fish and Wildlife Service, Anchorage, Alaska.

Wendell, F. E., R. A. Hardy, and J. A. Ames. 1986. An assessment of the accidental take of sea otters, *Enhydra lutris*, in gill and trammel nets. Marine Research Technical Report Number 54. California Department of Fish and Game, Sacramento, California.

Williams, T. M., and R. W. Davis (eds.). 1990. Sea otter rehabilitation program: 1989 *Exxon Valdez* oil spill. International Wildlife Research.

Chapter 4

Boat-Based Population Surveys of Sea Otters in Prince William Sound

Douglas M. Burn

INTRODUCTION

Sea otters (*Enhydra lutris*) in Alaska were commercially exploited from the time of Vitus Bering's voyage of 1741 until they were granted protection under the North Pacific Fur Seal Convention in 1911 (Kenyon 1969). They were extirpated throughout much of their range, with only a few small remnant populations surviving. It is believed that the present sea otter population in Prince William Sound (PWS) is derived from one such remnant population that persisted in the southwestern portion of the Sound (Lensink 1962). Surveys conducted in the 1960s and 1970s documented the northeastward expansion of sea otters into unoccupied areas of PWS (Pitcher 1975). By the mid-1980s, sea otters had recolonized most of the Sound, but likely had not reached carrying capacity in some areas (Irons et al. 1988). At the time of the *Exxon Valdez* oil spill (EVOS), the size of the PWS sea otter population was unknown but was believed to number between 5000 and 10,000 individuals. During the first few days following the spill, it was uncertain in which direction the oil would move and what proportion of the sea otter population would be injured.

The number of sea otter carcasses recovered could only produce a minimal estimate of the damage to the PWS sea otter population. Carcass recovery rates were determined, and were used to calculate an estimate of total spill-related mortality (Doroff and DeGange 1993). Another method to determine extent of damage was comparison of pre- and postspill population estimates. This study reports the results of nine postspill sea otter population surveys conducted in PWS between June 1989 and July 1991, and discusses their implications for damage of the PWS sea otter population as a result of the spill. The objectives of this study were to: (1) test that differences in sea otter densities were not significantly different between pre- and postspill surveys in oiled and unoiled areas in PWS; (2) estimate

61

the magnitude of any change between pre- and postspill sea otter population estimates in PWS; and (3) estimate postspill sea otter population size and monitor sea otter population trends in PWS.

METHODS

Study Area

In the southwest portion of PWS, the study area was bounded by a line extending from Cape Junken eastward to Montague Island (Fig. 4-1). Proceeding eastward, the northern shores of Montague, Hinchinbrook, and Hawkins Islands defined the southernmost extent of the study area. Prince William Sound contains numerous islands ranging in size from less than 1 km^2 to more than 250 km^2. The shoreline is highly convoluted, with numerous fiords, passes, and bays. Water depths within the study area varied from less than 2 m to more than 870 m.

Sampling Units

The study area was divided into three survey strata: shoreline, coastal, and pelagic. The shoreline stratum was based on shoreline transects surveyed by Irons et al. (1988) during May to August of 1984 and 1985, and was defined as the 200-m-wide strip immediately adjacent to the coastline. Within the study area, Irons et al. (1988) defined 742 shoreline transects with a total area of 822.3 km^2. Shoreline transects were of varying size, ranging from groups of rocks or small islands with less than 1 km of coastline, to sections of the mainland with over 25 km of coastline. The mean transect length was 6.57 km, with a sampled area of 1.11 km^2. Transect end points were often located at geographic features such as points of land or other landmarks to facilitate orientation in the field.

This survey was designed to count seabirds and sea otters simultaneously, so some of the sampling decisions in this survey were made primarily for seabird considerations. It was known that certain bird species occur in association with coastlines, while others occur farther from shore. Thus, waters outside the shoreline stratum were divided into sampling "blocks" based on a 5-minute latitude/longitude grid system. In an attempt to differentiate these blocks with respect to distance from shore, they were then stratified into two categories: coastal and pelagic. The coastal stratum consisted of those blocks located immediately adjacent to 1 km or more of shoreline, while the pelagic stratum consisted of those blocks adjacent to less than 1 km of shoreline. Where the grid intersected the coastline in such a way as to create an unmanageably small block, adjacent grid cells were pooled together into a longer or wider block. This classification scheme resulted in the creation of 207 coastal and 86 pelagic blocks, with total areas of 4524 km^2 and 3637 km^2, respectively.

⋙ ⋙ ⋙

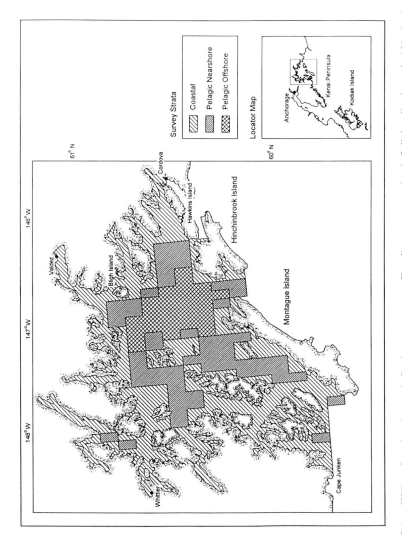

Figure 4-1. Prince William Sound study area, indicating survey strata. Shoreline stratum consisted of all shoreline located within the shaded areas.

Within each block, 200-m-wide strip transects were systematically placed along north–south meridians located 1 minute of longitude from the eastern and western block boundaries. The choice of these meridians was made to facilitate simultaneous aerial and boat-based sampling (aerial sampling was a component of a Natural Resource Damage Assessment Bird Study). For most blocks, these meridians resulted in the placement of two transects identified by the block designation, followed with an "E" or "W" to indicate if the transect was on the east or west side of the block. In some coastal blocks, one of the appropriate meridians may have fallen on land, thus only one transect was placed within the block. For those blocks that consisted of pooled grid cells as described above, a third transect was placed within the block if the appropriate meridian occurred over water. These additional transects were designated with an "E2" or "W2" subscript. Due to their intersection with the coastline, coastal transects ranged from hundreds of meters in length to the full 5-nautical-mile length of the block. Since pelagic blocks by definition did not intersect the coastline, transects were always paired and ran the entire 5-nautical-mile length of the block. An example of the three survey strata is presented in Figure 4-2.

After the first field season in June–August 1989, it was recognized that the pelagic stratum was not homogeneous with respect to sea otter distribution. Sea otters are benthic feeders that forage primarily in shallow subtidal areas (Riedman and Estes 1990). In PWS, sea otters have been observed to forage at mean depths of 7–28 m at various study sites (Garshelis 1983). Some pelagic blocks were located directly over shallow water, while others were located several kilometers distant. Assuming that sea otters occur in proximity to shallow water feeding areas, the pelagic stratum was poststratified into pelagic nearshore and pelagic offshore strata, based on distance from the 20-m bathymetric contour (contours in digital format for PWS were available in 20-m increments). The cutoff distance between the pelagic nearshore and pelagic offshore strata was 5 km from the 20-m contour. Under this new stratification, the pelagic nearshore strata had characteristics similar to the coastal strata (relatively close to shore or shallow water). However, these strata could not be pooled since the initial random samples were drawn separately from the coastal and pelagic strata.

Prespill Data

As stated earlier, the shoreline stratum in this study was based on a set of transects originally surveyed during May to August of 1984 and 1985 (Irons et al. 1988). Over the course of two field seasons, they surveyed virtually all the available shoreline transects within the PWS study area (708 out of the possible 742 transects). These data served as the prespill baseline for comparison with postspill surveys. It should be noted that changes in the PWS sea otter population

⋊⋉   ⋊⋉   ⋊⋉

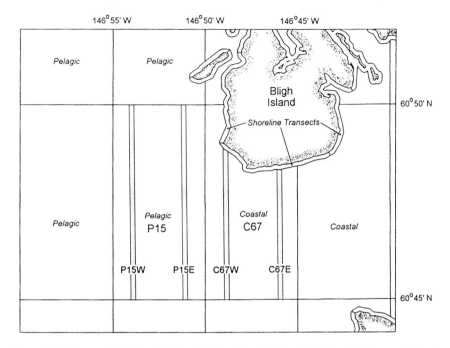

Figure 4-2. Area near Bligh Island in Prince William Sound, Alaska, illustrating shoreline, coastal, and pelagic survey strata used in this study. Coastal and pelagic transects are located along meridians 1' of longitude from eastern and western block boundaries.

between the Irons et al. (1988) survey and the time of the spill may have occurred. Waters beyond the shoreline stratum were not sampled during the prespill survey.

Field Methods

Sampling techniques in all postspill surveys duplicated those of the baseline study of Irons et al. (1988). Researchers used 8-m motor boats, with three crew members serving equally as operator and observers. Shoreline transects were surveyed from 100 m offshore at a cruising speed of 5–10 knots. One observer scanned the water from the vessel up to and including the shoreline, while another observer scanned the water from the vessel seaward an additional 100 m. The 100-m cutoff distance was visually estimated, with periodical calibration, using rangefinders or a float attached to a 100-m-long line. The coastal and pelagic transects were surveyed at a slightly faster cruising speed of 10–15 knots with two observers, one on each side of the boat, scanning the water from the trackline of the boat outward 100 m. In addition, the watercraft operator assisted with observations of animals directly ahead of the vessel. While the vessel was in motion, all marine mammals and birds sighted were recorded on standardized data sheets.

To insure consistency between years, an observer handbook was written after the first field season to familiarize new field personnel with the survey design and field methods. A transect guide was also developed to help field personnel locate transect end points based on geographic features. Surveys during the second and third years of the study used experienced personnel as boat team leaders and observers.

Survey Dates and Sample Sizes

Three replicates of the survey were conducted in June, July, and August 1989 to determine if a continued, ongoing effect of the spill was occurring. These replicates were repeated during June, July, and August 1990 for comparison with 1989, and to further examine variability within the field season. Due to reduced funding levels, a single survey was conducted in July 1991 to allow for comparisons with July 1989 and July 1990 results. July was considered the preferred month to survey for seabirds, minimizing the effect of certain species migrating into or out of the study area. Surveys were conducted in March 1990 and 1991 primarily to collect information on wintering seabird distribution and abundance. Allowing for inclement weather and mechanical failure, approximately 3 weeks were needed to complete each replicate of the survey.

Postspill surveys were initially conducted during summer 1989 as a simple random sample of approximately 25% of all shoreline transects and coastal and pelagic blocks. Due to logistic constraints, only the shoreline stratum was sampled during June 1989. All three strata were sampled in July and August 1989, and on each of the following surveys. Once the initial random sample of transects and blocks was chosen, each successive survey replicated the same sampling units to allow for comparison over time.

In order to complete surveys during March when daylight is a limiting factor and weather conditions are often less favorable, only a subset of the original set of shoreline transects and coastal and pelagic blocks was sampled. This subset consisted of approximately 14% of the shoreline transects and coastal blocks. The sample size of pelagic blocks ($n=25$) was not reduced during the March surveys. The magnitude of this reduction was based on an estimated time available for the survey of approximately 10 complete sampling days.

On the advice of peer reviewers, an additional 25 shoreline transects were added to the sample beginning with the fifth survey in June 1990, increasing the proportion sampled from 25% to 29%. These additional transects were randomly selected from western PWS. Sample sizes of the coastal and pelagic strata were not increased.

⋈ ⋈ ⋈

Oiling Classification

Classification of sampling units as oiled or unoiled was based on Alaska Department of Environmental Conservation overflight data collected at the time of the spill (ADEC 1989). Aerial observations were used to create a Geographic Information System (GIS) coverage depicting the movement of oil over the surface of the water. Since sea otters are mobile animals, those inhabiting areas adjacent to the path of the oil could have encountered oil during their normal movement patterns. Given this assumption, coupled with an inherent uncertainty as to the exact geographical extent of the surface oiling, a buffer zone of 5 km was added to the oiled zone boundary to represent an area within which sea otters might have been affected by oil (Fig. 4-3). Shoreline transects and coastal and pelagic blocks with any area located within this buffer zone (i.e., within 5 km of surface oil) were classified as oiled.

Analytical Methods

Sea otter density and abundance estimates for each survey strata were calculated using ratio estimator techniques (Cochran 1977). The following notation will be used to define the estimators density and abundance within a stratum in which i refers to the ith block within the stratum:

a_i=area of transects in the block,
A_i=area of the block (note: for shoreline stratum $a_i=A_i$),
A =total area of all blocks in the stratum,
n_i=number of sea otters observed in the block,
d_i=observed sea otter density in the block (=n_i/a_i),
B =number of blocks in the stratum, and
b =number of sampled blocks in the stratum.

The estimator of density for a stratum and its standard error (SE) are

$$\hat{D} = \frac{\sum_{i=1}^{b} A_i d_i}{\sum_{i=1}^{b}} \tag{1}$$

$$\hat{Se}(\hat{D}) = \frac{B}{A} \sqrt{\frac{\sum_{i=1}^{b} A_i^2 (d_i - \hat{D})^2}{b(b-1)} \frac{B-b}{B}} \tag{2}$$

⋈   ⋈   ⋈

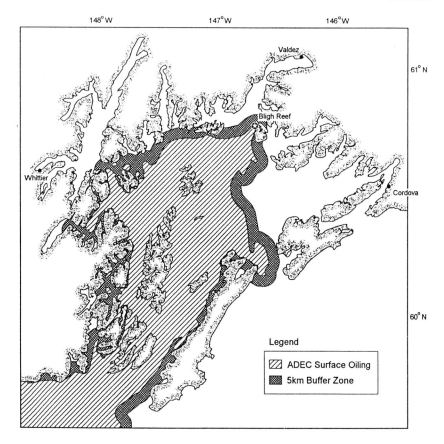

Figure 4-3. Extent of surface oiling in Prince William Sound, Alaska, following the *Exxon Valdez* oil spill. Data are from the Alaska Department of Environmental Conservation overflights. Sampling units located within 5 km of surface oiling were classified as oiled.

The finite population correction factor (B-b/B) was not used for the coastal and pelagic strata because the location of transects within a block were systematically placed. The estimator of abundance for a stratum and its standard error are

$$\hat{N} = A\hat{D} \qquad (3)$$

$$\hat{Se}(\hat{N}) = A \cdot \hat{Se}(\hat{D}) \qquad (4)$$

Abundance estimates for the entire PWS study area were calculated by summing the estimates for each strata. For the summer 1989 and 1990 field seasons, density and abundance estimates were also calculated using the mean of the three sea otter counts per transect from the June, July, and August surveys of that year.

Comparisons between pre- and postspill shoreline sea otter density estimates were made with a t test of the form

$$t = \frac{\hat{D}_2 - \hat{D}_1}{\hat{V}(\hat{D}_2 - \hat{D}_1)} \tag{5}$$

where:

$\hat{D}_1$	-	prespill density estimate,
$\hat{D}_2$	-	postspill density estimate, and
$\hat{V}(\hat{D}_2 - \hat{D}_1)$	-	variance of the difference between density estimates.

Degrees of freedom for this test were (n-1) for the smaller of the two sample sizes. The same test was used to compare sea otter density estimates in the coastal and nearshore pelagic strata between oiled and unoiled areas. A similar form of the t test was used to compare postspill abundance estimates of all survey strata combined. Effective degrees of freedom were calculated for each abundance estimate according to Satterthwaite's (1946) approximation as cited in Cochran (1977). Degrees of freedom for this test were (n-1) for the smaller of the two samples.

RESULTS

Survey Effort and Sea Otter Counts

During nine postspill surveys, a total of 2639 transects were sampled with 4791 sea otter sightings totaling 6469 individuals. The majority of sea otters counted (5975) were within the shoreline stratum (Table 4-1). Counts within the coastal and pelagic nearshore strata were lower and variable. With the exception of one sea otter sighted during July 1990, no otters were observed in the pelagic offshore stratum. Thus, density and abundance estimates are not presented for the pelagic offshore stratum.

Sea Otter Densities

Observed sea otter densities were highest in the shoreline stratum, followed by the coastal and pelagic nearshore strata (Table 4-2). Since the prespill surveys of Irons et al. (1988) were conducted over the course of a longer field season (May–August), mean summer shoreline density values were used for comparison.

>< >< ><

Table 4-1. Sample sizes (b) and numbers of sea otters counted (n) in oiled and unoiled areas of Prince William Sound, Alaska, before and after the *Exxon Valdez* oil spill. Pre-spill values are from Irons et al. (1988) data. Sample size in the shoreline stratum is the number of shoreline transects surveyed; in the coastal and pelagic nearshore strata, sample size is the number of blocks surveyed. (--- = no data)

Survey Date	Oiled areas						Unoiled areas					
	Shoreline		Coastal		Pelagic nearshore		Shoreline		Coastal		Pelagic nearshore	
	b	n	b	n	b	n	b	n	b	n	b	n
1984-1985	423	2191	---	---	---	---	285	1666	---	---	---	---
1989												
June	115	400	---	---	15	19	68	445	---	---	---	---
July	118	414	21	12	15	13	69	460	25	59	3	2
August	118	464	21	6	15	16	69	425	25	20	3	3
Summer[a]	118	430	21	9	15	16	69	445	25	40	3	3
1990												
March	61	173	15	16	15	5	38	216	14	14	3	9
June	133	219	20	10	14	15	78	305	24	44	3	0
July	134	384	21	12	15	7	78	253	25	61	3	1
August	134	411	21	8	15	7	78	388	25	38	3	4
Summer[a]	134	339	21	10	15	10	78	315	25	48	3	2
1991												
March	61	123	15	6	15	3	38	195	14	16	3	5
July	134	406	21	12	15	6	78	294	24	55	3	6

[a] Summer value calculated using the mean of transect counts from the June, July, and August surveys of that year.

Table 4-2. Estimated sea otter densities ($\hat{D}$) and associated standard errors ($\hat{S}e(\hat{D})$) in oiled and unoiled areas of Prince William Sound, Alaska, before and after the *Exxon Valdez* oil spill. Pre-spill values are from Irons et al. (1988) data. Density values are in units of otters/km². (--- = no data)

| Survey date | Oiled areas | | | | | | Unoiled areas | | | | | |
| | Shoreline | | Coastal | | Pelagic nearshore | | Shoreline | | Coastal | | Pelagic nearshore | |
	$\hat{D}$	$\hat{S}e(\hat{D})$	$\hat{D}$	$\hat{S}e(\hat{D})$	$\hat{D}$	$\hat{S}e(\hat{D})$	$\hat{D}$	$\hat{S}e(\hat{D})$	$\hat{D}$	$\hat{S}e(\hat{D})$	$\hat{D}$	$\hat{S}e(\hat{D})$
1984-1985	5.25	0.12	---	---	---	---	4.53	0.19	---	---	---	---
1989												
June	3.29	0.42	---	---	---	---	5.22	0.77	---	---	---	---
July	3.30	0.42	0.27	0.10	0.34	0.25	5.31	0.83	1.67	0.48	0.18	0.09
August	3.70	0.54	0.12	0.08	0.23	0.14	4.90	0.84	0.56	0.19	0.27	0.15
Summer [a]	3.43	0.38	0.20	0.06	0.29	0.17	5.14	0.61	1.13	0.30	0.22	0.09
1990												
March	2.60	0.23	0.53	0.27	0.09	0.03	4.35	0.75	0.61	0.22	0.80	0.41
June	1.58	0.39	0.23	0.16	0.29	0.13	3.16	0.67	1.29	0.47	0.00	0.00
July	2.76	0.47	0.26	0.11	0.13	0.11	2.62	0.38	1.65	0.82	0.09	0.09
August	2.95	0.49	0.19	0.08	0.13	0.07	4.02	0.58	1.01	0.31	0.36	0.09
Summer [a]	2.43	0.41	0.23	0.09	0.17	0.06	3.26	0.41	1.29	0.37	0.15	0.03
1991												
March	1.85	0.19	0.21	0.10	0.05	0.04	3.94	0.67	0.62	0.14	0.45	0.09
July	2.91	0.34	0.26	0.18	0.11	0.08	3.04	0.48	1.50	0.50	0.54	0.15

[a] Summer value calculated using the mean of transect counts from the June, July, and August surveys of that year.

71

Table 4-3. Estimated abundance ($\hat{N}$) and associated 95% confidence intervals (95%ci) of sea otters in oiled, unoiled and all areas of Prince William Sound, Alaska, before and after the *Exxon Valdez* oil spill. Pre-spill values are from Irons et al. (1988) data.

| | Oiled areas | | | | | | Unoiled areas | | | | | | All areas | |
| | Shoreline | | Coastal | | Nearshore pelagic | | Shoreline | | Coastal | | Nearshore pelagic | | | |
Survey date	$\hat{N}$	95%ci	$\hat{N}$	95%ci	$\hat{N}$	95%ci	$\hat{N}$	95%ci	$\hat{N}$	95%ci	$\hat{N}$	95%ci	$\hat{N}$	95%ci
1984-1985	2285	±128					1754	±143						
1989														
June	1430	±362	694	±500	688	±982	2020	±586	3293	±1857	76	±74	8240	±2280
July	1438	±360	304	±382	474	±554	2053	±631	1098	±749	113	±128	5497	±1283
August	1611	±460	499	±316	581	±382	1897	±635	2239	±1147	95	±74	6894	±1485
Summer[a]	1492	±320					1988	±465						
1990														
March	1130	±199	1361	±1338	182	±135	1685	±566	1203	±844	340	±340	5901	±1731
June	690	±331	577	±779	585	±520	1221	±511	2548	±1815	0		5621	±2131
July	1200	±397	650	±562	262	±442	1013	±291	3261	±3178	38	±74	6424	±3297
August	1284	±419	482	±384	255	±293	1554	±439	1991	±1208	151	±74	5717	±1438
Summer[a]	1059	±348	578	±433	354	±242	1263	±315	2546	±1449	63	±25	5881	±1602
1991														
March	804	±161	539	±524	109	±154	1524	±510	1234	±557	189	±74	4399	±948
July	1268	±287	664	±880	217	±308	1177	±362	2956	±1932	227	±128	6509	±2198

[a] Summer value calculated using the mean of transect counts from the June, July, and August surveys of that year.

72

Table 4-4. Distribution of sea otters in survey strata based on estimated abundances in oiled and unoiled areas of Prince William Sound, Alaska, during surveys conducted following the *Exxon Valdez* oil spill.

Survey date	Oiled areas			Unoiled areas		
	Shoreline	Coastal	Pelagic Nearshore	Shoreline	Coastal	Pelagic Nearshore
	percent	percent	percent	percent	percent	percent
1989						
July	51.0	24.6	24.4	37.9	60.7	1.4
August	67.5	12.7	19.8	61.0	35.3	3.7
Summer [a]	58.0	19.4	22.6	46.0	51.8	2.2
1990						
March	42.3	50.9	6.8	52.2	37.3	10.5
June	37.3	31.2	31.5	32.4	67.6	0.0
July	56.8	30.8	12.4	23.5	75.6	0.9
August	63.5	23.9	12.6	42.0	53.9	4.1
Summer [a]	53.2	29.0	17.8	32.6	65.6	1.6
1991						
March	55.4	37.1	7.5	51.7	41.9	6.4
July	59.0	30.9	10.1	27.0	67.8	5.2

[a] Summer value calculated using the mean of transect counts from the June, July, and August surveys of that year.

In the unoiled area, shoreline sea otter density in summer 1989 was approximately 14% greater than prespill density (t=0.941, df=68, P=0.35). In the oiled area, shoreline sea otter density declined approximately 35% during the same interval (t=-4.622, df=117, P<0.001). Surveys conducted in summer 1990 showed further declines in shoreline sea otter density in the oiled area to 54% below the prespill value. Shoreline otter density in unoiled areas also declined during the same period between 1989 and 1990. Mean summer 1990 density in the unoiled area was 28% lower than the prespill density (t=-2.779, df=77, P=0.007). Shoreline sea otter density within both oiled and unoiled areas did not appear to have changed between 1990 and 1991.

Within the coastal stratum, sea otter densities observed during postspill surveys in June, July, and August were consistently higher in the unoiled area. The difference was significant at the P<0.05 level for all but the July 1990 survey. Density estimates in the pelagic nearshore stratum were not significantly different between oiled and unoiled areas.

Sea Otter Abundance

Although sea otter densities were lower in coastal and nearshore pelagic strata than in the shoreline stratum, given their large total areas, these strata contained a considerable number of otters (Table 4-3). In some instances, these strata accounted for over 50% of the total estimated population (Table 4-4). The proportion of otters within each of the three survey strata varied from survey to survey. In some instances, changes in density within one stratum were offset by changes in other strata. For this reason, monitoring the sea otter population over the course of this study can best be done by comparisons between abundance estimates of all survey strata combined (Fig. 4-4). All survey strata were sampled in July and August 1989; June, July, and August 1990; and July 1991. Since July was the only month all strata were surveyed in all years, comparisons between these data points were used to assess population trends following the spill.

Within the oiled area, the July 1990 estimate was 654 fewer sea otters than the July 1989 estimate (2165 versus 2819). Given their large variances, the estimates were not significantly different (t=-0.9, df=124, P=0.37). The July 1991 abundance estimate for the oiled area was virtually identical to the July 1990 estimate (2165 versus 2149). Within the unoiled area, the July 1990 estimate was 1110 fewer sea otters than the July 1989 estimate (4312 versus 5422). Again, due primarily to the large variances, these estimates were not significantly different (t=-0.58, df=34, P=0.57). Similar to the oiled area, the July 1990 and 1991 estimates were almost identical (4312 versus 4360).

⋙   ⋙   ⋙

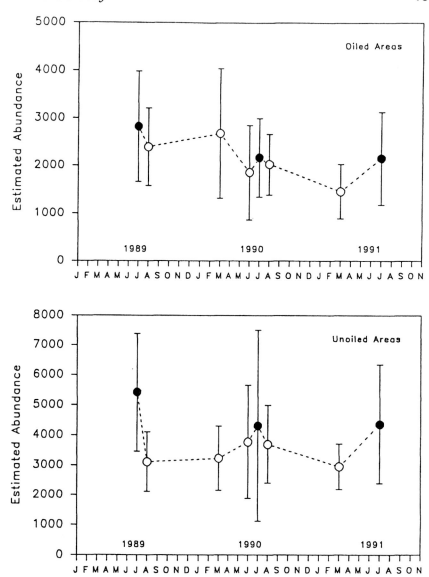

Figure 4-4. Estimated abundance of sea otters in oiled and unoiled areas of Prince William Sound, all survey strata combined. Error bars represent 97% confidence intervals. July estimates indicated for comparison by filled circles.

DISCUSSION

Based on comparisons of prespill and summer 1989 survey data, the 35% decline in shoreline sea otter density within the oiled area suggested a significant first-year effect of the oil spill on the PWS sea otter population. This result is not surprising, given that over 400 sea otter carcasses had been retrieved from PWS prior to the first postspill survey in June 1989 (DeGange and Lensink 1990). Other studies suggested that the number of carcasses recovered may have represented only 20% of the total mortality (Doroff and DeGange 1993). Based on 424 carcasses recovered, an estimated 20% carcass recovery rate, and 89 otters that died in rehabilitation centers, Doroff and DeGange (1993) estimated an acute loss of 2209 sea otters from PWS due to the spill.

Further declines in shoreline sea otter density within the oiled area between 1989 and 1990 suggested a continuing oil effect. However, this decline was mirrored by a decline in shoreline sea otter density within unoiled areas of PWS. Was there a Sound-wide decline in the sea otter population between the summers of 1989 and 1990? Abundance estimates of all survey strata combined for July 1989 and July 1990 were not significantly different in either the oiled or unoiled areas. Due to the large variance of the abundance estimates, changes in the sea otter abundance, if real, would have had to be very large to be detected with this survey design.

Other damage assessment studies suggested an ongoing effect of the oil spill on sea otters in PWS. The age-class structure of dying sea otters, based on carcasses recovered during the spill year (defined as 1989) and postspill (defined as 1990 and 1991) within the oiled area was significantly different from the prespill age-class structure (Monson 1993). Specifically, the proportion of "prime-age" animals (2 to 8 years old) in the spill year and postspill samples was higher than that observed prespill. This result suggested that some abnormal mortality had occurred within the oiled area beyond the first year of the spill.

Reasons for a possible decline in sea otter abundance in unoiled areas of PWS may be less obvious. Results of radiotelemetry studies conducted to monitor the fate of sea otters released back into the wild from the rehabilitation centers indicated that these individuals exhibited relatively low survivorship when compared to radio implanted sea otters from other study groups (Monnett et al. 1990). Furthermore, following the release of these rehabilitated sea otters, other study groups of otters in eastern PWS (an unoiled area) that had been radio-implanted prior to the spill also exhibited reduced survivorship (Monnett and Rotterman 1993). Monnett and Rotterman (1993) have suggested that the release of rehabilitated otters deleteriously affected sea otters in eastern PWS, perhaps through the introduction of disease, resulting in unusually high mortality in the wild sea otter population.

The variability in the proportion of sea otters within the shoreline stratum among surveys adds an element of uncertainty to the use of shoreline density values as an

✕✕ ✕✕ ✕✕

index of the population. However, it seems unlikely that the 35% decline in shoreline sea otter density in the oiled area between the prespill surveys and summer 1989 was due to a redistribution of otters among strata. If one assumes that the net loss of sea otters in the shoreline stratum in the oiled area was offset by an increase in the coastal or nearshore pelagic strata (as may have occurred in unoiled areas between the summers of 1989 and 1990), it would require the prespill abundance of these strata to have been almost zero.

Sea otter density within the coastal stratum of the oiled area was significantly lower than in unoiled areas during the postspill surveys. Lacking prespill data for all but the shoreline stratum, there is no method to determine if this difference represents injury due to the spill, or merely a reflection of differences in habitat or population status between the two areas. However, if sea otter densities within the coastal stratum were homogeneous throughout the Sound prior to the spill, this difference between oiled and unoiled coastal strata would represent the loss of a considerable number of otters from the oiled coastal stratum. Alternatively, this difference could also have been the result of a shift in sea otter distribution. Continued monitoring of the sea otter population within the coastal stratum of the oiled area may eventually provide an indication of what the prespill density may have been.

Accepting that the decline in shoreline sea otter density between prespill and mean summer 1989 estimates represented a significant oil effect, results of studies on sighting probability, carcass recovery rates, and the age structure of the recovered carcasses were combined with these survey data to calculate an estimate of the initial first-year injury to the PWS sea otter population (Garrott et al. 1993). This exercise produced a loss estimate of approximately 2800 sea otters for PWS. This result is comparable to the estimate of 2209 sea otters lost based on carcass recovery rates (Doroff and DeGange 1993).

The long-term effects of the spill on sea otters in the western portion of PWS are unknown. Two key factors that will influence potential long-term effects on sea otters are the impact of the spill on the populations of sea otter prey items (primarily mussels and clams), and continued exposure of sea otters to hydrocarbons through their prey. Either of these factors could have a profound impact on the recovery of the sea otter population in the oiled area of PWS.

Continued monitoring of the PWS sea otter population will yield postspill population trends, but due to the uncertainty involved with using shoreline sea otter abundance as an index of the population, the issue of recovery may be difficult to address. A common limitation of many of the damage assessment studies was the quality of available prespill data (Spies 1993). Study design, methodology, and the amount of time between pre- and postspill data points are all potential sources of uncertainty. While it was a relatively easy matter in this study to replicate the sampling methodology of the shoreline stratum of Irons et al. (1988) there is no

❧   ❧   ❧

way to compensate for lack of prespill data in other strata. While sea otter density within the shoreline stratum may not be an ideal index of the total population, it is the only statistic available for comparison with a prespill value. An increase in shoreline sea otter density within the oiled area accompanied by proportional increases in the coastal and nearshore pelagic strata may indicate a real increase in the population rather than redistribution, and may be the best means of gauging recovery of the population.

In order to improve the precision of abundance estimates, modifications to the survey design are necessary. Keeping the shoreline stratum intact to allow for comparison with prespill data, I suggest a redesign of the coastal and pelagic strata with a more biologically meaningful basis. For sea otters, proximity to shallow water feeding areas should be a useful criterion. Use of a GIS with bathymetric data layers would aid in this objective. Sampling units (blocks) within these areas should be smaller than those of the current design, and have their transects placed parallel to one another oriented perpendicular to the general direction of the coastline. The current survey design samples a relatively small proportion (approximately 2%) of the total area of the coastal and pelagic strata. An increase in the sample sizes in these strata would likely reduce the variance of the estimates. This could be accomplished by extending the survey window beyond the present 3-week period, or by adding additional survey vessels. Based on the results from the pelagic offshore stratum, it would seem that this area of the Sound would not need to be sampled for sea otters. However, it may be necessary to continue sampling this area in consideration of some seabird species.

CONCLUSIONS

Shoreline sea otter densities in the unoiled area increased 14% between prespill surveys conducted in 1984–1985 and 1989, while densities in the oiled area declined 35%. The mean summer 1989 density estimate in the oiled area was significantly lower than prespill density. Based on these and related data, a cooperative effort to quantify total injury to the PWS sea otter population estimated that approximately 2800 otters were initially killed by the spill (Garrott et al. 1993). Shoreline sea otter densities from additional surveys conducted in June, July, and August 1990 suggested a decline between 1989 and 1990 in both oiled and unoiled areas of the Sound. However, abundance estimates of all survey strata combined for these areas were not significantly different between July 1989, 1990, and 1991. As a measure of recovery, the population trend of all strata combined should be considered along with future comparisons with prespill shoreline sea otter density within the oiled area.

⋈   ⋈   ⋈

ACKNOWLEDGMENTS

I thank the following persons for their assistance in preparation of this report: T. Jennings, B. Boyle, R. Slothower, G. Balogh, and H. Buckholtz provided GIS support. D. Bowden and S. Klosiewski provided statistical advice. Damage assessment surveys were designed by S. Klosiewski. He and K. Laing served as project coordinators and were instrumental in getting this effort into the field. I also thank D. Irons for providing the prespill data set for comparative analysis. I thank B. Ballachey, D. Bowden, T. DeGange, D. DeMaster, T. Evans, R. Garrott, K. Oakley, and three anonymous reviewers for providing comments on the manuscript.

REFERENCES

Alaska Department of Environmental Conservation (ADEC). 1989. *Exxon Valdez* oil spill: Cumulative oiling on water through June 20, 1989. ARC/INFO Data Files.

Batten, B. T. 1990. Press interest in sea otters affected by the T/V *Exxon Valdez* oil spill: a star is born. Pages 32–40 *in* K. Bayha and J. Kormendy (eds.), Sea otter symposium: Proceedings of a symposium to evaluate the response effort on behalf of sea otters after the T/V *Exxon Valdez* oil spill into Prince William Sound. Anchorage, Alaska, 17–19 April 1990. U.S. Fish & Wildlife Service Biological Report 90(12).

Cochran, W. G. 1977. Sampling techniques. John Wiley and Sons, Inc., New York, New York. 428 p.

DeGange, A. R., and C. J. Lensink. 1990. Distribution, age, and sex composition of sea otter carcasses recovered during the response to the T/V *Exxon Valdez* oil spill. Pages 124–129, *in* K. Bayha and J. Kormendy (eds.), Sea otter symposium: Proceedings of a symposium to evaluate the response effort on behalf of sea otters after the T/V *Exxon Valdez* oil spill into Prince William Sound. Anchorage, Alaska, 17–19 April 1990. U.S. Fish & Wildlife Service Biological Report 90(12).

Doroff, A. M., and A. R. DeGange. 1993. Experiments to determine drift patterns and rates of recovery of sea otter carcasses following the *Exxon Valdez* oil spill. Marine Mammal Study No. 6. Final Report. U.S. Fish and Wildlife Service. Anchorage, Alaska.

Garrott, R. A., L. L. Eberhardt, and D. M. Burn. 1993. Mortality of sea otters in Prince William Sound following the *Exxon Valdez* oil spill. Marine Mammal Science 9:343–359.

Garshelis, D. L. 1983. Ecology of sea otters in Prince William Sound, Alaska. Ph.D. thesis, University of Minnesota, Minneapolis, Minnesota. 321 p.

Irons, D. B., D. R. Nysewander, and J. L. Trapp. 1988. Prince William Sound sea otter distribution in relation to population growth and habitat type. Unpublished Report, U.S. Fish and Wildlife Service. Anchorage, Alaska. 31 p.

Kenyon, K. W. 1969. The sea otter in the eastern Pacific Ocean. North American Fauna 68. 352 p.

Lensink, C. J. 1962. The history and status of sea otters in Alaska. Ph.D. thesis, Purdue University, Lafayette, Indiana. 186 p.

Monnett, C., and L. M. Rotterman. 1993. The efficacy of the T/V *Exxon Valdez* oil spill rehabilitation program and the possibility of disease introduction into recipient sea otter populations. Page 273, *in Exxon Valdez* oil spill symposium, 2–5 February 1993, Anchorage, Alaska. (Available, Oil Spill Public Information Center, 645 G Street, Anchorage, Alaska 99501.)

 ⋈   ⋈   ⋈

Monnett, C., L. M. Rotterman, C. Stack, and D. Monson. 1990. Postrelease monitoring of radio-instrumented sea otters in Prince William Sound. Pages 400–409, *in* K. Bayha and J. Kormendy (eds.),
Sea otter symposium: Proceedings of a symposium to evaluate the response effort on behalf of sea
otters after the T/V *Exxon Valdez* oil spill into Prince William Sound. Anchorage, Alaska, 17-19
April 1990. U.S. Fish & Wildlife Service Biological Report 90(12).

Monson, D. H. 1993. Age distributions of sea otters dying in Prince William Sound, Alaska, following
the *Exxon Valdez* oil spill. Marine Mammal Study No. 6. Final Report. U.S. Fish and Wildlife
Service. Anchorage, Alaska.

Pitcher, K. W. 1975. Distribution of sea otters, Steller sea lions, and harbor seals in Prince William
Sound, Alaska. Division of Game, Alaska Department of Fish and Game unpublished report. 76 p.

Riedman, M. L., and J. A. Estes. 1990. The sea otter (*Enhydra lutris*): behavior, ecology, and natural
history. U.S. Fish and Wildlife Service, Biological Report 90(14). 126 p.

Satterthwaite, F. E. 1946. An approximate distribution of estimates of variance components. Biomet-
rics 2:110–114.

Spies, R. B. 1993. So why can't science tell us more about the effects of the *Exxon Valdez* oil spill?
Pages 1–5, *in Exxon Valdez* oil spill symposium, 2–5 February 1993, Anchorage, Alaska. (Available, Oil Spill Public Information Center, 645 G Street, Anchorage, Alaska 99501.)

Chapter 5

An Intersection Model for Estimating Sea Otter Mortality along the Kenai Peninsula

James L. Bodkin and Mark S. Udevitz

INTRODUCTION

One of the primary objectives of state and federal resource agencies following the *Exxon Valdez* oil spill (EVOS) was to estimate the mortality of marine birds and mammals. The recovery of 781 carcasses after the spill (Doroff et al. 1993) indicated extensive acute mortality of sea otters (*Enhydra lutris*). Causes of mortality included hypothermia resulting from oiled pelage and interstitial pulmonary emphysema, gastric erosion and hemorrhage, hepatic and renal lipidosis, and centrilobular hepatic necrosis (Lipscomb et al. 1993, Chapter 16). Estimating the magnitude of acute sea otter mortality resulting from the spill beyond the number of recovered carcasses was difficult because baseline population data on sea otters throughout the spill area did not exist. This was particularly true of otters along the Kenai Peninsula (Fig. 5-1).

Garrott et al. (1993) and Doroff et al. (1993), using different approaches, estimated the numbers of sea otters killed from acute exposure to oil from the EVOS. However, each approach had significant limitations. Garrott et al. (1993) compared the number of animals in Prince William Sound (PWS) before and after the spill. This method required accurate pre- and postspill population estimates for each affected area. Doroff et al. (1993) assumed that carcasses recovered during the spill represented a proportion of the total mortality. Their method required accurate estimates of the carcass recovery rate for each affected area.

Along the Kenai Peninsula, a comparison of pre- and postspill survey data did not detect a significant loss of sea otters and data on carcass recovery rates were not available. Therefore, we examined a third approach for estimating the loss of sea otters along the Kenai Peninsula based on their exposure to oil and the relation between exposure and sea otter mortality. We developed an intersection model to integrate parameters estimated from three distinct data sets that resulted from

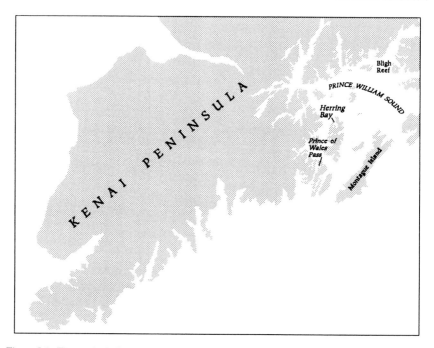

Figure 5-1. The survival of sea otters was estimated at two sites in Prince William Sound, Herring Bay and Prince of Wales Pass. Estimated survival rates were based on the recovery of carcasses and the survival of otters captured live and treated at rehabilitation facilities.

the EVOS: (1) the distribution, amount, and movements of spilled oil; (2) the distribution and abundance of sea otters along the Kenai Peninsula; and (3) the estimates of site-specific sea otter mortality relative to oil exposure from otters captured in PWS for rehabilitation and from collected carcasses. In this chapter, we describe the data sets and provide examples of how they can be used in the model to generate acute loss estimates. We also examine the assumptions required by the model and provide suggestions for improving and applying the model.

METHODS

Oil Movements

The On-Scene Spill Model (OSSM), a generalized computer model developed by the National Oceanic and Atmospheric Administration (Torgrimson 1984), was used to describe the distribution of oil particles as they traveled through PWS and along the Kenai and Alaska Peninsulas. The OSSM model output was iteratively

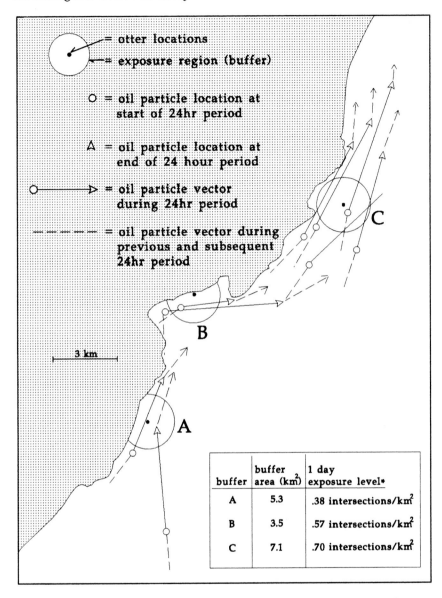

Figure 5-2. A hypothetical example of the calculation of intersection densities of three sea otter locations during one 24-hour period. Arrows indicate direction of oil particle movements.

adjusted based on the observed and computed distribution and movements of oil in a hindcast trajectory analysis (Galt et al. 1991). The hindcast trajectory analysis traced the movement of 10,000 particles of oil, each of which represented about 4169 liters, from their origin at Bligh Reef on 24 March through 23 May 1989 Depending on the thickness of the oil on the water, the sea surface covered by one particle (4169 liters) could range from 1.2 to 85 km^2. (Our model required only the assumption that the area represented by each particle was the same and that area remained constant over time. The validity and consequences of this assumption are discussed below.) We used the location of each modeled particle at 3-hour intervals to represent the distribution of oil over time along the Kenai Peninsula (Fig. 5-2). We defined a continuous path for each of the 10,000 oil particles by the vectors between the locations of each oil particle at 3-hour intervals. As oil evaporated, sank, or became beachcast, the number of oil particles decreased, but each remaining particle continued to represent the constant oil volume of 4169 liters.

Sea Otter Abundance and Distribution

The abundance and distribution of sea otters in nearshore and offshore habitat along the Kenai Peninsula at the time the oil passed through was estimated with data from a helicopter survey (DeGange et al. 1993). The survey combined a strip count of otters along the coastline and line transect counts running perpendicular to the shoreline out to the 50-fathom (approximately 100 m) depth contour. The location of each otter or group of otters was recorded on navigational charts (1:82,000). The counts were corrected for group size and visibility bias to estimate total population size (DeGange et al. 1993). Observed locations of otters at the time of the survey were used to estimate the distribution of the animals relative to exposure for our model.

Sea Otter Mortality

Data for relating exposure levels to oiling and subsequent mortality of otters were collected in two areas of PWS (Fig. 5-1). One area was Herring Bay (HB), (60°28'N, 147°45'W) on the north end of Knight Island, where heavy oiling persisted over time, all captured otters were oiled, and 22 oiled sea otter carcasses were recovered (Table 5-1). The second site included the northeast third of Prince of Wales Pass (PWP) including Iktua Bay (60°06'N, 148°00'W) between Evans and Bainbridge Islands. This area received less oil than Herring Bay and the oil passed through during a shorter period of time. Most captured otters were either lightly oiled or not oiled, and only one carcass was recovered.

During the first 3 weeks of April, attempts were made to capture all otters in these areas irrespective of the presence or degree of oiling (Bodkin and Weltz 1990), and to recover carcasses. Each otter was subjectively classified by one of

≫⊃ ≫⊃ ≫⊃

Table 5-1. Mortality rates of sea otters by degree of oiling in Herring Bay and Prince of Wales Pass, western Prince William Sound, and the Kenai Peninsula. Rates in Herring Bay and the Prince of Wales Pass include oiled carcasses collected between 1 and 15 April 1989 but do not include three carcasses with unknown oiling status. (H=heavy, M=moderate, L=light, N=none). Carcasses with undetermined oiling condition were not included.

	Degree of oiling				
	H	M	L	N	Total
Herring Bay					
Captured live (survived >35 days)	1	2	1	0	4
Captured live (survived ≤ 35 days)	4	2	1	0	7
Number of recovered carcasses	10	10	2	0	22
Total	15	14	4	0	33
Mortality rate ($\bar{x}=0.88$)	.93	.86	.75	nd	.88
Prince of Wales Pass					
Captured live (survived >35 days)	0	2	12	7	21
Captured live (survived ≤ 35 days)	0	1	7	3	11
Number of recovered carcasses	0	0	1	0	1
Total	0	3	20	10	33
Mortality rate ($\bar{x}=0.36$)	nd	.33	.40	.30	.36
Combined Mortality ($\bar{x}=0.62$) (both sites)	.93	.76	.46	.30	.62
All Western Prince William Sound					
Captured live	50	14	44	10	118
Percent of total	.42	.12	.37	.08	
Mortality rate	.76	.50	.31	.70	.58
Kenai Peninsula					
Captured live	3	19	70	32	124
Percent of total	.02	.15	.56	.26	
Mortality rate	.00	.11	.11	.12	0.1

us (JLB) into one of four categories based on the quantity of oil observed on its pelage at the time of capture. The categories were

no oiling—oil not visually or tactically evident on the pelage;

light oiling—oil not easily visible or detectable, or a small proportion (%) of the pelage showed visible oil;

moderate oiling—oiling visible on about 25–75% of the pelage; and

heavy oiling—oil visible on all or nearly all of the pelage.

⋈   ⋈   ⋈

A similar subjective classification was used for all animals brought to rehabilitation centers (Lipscomb et al. 1993). No quantitative means of assessing the degree of oiling were available and in the case of dead otters, postmortem oiling was possible.

While the capture of live animals was under way, the carcasses of dead otters were also collected. The date, location, and degree of oiling of each carcass were recorded. All carcasses collected were relatively fresh. With the exception of five nonoiled animals that were released after capture in Prince of Wales Pass, all captured otters were transported to rehabilitation centers where they were cleaned and held. We assume that the five nonoiled animals survived.

To estimate oiling and mortality rates for otters in each capture area, we used data from all captured live otters and from carcasses. The proportion of animals in each degree of oiling category was estimated by capture area (Table 5-1). Mortality rates were estimated by capture area and by category of oiling (Table 5-1). Pups born at the rehabilitation facilities, otters with an undetermined oiling status, and otters with obvious non-oil-related pathology (e.g., gunshot wounds) (Lipscomb et al. 1993) were excluded from the calculations of oiling and mortality rates. We assumed that sea otters that were able to survive more than 35 days in captivity did not die as an immediate result of the spill.

Oil Exposure

To estimate exposure (the amount and persistence of oil), an exposure region was defined for each otter or group of otters as a circle (area=7.1 km^2 with a 1.5-km radius) centered at the otter's observed location during the helicopter survey (Fig. 5-2). This radius represented the average distance sea otters moved between successive radio relocations recorded between 18 and 36 hours apart in California (Ralls et al. 1988). These data included movements of adult and subadult male and female sea otters ($n=38$).

Exposure to oil was estimated at each location of a Kenai Peninsula otter by summing the number of oil particles that were present in an exposure region during each 24-hour interval, summing over the time period of the spill, and dividing by the area of the exposure region (exposure regions were usually less than 7.1 km^2 because most otters were observed less than 1.5 km from a shoreline; Fig. 5-2, Table 5-2). For example, 10 particles intersecting one complete exposure region of 7.1 km^2 in 1 day would result in an exposure level of 1.4 intersections/km^2. The same exposure level would be obtained from one particle remaining inside that same exposure region for 10 days. (This method of estimating exposure weights quantity and persistence of oil in an area equally. Additional information on the effects of oil on otters and their response to exposure may suggest improved measures of exposure.) The proportion of otters observed at each location was used to estimate the proportion of the total Kenai Peninsula population with that location's level of

⋙   ⋙   ⋙

Table 5-2. Estimated exposure of discrete hypothetical sea otter exposure regions (3.0 km diameter) in western Prince William Sound and at 131 observed locations of otters along the Kenai Peninsula 24 March - 23 May 1989.

Exposure area	Number of exposure regions	Mean exposure[a]	SE	Max	Min
Herring Bay	10	226	18	316	158
Prince of Wales Pass	10	51	12	101	13
Kenai Peninsula	131	6	0.7	32	0.16

[a] Intersections/km^2.

exposure. The range of exposure levels in HB and PWP was estimated by calculating the exposure in intersections/km^2 for 10 randomly distributed exposure regions of 1.5-km radius in each of the two capture areas.

Examples of Model-Generated Loss Estimates

The mortality data from HB and PWP provide only two independent observations for estimating the relation between exposure and mortality. This sample size limited consideration to only the simplest functional forms for the relation. We used two different sets of assumptions about the functional form to derive two different relations between exposure and mortality. Because of the limited amount of data and problems with the type of data that could be obtained from rehabilitation centers (discussed below), neither of these relations is likely to be realistic, but we used these relations as examples to illustrate how the model can generate loss estimates.

Each exposure-mortality relation provided a separate estimate of the mortality rate associated with the estimated exposure level at each location where otters were observed from the helicopter along the Kenai Peninsula. The total number of otters estimated to have received one of the estimated exposure levels was multiplied by the mortality-rate estimate for that exposure level and these numbers were then summed to obtain an estimate of the total number of otters lost. A separate estimate of total loss was obtained with each of the examples of the mortality-exposure relation (Fig. 5-3).

In Example 1, we assumed the mortality-exposure relation could be approximated as a straight line through the mean mortality rates of otters in HB and PWP at the mean exposure levels for those locations (Table 5-3, Fig. 5-3). Symbolically, the relation is given by

$$Y_i = a + bX_i, \qquad X_i \leq X_q$$
$$Y_i = 1.0, \qquad X_i \geq X_q,$$

>< >< ><

Bodkin/Udevitz

Table 5-3. Estimated or assumed mortality rates, proportions of otters in oiling categories, and calculated weighted mortality rates (Example 2) used in the mortality estimate examples. PWP=Prince of Wales Pass and HB=Herring Bay.

Example 1

Location	Mean exposure	Mean mortality rate
PWP	51	0.36
HB	226	0.88

Example 2

Oiling category	Estimated mortality rates (Y)	% of otters in each oil category at PWP (P_{PWP})	Mortality by oil category at PWP $(Y\ P_{PWP})$	% of otters in each oil category at HB (P_{HB})	Mortality by oil category at HB $(Y\ P_{HB})$
H	0.93	0.00	0.00	0.45	0.42
M	0.76	0.09	0.07	0.42	0.32
L	0.46	0.61	0.28	0.12	0.06
N	0.3	0.3	0.09	0	0
			$\Sigma\ 0.44$		$\Sigma\ 0.80$

where Y_i is the mortality rate at location i, X_i is the exposure level at location i (in intersections/km^2), and X_q is the exposure level that resulted in 100% mortality. We estimated the parameters a and b by fitting a straight line through the mean exposure and mortality values at HB and PWP. We estimated X_q by extrapolating the line to the point where mortality (Y) was equal to 1.0.

In Example 2, we used the survival of sea otters from HB and PWP by degree of oiling to approximate the exposure-mortality relation (Tables 5-1 and 5-3). The relation was calculated in two steps. First, we estimated the proportion of otters in each oiling category resulting from each level of exposure. We then multiplied these proportions by mortality rates estimated specifically for each oiling category

⋈ ⋈ ⋈

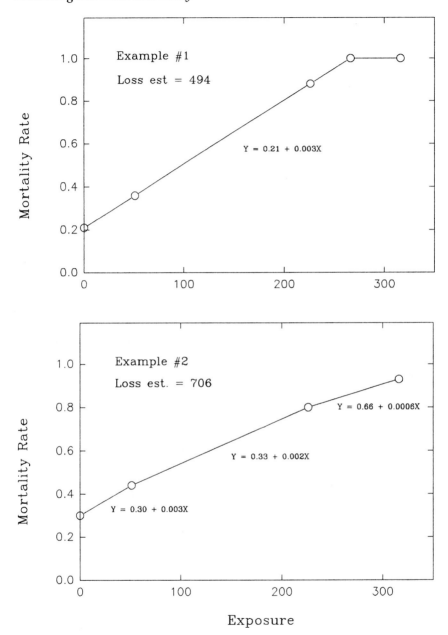

Figure 5-3. Two examples of the application of exposure-dependent mortality rates for the estimation of total acute sea otter mortality along the Kenai Peninsula. Exposure measures the number of oil particles that pass within 1.5 km of an observed otter location between 24 March and 23 May 1989.

and summed the products to obtain an overall mortality rate for each level of exposure. The first step is based on the assumption that the probability of an otter being in any particular degree of oiling category was a piecewise linear function of exposure. We assumed that: (1) none of the otters were oiled at locations without exposure, and (2) there was some exposure level above which all of the otters were heavily oiled. For locations with intermediate levels of exposure, we assumed that

$$P_{ij} = a_{j1} + b_{j1}X_i, \qquad 0 \leq X_i \leq x_{(PWP)}, \qquad j=1,\cdots,4$$

$$P_{ij} = a_{j2} + b_{j2}X_i, \qquad x_{(PWP)} \leq X_i \leq x_{(HB)}, \qquad j=1,\cdots,4$$

$$P_{ij} = a_{j3} + b_{j3}X_i, \qquad x_{(HB)} \leq X_i \leq X_H, \qquad j=1,\cdots,4,$$

where j indexes the oiling categories (1=none, 2=light, 3=moderate, 4=heavy), P_{ij} is the probability of an otter at location i (PWP or HB) being in category j, $x_{(PWP)}$ is the mean exposure level at PWP, $x_{(HB)}$ is the mean exposure level at HB, and X_H is the exposure level that resulted in all otters being heavily oiled. We estimated the parameters a_{jk} and b_{jk}, j=1,4, k=1,3, by fitting straight lines between the points $(0, P_{0,j})$, $(x_{(PWP)}, P_{(PWP),j})$, $(x_{(HB)}, P_{(HB),j})$, and $(X_H, P_{H,j})$, separately for each j=1,4. $P_{(PWP),j}$ and $P_{(HB),j}$, j=1,4 were estimated by the proportion of otters in category j from PWP and HB, respectively. By assumption, $P_{0,1}=1$; $P_{0,j}=0$, j=2,4; $P_{H,j}=0$, j=1,3; and $P_{H,4}=1$. We estimated X_H by extending the line through $(x_{(PWP)}, P_{(PWP),4})$ and $(x_{(HB)}, P_{(HB),4})$ to the point where $P_{i,4}=1$. In the second step, we estimated overall mortality rates as weighted sums specific for each oiling category; weights were equal to the probabilities for each of the oiling categories. Thus, the mortality functions had the form

$$Y_i = y_1 P_{i1} + y_2 P_{i2} + y_3 P_{i3} + y_4 P_{i4},$$

where y_j, j=1,4, are the category-specific mortality rates. We estimated each of the y_j, j=1,4, by the mortality rates for the respective categories in the combined HB and PWP data sets.

RESULTS

Oil Movements

The OSSM model indicated that oil first entered the waters along the Kenai Peninsula on about 30 March 1989. The quantity of oil leaving PWS and entering the Kenai Peninsula diminished through late April. According to Galt et al. (1991), about 25% of the spilled oil left PWS and traveled along the Kenai Peninsula. Local

physiography and climatology resulted in incomplete coverage of oil on Kenai Peninsula beaches; the heaviest oiling occuring along prominent headlands.

Sea Otter Abundance and Distribution

A total of 351 groups totaling 1114 sea otters were detected from helicopters during surveys along coastal and offshore transects along the Kenai Peninsula during April 1989. Ninety-seven percent of the individual otter sightings were detected on the coastal transects. Based on these counts, DeGange et al. (1993) estimated a population size of 1275 sea otters (SE=26) in 778 km^2 of coastal habitat and 1055 sea otters (SE=215) in 3353 km^2 of offshore habitat. The estimated total number of sea otters on the Kenai Peninsula was 2330 (SE=217).

Sea Otter Oiling and Mortality

We captured or handled 43 live sea otters, 11 from HB and 32 from PWP, and recovered 23 carcasses, 22 from HB and 1 from PWP (Table 5-1). Rates of oiling, degree of oiling, and estimated mortality rates at HB and PWP, and comparisons between sea otters captured throughout western PWS and at the Kenai Peninsula are presented in Table 5-1. The proportion of heavily and moderately oiled sea otters captured in HB was 82%, whereas only 14% of the otters captured in PWP were in these categories. Twenty-two carcasses were recovered from HB (91% of which were heavily or moderately oiled) but only one carcass was recovered from PWP and it was lightly oiled. The mean mortality rates were 0.88 (29 of 33) at HB and 0.36 (12 of 33) at PWP. By the end of April 1993 only two live otters could be found in HB, at which time capture efforts were discontinued.

Oil Exposure

OSSM oil particles intersected 131 of the 351 sea otter exposure regions along the Kenai Peninsula on one or more days between 24 March and 23 May 1989 (Table 5-2). These 131 exposure regions represented an estimated 1211 of the estimated 2330 (52%) sea otters along the Kenai Peninsula. Potential exposure levels of this group averaged 6.1 intersections/km^2 (SE=0.75; range=0.17–32). Mean exposure levels at HB and PWP were 226 and 51 intersections/km^2, respectively (Table 5-2). Because 75% of the oil did not leave PWS and the oil that moved along the Kenai Peninsula tended to be offshore, we did not obtain the high levels of oil exposure at otter locations along the Kenai Peninsula that were measured in PWS. However, heavily and moderately oiled animals were captured along the Kenai Peninsula (Table 5-1).

⋙ ⋙ ⋙

Examples of Model-Generated Loss Estimates

Estimated mortality rates in Example 1 ranged from 0.21 at an exposure level of $X_i=0$ intersections/km^2 to 1.0 at exposure levels greater than or equal to $X_i=266$ intersections/km^2 (Fig. 5-3). At the Kenai Peninsula, estimated mortality rates attained a maximum value of 0.30 at the maximum estimated exposure of $X_i=32$ intersections/km^2. This example generated a total mortality estimate of 494 otters at the Kenai Peninsula.

In Example 2, estimated mortality rates ranged from 0.30 at $X_i=0$ to 0.93 at $X_i \geq X_H = 316$ (Fig. 5-3). Although the mortality functions of Examples 1 and 2 crossed, estimated rates were higher in Example 2 than in Example 1 over the full range of the estimated exposure levels at the Kenai Peninsula. The maximum mortality rate estimated in Example 2 at the Kenai Peninsula was 0.39 at $X_i=32$. Example 2 generated a total mortality estimate of 706 otters along the Kenai Peninsula.

DISCUSSION

Several assumptions were required for the use of the data sets with the intersection model in addition to the assumptions of the model itself. The most important data assumptions were that: (1) the OSSM model accurately reflected the spatial and temporal distribution of spilled oil; (2) estimates of sea otter distribution and abundance from the helicopter survey were unbiased; (3) sea otter collection methods at the two capture sites were not selective; and (4) mortality rates of captured otters were not affected by the capture or the rehabilitation process.

Four key assumptions of the intersection model were (1) the relation between exposure to oil and mortality as measured by intersections/km^2 can be approximated by a piecewise linear function; (2) the relative exposure level measured in the region around each observed otter location reflected the relative exposure of those otters to oil; (3) the relation between exposure and mortality at the capture sites in PWS was the same as the relation at the Kenai Peninsula; and (4) the distribution and abundance of sea otters remained constant throughout the period of potential exposure.

Assumptions for Data

We did not have information to address all of the assumptions associated with estimates of oil and sea otter abundance and distribution. Bodkin and Weltz (1990) provided data that support the assumption of unbiased capture of live otters in HB and PWP. Most potential biases in carcass retrieval would result in fewer rather than more retrieved carcasses. This would have the potential effect of reducing

><> ><> ><>

mortality rates and, thereby, loss estimates. However, inclusion of oiled but non-spill-related carcasses would increase mortality rates and loss estimates. Because the carcasses collected in HB and PWP were fresh, it is most likely that mortality occurred at or near these areas and was, in fact, spill related.

Our estimates of mortality-exposure relations from the PWP and HB data required the assumption that rehabilitation did not affect survival rates. If this assumption was valid, we would expect little or no mortality among nonoiled otters. However, mortality rates of lightly oiled and nonoiled sea otters in PWS were 0.40 and 0.30, respectively, suggesting that the rehabilitation treatment had a negative effect on nonoiled and lightly oiled sea otters. Also, it may be reasonable to assume that all heavily and moderately oiled otters would die (Costa and Kooyman 1982) and that any survival among the animals in these oiling categories could be attributed to rehabilitation. Of the 86 heavily or moderately oiled sea otters captured in either PWS or the Kenai Peninsula (Table 5-1), 42 survived >35 days, resulting in a mortality rate of 0.55. This suggests that rehabilitation may have had a positive effect on the survival rate of heavily and moderately oiled sea otters. It would be a straightforward procedure to use our approach to generate loss estimates based on the assumption that there was no mortality of nonoiled or nonexposed otters, but without additional data, there is no basis for determining the magnitude of the corresponding adjustment required for the heavily and moderately oiled otters.

Assumptions for the Model

Our data indicated that mortality and the proportions of heavily oiled otters were increasing functions of exposure, but we were unable to evaluate the shape of these functions with the two data points available from HB and PWP. In order for our estimated relative exposure levels to have strictly corresponded to the actual relative exposure of otters to oil, the surface area represented by each OSSM oil particle would have to have been the same for all particles and would have had to remain constant over time. Also, sea otter movements and responses to potential exposure would have to have been the same for all otters. In fact, the surface area represented by an oil particle is likely to have changed with time. We did not have information to assess how sea otter behavior may have varied over the area affected by the spill. Failure to meet our assumption of similarity in the exposure-mortality function between PWS and the Kenai Peninsula may have biased the loss estimates to the extent that the exposure-mortality function varied over time. Potential changes may have resulted from cumulative effects of exposure to oil or changes in the quality (e.g., weathering) of spilled oil over time in addition to the effects of the change in surface area represented by each particle. The model also required the assumption that sea otters did not move in response to oil. We had no data to evaluate the validity or the effects of violating this assumption. During the capture of otters in PWS, we observed otters in apparently unoiled areas in close proximity to oiled areas; whether these observations were of avoidance or of good fortune is unknown.

Loss Estimates

Our examples provided loss estimates lower than the estimated loss of 868 sea otters provided by Doroff et al. (1993) and higher than the postspill difference of 183 sea otters provided by DeGange et al. (1993). Both of our examples resulted in loss estimates that were greater than the 167 carcasses recovered from the Kenai Peninsula. The mortality-exposure relationships used in our examples are unrealistic because they assigned nonzero mortality rates to nonoiled otters. However, reducing the mortality rate of nonoiled otters in the model would result in unrealistically low loss estimates. This suggests that mortality rates for lightly oiled otters are likely to have been greater than assumed in our examples.

CONCLUSIONS

The intersection approach seemed to provide an adequate estimate of relative exposure to oil for a case where suitable data on oil abundance and movement and on the abundance and distribution of sea otters were available. Our examples of loss estimates generated with the model demonstrated its sensitivity to the assumed relation between exposure to oil and mortality rates. The mean exposure values in HB and PWP were 38 to 8.5 times greater than those at the Kenai Peninsula (Table 5-2). Refinement of the model should include better definition of the exposure-mortality relation, particularly at low exposure levels. Mortality rates of otters subjected to rehabilitation should not be used to estimate this relation unless the effects of the interaction between rehabilitation and degree of oiling on mortality can be quantified. Further development of the model should also focus on obtaining an estimate of the precision of loss estimates. It is likely that bootstrap procedures (Efron 1982) will be required because of the complex nature of the loss estimator.

The need for a theoretical approach to estimate acute losses of sea otters arose because complete baseline data on sea otter abundance before the spill were lacking. This limited our ability to quantify the acute mortality of sea otters from the spill beyond the number of recovered carcasses. Direct estimates of losses from well designed pre- and postspill surveys would be preferable to those provided by models such as the intersection. Therefore, we recommend development and implementation of rigorous survey protocols in areas shared by sea otters and oil recovery, storage, and transportation. The most effective uses of the intersection model may be predictive modeling of the effects of future spills and quantifying exposure in a spill area.

≫ ≫ ≫

ACKNOWLEDGMENTS

This work was supported by the Alaska Fish and Wildlife Research Center and the *Exxon Valdez* Oil Spill Trustee Council. B. Ballachey, D. Bruden, L. Pank, and C. Robbins made significant contributions to the project. B. Ballachey, L. Holland-Bartels, K. Oakley, E. Rockwell, B. Rothschild, and four anonymous reviewers provided helpful critism that resulted in improvements to the manuscript.

REFERENCES

Bodkin, J. L., and F. Weltz. 1990. A summary and evaluation of sea otter capture operations in response to the *Exxon Valdez* oil spill, Prince William Sound Alaska. Pages 61–69, *in* K. Bayha and J. Kormendy (eds.), Sea otter symposium: Proceedings of a symposium to evaluate the response effort on behalf of sea otters after the T/V *Exxon Valdez* oil spill into Prince William Sound, Anchorage, Alaska, 17–19 April 1990. U.S. Fish and Wildlife Service, Biological Report 90(12).

Costa, D. P., and G. L. Kooyman. 1982. Oxygen consumption, thermoregulation, and the effect of fur oiling and washing on the sea otter, *Enhydra lutris*. Canadian Journal of Zoology 60:2761–2767.

DeGange, A. R., D. C. Douglas, D. H. Monson, and C. Robbins. 1993. Surveys of sea otters in the Gulf of Alaska in response to effects of the *Exxon Valdez* oil spill. Marine Mammal study 6, final report. U.S. Fish and Wildlife Service, Anchorage, Alaska.

Doroff, A. M., A. R. DeGange, C. Lensink, B. E. Ballachey, J. L. Bodkin, and D. Bruden. 1993. Recovery of sea otter carcasses following the *Exxon Valdez* oil spill. Pages 285–288, *in Exxon Valdez* oil spill symposium, 2-5 February 1993, Anchorage, Alaska. (Available, Oil Spill Public Information Center, 645 G Street, Anchorage, Alaska 99501.)

Efron, B. 1982. The jackknife, the bootstrap and other resampling plans. Society for Industrial and Applied Mathematics, Philadelphia. 92 p.

Galt, J. A., G. Y. Watabayashi, D. L. Payton, and J. C. Peterson. 1991. Trajectory analysis for the *Exxon Valdez*: Hindcast study. Pages 629–634, *in* Proceedings 1991 International Oil Spill Conference, 4–7 March 1991, San Diego, California, American Petroleum Institute, Washington, D.C.

Garrott, R. A., L. L. Eberhardt, and D. M. Burn. 1993. Mortality of sea otters in Prince William Sound following the *Exxon Valdez* oil spill. Marine Mammal Science 9(4):343–3359.

Lipscomb, T. P., R. K. Harris, A. H. Rebar, B. E. Ballachey, and R. J. Haebler. 1993. Pathological studies of sea otters. *In* Assessment of the magnitude, extent and duration of oil spill impacts on sea otter populations in Alaska. Final Report, U.S. Fish and Wildlife Service, Alaska Fish and Wildlife Research Center, Anchorage, Alaska.

Ralls, K., T. Eagle, and D. B. Siniff. 1988. Movement patterns and spatial use of California sea otters. *In* D. B. Siniff and K. Ralls (eds.), Population status of California sea otters. U.S. Fish and Wildlife Service, Minerals Management Service, Contract No. 14-12-001-30033.

Torgrimson, G. M. 1984. The on-scene spill model. U.S. Department of Commerce, NOAA Technical Memorandum NOS-OMA-12. Rockville, Maryland. 100 p.

Chapter 6

Impacts on Distribution, Abundance, and Productivity of Harbor Seals

Kathryn J. Frost, Lloyd F. Lowry, Elizabeth H. Sinclair,
Jay Ver Hoef, and Dennis C. McAllister

INTRODUCTION

Oil that spilled from the T/V *Exxon Valdez* after it struck Bligh Reef impacted some of the largest harbor seal (*Phoca vitulina richardsi*) haulout sites in Prince William Sound (PWS). The potential effects of the spill on harbor seals were of particular concern because their numbers had declined substantially since the 1970s (Pitcher 1986, 1989), and because newborn pups and their mothers were likely to be exposed to oil contamination and disturbance from spill response and clean-up activities.

As part of the Natural Resources Damage Assessment program, a study was designed to investigate and quantify, if possible, the effects of oil and the disturbance associated with cleanup on the distribution, abundance, and health of harbor seals. Two of the objectives were to measure the effects of the *Exxon Valdez* oil spill (EVOS) on the abundance of harbor seals in PWS and to determine whether production and survival of pups were affected.

It is not practical to determine the exact number of harbor seals inhabiting an area such as PWS, which has over 4800 km of coastline consisting of many bays, fiords, islands, and offshore rocks. For this reason, the Alaska Department of Fish and Game (ADF&G) established an aerial survey trend-count route consisting of 25 selected haulout sites (trend sites) in eastern and central PWS, and made repetitive counts of seals on those sites during the annual molt in 1983, 1984, and 1988 (Calkins and Pitcher 1984; Pitcher 1986, 1989). Because some of those sites were oiled by the EVOS and others were not, counts of molting seals at those locations after the spill could be compared with historical data to provide a measure of impact on overall abundance. Additional counts were made at the trend sites during pupping to determine whether the spill had measurable impact on pup production.

METHODS

Aerial surveys were flown in PWS in 1989–1992 along the previously established trend-count route. The route included seven sites that were contaminated during the EVOS and 18 unoiled sites that were north, east, and south of the primary area impacted by oil (Fig. 6-1). Sites were classified as oiled or unoiled based on our direct observations (Lowry et al., Chapter 12), and maps provided by the Alaska Department of Environmental Conservation. Sites classified as oiled were those that were significantly and persistently oiled as a result of the EVOS.

We surveyed haulout sites during both pupping and molting, the periods when maximum numbers of seals haul out (Pitcher and Calkins 1979; Calambokidis et al. 1987). Our methods were identical to those used in previous trend-count surveys (Pitcher 1986, 1989). Surveys were flown within 2 hours before or after daylight low tides. Haulout sites were surveyed from a single engine fixed-wing aircraft (Cessna 180 or 185) at an altitude of 150–300 m. Visual counts of seals were made at each site, usually using 7-power binoculars. For groups of 40 or more seals, color photographs were taken using a hand-held 35-mm camera with a 70- to 210-mm zoom lens and high-speed slide film (ASA 400). Seals were counted from projected images. During June surveys, separate counts were made of pups and nonpups. Pups were identified by their relatively smaller size and orientation to their mothers.

To estimate variability in seal counts, we attempted to obtain 7–10 counts of seals at each site within a survey period. In practice, the number of counts was almost always less because of inclement weather, timing of low tides, and the length of the pupping and molting periods. If a particular count was considered invalid (e.g., if the haulout site was empty and a boat was observed nearby), it was excluded from the analysis.

Data were initially analyzed using the simple mean, the trimean (Rosenberger and Gasko 1983), and the maximum count per site per year. The three types of analyses yielded similar results, but simulations indicated that comparisons using the simple mean had somewhat more statistical power than comparisons using the trimean or maximum value (Frost and Lowry 1993). Therefore, the replicate counts of seals at each site in each year were averaged and the mean counts were used in all further analyses.

We tried two general types of statistical techniques for comparing changes in seal counts at oiled and unoiled sites over time; analysis of variance (ANOVA) (Winer 1971; Neter and Wasserman 1974) and categorical models (Agresti 1990). Although significant differences were found between oiled and unoiled sites using

⋈ ⋈ ⋈

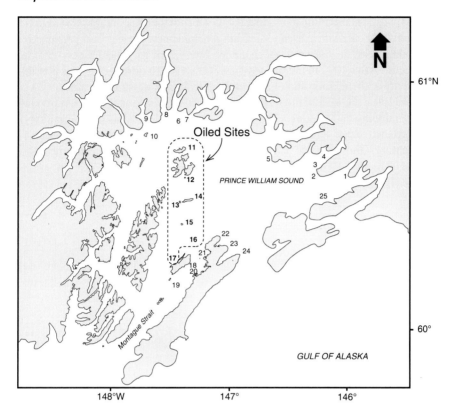

Figure 6-1. Map of Prince William Sound showing locations of trend-count sites and the sites that were affected by the *Exxon Valdez* oil spill (bold numbers). (1-Sheep Bay, 2-Gravina Island, 3-Gravina Rocks, 4-Olsen Bay, 5-Porcupine Point, 6-Fairmount Island, 7-Payday, Olsen Island, 9-Point Pellew, 10-Little Axel Lind Island, 11-Storey Island, 12-Agnes Island, 13-Applegate Rocks, 17-Green Island, 18-Channel Island, 19-Little Green Island, 20-Port Chalmers, 21-Stockdale Harbor, 22-Montague Point, 23-Rocky Bay, 24-Schooner Rocks, 25-Canoe Passage.)

ANOVA (Frost 1991), this technique was rejected because it weights data from each site equally, which was inappropriate since the actual counts differed greatly among sites. Categorical models are more suited for testing hypotheses that compare proportional changes among sample groups because they give more weight to sites with larger numbers of animals. However, these models assume independent binary data or Poisson-distributed count data, and our data were mean counts per site. We therefore used bootstrapping (Efron and Tibshirani 1986) to account for the variability in the mean count per site, as well as variability among sites. The bootstrap method resampled with replacement from the actual daily counts at each haulout site to produce a data set with the same sample size (number of counts) for each site in each year. This generated new mean counts per year that were then used in the categorical model framework to generate new contrast estimates. After 1000 iterations, yielding 1000 contrast estimates, it was possible to determine the relationship of zero to the distribution of contrast values.

A loglinear categorical model (Agresti 1990) was used to analyze nonpup counts during pupping and molting. The log link, $\ln(m)$, is defined as the natural log of the expected mean number of seals per site, and is modeled as a function of explanatory variables, year, and oil group. A model linking the expected mean seal counts, m, in the ith year and jth oil group to effects for α_i (year) and β_j (oil group) can be written as:

$$\ln(m_{ij}) = \mu + \alpha_i + \beta_j + \alpha\beta_{ij} \tag{1}$$

where μ is the intercept and $\alpha\beta_{ij}$ is an interaction term.

The null hypothesis, that the oiled:unoiled ratio of harbor seal counts was unchanged between prespill years and the year of the EVOS can be written as:

$$H_0: \frac{m_{89\ oil}}{m_{89\ no\ oil}} = \left(\frac{m_{83\ oil}}{m_{83\ no\ oil}} x \frac{m_{84\ oil}}{m_{84\ no\ oil}} x \frac{m_{88\ oil}}{m_{88\ no\ oil}}\right)^{1/3} \tag{2}$$

where the right hand side of the equation is the geometric mean of the ratios for the three prespill count years. Equation (2) is completely additive at the log scale, and the hypothesis can be tested using the contrast:

$$C1: [\ln m_{89\ oil} - 1/3(\ln m_{83\ oil} + \ln m_{84\ oil} + \ln m_{88\ oil})]$$
$$-[\ln m_{89\ no\ oil} - 1/3(\ln m_{83\ no\ oil} + \ln m_{84\ no\ oil} + \ln m_{88\ no\ oil})] \tag{3}$$

In order for (3) to be nonzero, the interaction term in (1) is required. In the categorical model framework, m_{ij} is estimated by:

$$\hat{m}_{ij} = \sum_k Y_{ijk}$$

where Y_{ijk} is the mean count for site k in year i and oil group j. One thousand bootstrap samples of Y_{ijk} determined the distribution of (3). Under the null

⋈ ⋈ ⋈

hypothesis H_0, we would expect the contrast (3) to be zero, indicating no change in the relationship of oiled to unoiled counts. A negative contrast value would indicate a decline had occurred in the oiled areas relative to the unoiled areas in the year of the EVOS.

We also examined the oiled:unoiled ratio of expected mean seal counts for the spill year and postspill years to examine the relative changes in seal numbers in years following the EVOS. Similar to (3), the hypothesis may be tested using the contrast:

$$C2:[\ln m_{89\ oil}-1/3(\ln m_{90\ oil}+\ln m_{91\ oil}+\ln m_{92\ oil})]$$
$$-[\ln m_{89\ no\ oil}-1/3(\ln m_{90\ no\ oil}+\ln m_{91\ no\ oil}+\ln m_{92\ no\ oil})] \quad (4)$$

If seal counts in oiled and unoiled areas changed in the same way during 1989–1992, we would expect (4) to be zero. If a relatively greater increase in numbers occurred in oiled areas, (4) would be negative.

Spill-year and postspill-year comparisons of pup production were conducted using a logit categorical model (Agresti 1990). If π is the expected ratio of mean percentage of pups per site, then the logit link is defined to be

$$\ln\frac{\pi}{1-\pi}\equiv\ln(r),$$

where r is the expected pup:nonpup ratio. Next, r is modeled as a function of explanatory values, year, and oil group. The expected pup:nonpup ratio (r_{ij}) in the ith year and the jth oil group is linked to effects for α_i (year) and β_j (oil group) as,

$$\ln(r_{ij})=\mu+\alpha_i+\beta_j+\alpha\beta_{ij} \quad (5)$$

where μ is the intercept and $\alpha\beta_{ij}$ is the interaction term. The null hypothesis of no change in the oiled:unoiled ratio of harbor seal pup counts, adjusted for the number of nonpups, between the spill year and postspill years is written as:

$$H_0:\frac{r_{89\ oil}}{r_{89\ no\ oil}}=(\frac{r_{90\ oil}}{r_{90\ no\ oil}}x\frac{r_{91\ oil}}{r_{91\ no\ oil}}x\frac{r_{92\ oil}}{r_{92\ no\ oil}})^{1/3} \quad (6)$$

where, as in (2), the right-hand side of (6) is the geometric mean, but in this case it is a ratio of ratios for the last 3 years. Equation (6) is completely additive at the log scale, and is written and tested as the contrast:

$$C3 : [lnr_{89, \ oil} - 1/3(lnr_{90, \ oil} + lnr_{91, \ oil} + lnr_{92, \ oil})]$$
$$- [lnr_{89, \ no \ oil} - 1/3(lnr_{90, \ no \ oil} + lnr_{91, \ no \ oil} + lnr_{92, \ no \ oil})] \qquad (7)$$

In order for (7) to be nonzero, the interaction term in (5) is required. In the categorical model framework, r_{ij} is estimated by:

$$\hat{r}_{ij} = \frac{\displaystyle\sum_k X_{ijk}}{\displaystyle\sum_k Y_{ijk}}$$

where Y_{ijk} is the mean nonpup count and X_{ijk} is the mean pup count for site k in year i and oil group j. One thousand bootstrap samples of X_{ijk} and Y_{ijk} determined the distribution of (7). If no change in birth rate (or neonatal survival) occurred from 1989 to postspill years, (7) would be zero. An increase in birth rate in the oiled area in postspill years would result in a negative contrast value.

The number of seals missing in the oiled trend-count area was estimated by multiplying the expected ratio of seals at oiled and unoiled sites by the observed number of seals at unoiled sites in 1989, and then subtracting from this the observed number of seals at the oiled sites in 1989. The formula used was:

$$\text{missing} = (\frac{\hat{m}_{83, \ oil}}{\hat{m}_{83, \ no \ oil}} x \frac{\hat{m}_{84, \ oil}}{\hat{m}_{84, \ no \ oil}} x \frac{\hat{m}_{88, \ oil}}{\hat{m}_{88, \ no \ oil}})^{1/3} \hat{m}_{89, \ no \ oil} - \hat{m}_{89, \ oil} \qquad (8)$$

Equation (8) follows directly from (2). Likewise, the number of pups missing in 1989 was estimated from the ratios of pups to nonpups in oiled areas by,

$$\text{missing} = (\frac{\hat{m}_{90, \ pup}}{\hat{m}_{90, \ nonpup}} x \frac{\hat{m}_{91, \ pup}}{\hat{m}_{91, \ nonpup}} x \frac{\hat{m}_{92, \ pup}}{\hat{m}_{92, \ nonpup}})^{1/3} \hat{m}_{89, \ nonpup} - \hat{m}_{89, \ pup} \qquad (9)$$

Bootstrapping was used to estimate 95% confidence intervals for the numbers of missing seals and pups.

⋙   ⋙   ⋙

Table 6-1. Mean counts and annual percent change in numbers for oiled and unoiled sample groups based on 25 trend-count haulout sites in Prince William Sound, surveyed in August-September 1984, 1988, and 1989-1992. Percent change shown for 1988 is the average annual rate of decline from 1984-1988. Data for 1983-1988 are from Pitcher (1986, 1989, and unpublished). (--- = no data)

| | Oiling Category | | | | | |
| | Oiled (n=7) | | Unoiled (n=18) | | All (n=25) | |
Year	mean	annual % change	mean	annual % change	mean	annual % change
1983	743	---	868	---	1611	---
1984	675	-9	1121	+29	1796	+11
1988	418	-11	639	-13	1057	-12
1989	239	-43	568	-11	807	-24
1990	276	+15	504	-11	780	-3
1991	290	+5	631	+25	921	+18
1992	276	-5	493	-22	769	-16

RESULTS

Molt-Period Counts

Molt-period aerial surveys were conducted during 3–16 September 1989, 28 August–11 September 1990, 22 August–1 September 1991, and 27 August–6 September 1992. In addition, data were available from molt-period surveys conducted by ADF&G at these same sites in August-September 1983, 1984, and 1988 (Pitcher 1986, 1989). In most years 8–10 counts were made at each site (Appendix 6-A).

Between 1984 and 1988 the proportional decline at the two groups of sites was similar, with an 11% average annual decline at oiled sites and a 13% decline at unoiled sites. However, between 1988 and 1989 the average number of seals at oiled sites declined 43%, compared to 11% at unoiled sites (Table 6-1). The contrast value from equation (3) was negative, indicating a significant decline had occurred in the oiled area relative to the unoiled area in the year of the EVOS (C1=-0.45, P=0.002; Fig. 6-2A).

Overall, postspill molt-period counts of seals at trend-count sites have shown relatively little change since 1989. In 1990 there was an increase in the number of seals at oiled sites and a decline at unoiled sites (Table 6-1). Counts in 1991 increased in both sample groups, but were offset by declines of similar magnitude in 1992. Weather was exceptionally good in 1991, perhaps resulting in a somewhat higher proportion of seals being hauled out than in other years. In 1992 counts at

✂   ✂   ✂

Table 6-2. Mean counts of harbor seals and harbor seal pups in oiled and unoiled sample groups based on 25 trend-count haulout sites in Prince William Sound surveyed during June 1989-1992. Data for 1992 are from the National Marine Mammal Laboratory (unpublished).

	Oiled (n=7)			Unoiled (n=18)			Combined (n=25)		
	non-pups	pups	pups/100 non-pups	non-pups	pups	pups/100 non-pups	non-pups	pups	pups/100 non-pups
1989	279	72	26	471	98	21	750	170	23
1990	296	99	34	430	72	17	726	171	24
1991	317	111	35	302	56	18	619	167	27
1992	248	92	37	268	55	21	516	147	28

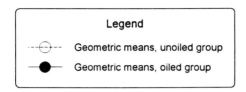

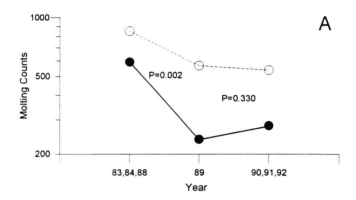

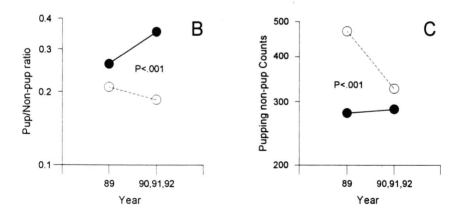

Figure 6-2. Graphs showing the results of categorical analyses of seal counts at oiled and unoiled sites in Prince William Sound. Figure A shows comparison of the geometric means of counts of all seals made during the molt. Figure B shows comparisons of geometric means of the ratio of pups to nonpups counted during pupping. Figure C shows the geometric means of counts of nonpup seals made during pupping.

oiled sites were 34% lower than they were before the spill in 1988, while counts at unoiled sites were 23% lower. Statistical analysis of spill-year and postspill data showed that the EVOS-related reduction in counts at oiled sites was confined to the year of the spill, and that changes in seal numbers in oiled areas as compared to unoiled areas were similar in 1990–1992 (C2=-0.16, P=0.330; Fig. 6-2A).

Pupping-Period Counts

No pupping-period counts of seals were made on the PWS trend-count route prior to 1989. After the EVOS, pupping-period aerial surveys were conducted during 8–27 June 1989, 7–15 June 1990, 6–20 June 1991, and 14–20 June 1992. In 1989–1991, 6–10 counts at each site were suitable for use in the analysis, while in 1992 there were 3–4 counts per site (Appendix 6-B).

Pup production was lower in the oiled area in 1989 than it was in postspill years. In 1989, there were 26 pups/100 nonpups at the oiled sites compared to 34–37 pups/100 nonpups in 1990–1992 (Table 6-2). At unoiled sites, pup production was similar during 1989–1992, ranging from 17 to 21 pups/100 nonpups. Contrast C3 was negative, indicating a significant increase in pup production at oiled sites, compared to unoiled sites, in 1990–1992 (C3=-0.43, P<0.001; Fig. 6-2B).

Mean counts of nonpups indicated a small increase (+6%) at oiled sites between pupping-period surveys in 1989 and 1990 (Table 6-2). Counts of nonpups in the oiled area increased again in 1991, then decreased in 1992, for an overall decrease of 11% between June 1989 and 1992. At unoiled sites, the mean counts declined steadily from 1989 through 1992. In 1992 there were 43% fewer nonpup seals at the unoiled sites than there were in 1989. In the trend-count area as a whole, there were 31% fewer nonpup seals counted in 1992 than in 1989. Statistical analysis indicated that the decline in nonpup seals during pupping has been significantly greater at unoiled sites than at the oiled sites (C2=-0.39, P<0.001; Fig. 6-2C).

Seals Missing After the EVOS

The expected number of seals on oiled sites in the trend-count area in 1989 [from Equation (8)] was 374. When the actual number of seals counted in the oiled area (239) was subtracted from the expected number, this showed that 135 (95% CI=43-209) more seals were missing from the oiled sites in 1989 than would have been expected based on the trend indicated by historical data.

The area included in trend-count surveys, and for which we had prespill data, did not include all harbor seal haulout sites in PWS that were impacted by the EVOS. No systematic aerial survey data were collected in 1989 for oiled haulout sites outside the trend-count area. To estimate the number of seals in these areas, we summed the maximum counts obtained at haulout sites during our small boat operations in May–July 1989 (Lowry et al., Chapter 12). This total (296 seals) is conservative because not all oiled areas were counted, and some areas were counted

✂ ✂ ✂

in inappropriate weather and tide conditions. The number of seals missing in other oiled areas was calculated as:

Missing in other oiled areas of PWS =
Missing$_{oiled\ trend}$ x (Seals$_{oiled\ other\ PWS}$ / Seals$_{oiled\ trend\ 1989}$).

Substituting values from above results in an estimate of 167 seals missing in oiled areas outside the trend-count area. Our estimate of the total number of seals missing in PWS due to the EVOS is the total of those missing in the trend-count area plus those missing outside the trend-count area, or 302 missing seals. Mortality of pups resulting from the EVOS was calculated from Equation (9). When the actual number of pups counted in the oiled area in 1989 (72) was subtracted from the expected number (98), this indicated that there were 26 fewer seal pups (95% CI=8-41) than expected based on postspill pupping rates.

DISCUSSION

Mortality of Seals Due to the EVOS

Relatively few harbor seal carcasses were found following the EVOS despite extensive search efforts by scientists and other people working in the area (Zimmerman et al., Chapter 2; Spraker et al., Chapter 17). This was not surprising, since dead seals usually do not float. Animals that died at sea probably sank and carcasses on haulout sites would have been washed off by daily 2- to 3-m tides. For these reasons, we did not consider the number of dead seals found in oiled areas after the EVOS to be a useful indication of the number of seals that died because of the spill. We therefore used a method that compared prespill counts to postspill counts to measure the spill's impact on harbor seals. A similar method was used to estimate mortality resulting from an outbreak of phocine distemper in the North Sea in 1988 when thousands of seals died (Thompson and Miller 1992). Based on their studies, Thompson and Miller (1992) concluded that observing changes in the number of seals surviving provided a more reliable method for estimating mortality than collections of carcasses.

Data from PWS aerial surveys conducted during the molt in 1983, 1984, and 1989–1992 clearly indicate that counts of harbor seals decreased more in oiled areas than in unoiled areas in the year of the EVOS (Fig. 6-3). In 1989, there were fewer seals present at the seven oiled sites along the trend-count route than were present at those sites in 1988, and fewer than would have been expected based on prespill data that showed an ongoing decline of similar magnitude in oiled and unoiled areas. The 135 missing seals represent an estimated EVOS-related loss of approximately 36% of the seals using the oiled haulout sites.

⋈ ⋈ ⋈

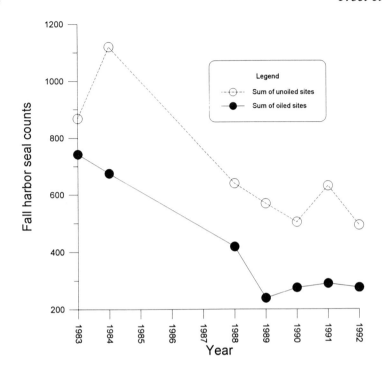

Figure 6-3. The overall trend in mean counts of harbor seals in Prince William Sound based on aerial surveys conducted during the molt in 1983 through 1992.

Our estimate of the number of seals missing in oiled areas could be biased upward if it was more difficult to see and count oiled seals in oiled areas during molt surveys in 1989. We do not think this was a problem for two reasons. First, we did not have any difficulty counting oiled seals in oiled areas during June 1989 pupping surveys. From survey altitudes oiled seals looked like normal dark-phase seals, and they were readily distinguishable from the substrate and algae on which they were lying. Second, and most important, small boat observations conducted on 4 September 1989 showed that most nonpup seals in oiled areas had molted and appeared unoiled (Lowry et al., Chapter 12). Therefore, there should have been very few oiled seals to be counted on the haulout sites when aerial surveys were conducted during 3–16 September 1989.

The slight increase in the number of seals at oiled trend-count sites between 1989 and 1990 (Fig. 6-3) could be interpreted as suggesting that some short-term displacement of seals occurred in 1989. The trend-count route we surveyed included major seal haulout sites to the north, east, and south of the sites affected by the spill (Fig. 6-1). Many of those haulout sites were a considerable distance away from the area affected by EVOS response and clean-up activities. If displacement had occurred, the number of seals in the unoiled part of the trend-count route should have increased in 1989. Instead, seal counts at unoiled sites declined by 11%, which was similar to the rate of decline for 1984-1988. In any event, the mean counts at oiled sites in 1990 were still 34% lower than the 1988 prespill counts, as compared to a 21% difference at unoiled sites (Table 6-1).

Several additional lines of evidence suggest that seals did not move away from oiled areas in the months immediately following the EVOS. Oiled seals were very lethargic and reluctant to enter the water, and they showed little response to human presence (Lowry et al., Chapter 12). We conducted small boat observations in May–June 1989 throughout most of PWS, and although we examined hundreds of seals we saw no oiled seals at unoiled sites that were more than a few kilometers away from oiled sites. Heavily oiled areas that were subjected to intense human activity were not abandoned by seals. Counts made in one such area, Herring Bay, were very similar in mid-May and mid-September. When clean-up crews were working on Applegate Rocks in May, many seals remained in the area on offshore rocks and in the water nearby (Lowry, unpublished observations).

Available data suggest that in normal circumstances the short-term movements of harbor seals in the Gulf of Alaska are mostly local. Radiotagged seals at Tugidak Island showed considerable fidelity to a particular haulout site, and movements to other haulout sites were usually to the nearest adjacent location (Pitcher and McAllister 1981). Preliminary analysis of data from studies that attached satellite-linked transmitters to seals at oiled sites in central PWS suggests that seals had a strong tendency to return to the tagging site after making trips to sea (Frost and Lowry, unpublished data).

We know of no information that indicates that seals moved out of oiled parts of PWS after the EVOS. We therefore conclude that most of the seals that were missing at oiled sites had died.

Our calculations of nonpup mortality based on prespill and postspill data assume that all EVOS-caused mortality took place before the 1989 molt-period survey. We think this assumption is valid because histopathologic findings indicated that most mortality would probably have occurred within the first 1–2 months after the spill (Spraker et al., Chapter 17), and comparisons of molting counts in 1989 with 1990–1992 showed no significant difference in the change in numbers at oiled and unoiled sites. If additional EVOS-caused mortality occurred after mid-September 1989, it was not detected by our surveys.

Impacts of the EVOS on Pup Production and Survival

Because pupping-period surveys were not conducted in PWS prior to 1989, our evaluation of the impact of the EVOS on pup production was based on a comparison of 1989 data with data from subsequent years. Since the proportion of pups varies considerably by site, it cannot be assumed that production of pups should be the same in oiled and unoiled areas. However, we think it is reasonable to assume that within a given area productivity should be generally similar from year to year, and any environmentally caused changes would affect nearby haulout sites similarly. Our surveys showed that pup production was similar at unoiled trend-count sites in all years. The relatively low proportion of pups in oiled areas in 1989, plus the 12 dead fetuses and pups collected from oiled areas (Spraker et al., Chapter 17), strongly suggest that the EVOS did cause the deaths of harbor seal pups. Twenty-six of the 98 pups that should have been born at oiled trend sites were missing, which equals an EVOS-related pup mortality of 26%.

A similar level of pup mortality was estimated for Herring Bay based on recovered carcasses (Frost and Lowry 1993). Herring Bay is an important pupping location and was one of the most heavily oiled areas in PWS. In 1989 our observations indicated that a minimum of 31 seal pups were born there. Between May and July five pups were found dead and two other prematurely born pups died in captivity. While we cannot be sure that we counted all the pups born or found all those that died, these figures suggest that about 23% of the pups that were born in Herring Bay in 1989 died within the first 2 months of life.

Natural preweaning mortality rates for harbor seals have been estimated to be 12% on Sable Island, Nova Scotia (Boulva 1971) and 7% at Double Point, California (Allen 1980). Steiger et al. (1989) reported 12–18% neonatal mortality for sites in the inland waters of Washington and considered these rates to be unusually high. Bigg (1969) estimated mortality in British Columbia to be about 21% for the entire first year of life. There are no baseline data on harbor seal pup mortality for PWS. However, the 23–26% mortality observed in Herring Bay and estimated for oiled trend-count areas appears to be unusually high, and is probably an indication of the impact of the EVOS on early survival of pups.

Current Status of Harbor Seals in the PWS Area

In the mid-1970s harbor seals were abundant in PWS and the Gulf of Alaska and the population was considered healthy (Pitcher and Calkins 1979). Approximately 4000 seals were counted in PWS during June 1973 (Pitcher and Vania 1973), and a minimum population of 13,000 was later estimated from harvest records based on bounty payments (Calkins et al. 1975). The next complete surveys of PWS were in 1991 when a minimum of 2500 seals was counted during molt-period surveys (NMFS, unpublished data).

The first trend-count surveys in PWS were flown during the molt period in 1983. While these counts are useful as a general reference point, they were not of the same quality as later surveys (J. Lewis, ADF&G, Anchorage, Alaska, personal communication). Trend-count surveys in 1984 and 1988–1992 (Pitcher 1986, 1989; Table 6-3) indicated a 41% decline in mean counts from 1984 to 1988. By 1992,

Table 6-3. Mean counts of harbor seals hauled out in Prince William Sound and at Tugidak Island, 1976-1992. Counts at Tugidak Island were all done during the molt. Pupping counts in Prince William Sound include only seals older than pups. Molting counts include all seals. Tugidak Island data are from Pitcher (1990, 1991) and the National Marine Mammal Laboratory (unpublished). (--- = no data)

	Prince William Sound		Tugidak Island
Year	Pupping	Molting	
1976	---	---	6919
1977	---	---	6717
1978	---	---	4839
1979	---	---	3836
1982	---	---	1575
1983	---	1611	---
1984	---	1796	1390
1986	---	---	1270
1988	---	1057	1014
1989	750	807	---
1990	726	780	960
1991	619	921	---
1992	516	769	571

molt-period counts were 57% lower than in 1984. Counts of the same trend-count sites during pupping indicate a continuous decline since they were begun (Table 6-3) with 1992 counts 31% lower than in 1989. At this time it is not clear whether counts during the molt or during the pupping season provide the best indication of the overall trend in numbers for this area.

An extensive series of trend counts (Table 6-3) for another site within the EVOS area, Tugidak Island, indicates a major decline from almost 7000 animals in 1976 to less than 600 in 1992 (Pitcher 1990, 1991; National Marine Mammal Laboratory, unpublished data). This represents a 92% overall decline from 1976 through 1992 at what was once the largest harbor seal haulout in the world. Similarly, counts of harbor seals in Aialik Bay (northern Gulf of Alaska) during pupping in June 1980 and 1989 suggested a decline of 82% in total seals and 77% in the number of pups (Hoover-Miller 1989). Aialik Bay was in the area impacted by the EVOS, but because the two counts were 9 years apart no conclusions could be drawn about how the EVOS may have affected harbor seals in this area.

Harbor seals are clearly much less numerous in PWS and adjacent parts of the Gulf of Alaska than they once were. Although the total number of seals in the area is not known, available counts suggest that the declines are on the order of 60–90%.

⋈ ⋈ ⋈

Results from this study indicate that the EVOS has contributed to the recent declines in seal abundance in PWS. Other factors that may have contributed to the decline are poorly understood (Pitcher 1990; Sease 1992), but there are several human activities in PWS that may also affect harbor seals. PWS supports a large commercial fishery for salmon and other smaller fisheries for shellfish, groundfish, and herring. These fisheries may kill some seals through net entanglement and shooting, and may influence the availability of their prey (Sease 1992). Tourism is growing rapidly, bringing with it increased vessel traffic in areas that were once remote and relatively undisturbed habitat. The logging industry has increased greatly, causing habitat changes in nearshore areas that may be important to harbor seals or their prey. Coastal native residents harvest harbor seals for subsistence (Stratton and Chisum 1986; Stratton 1990), and these removals may have had some influence on seal numbers and may affect population recovery.

Until the factors that are affecting harbor seals are identified and, if possible, controlled, it cannot be predicted whether or when seal numbers may recover.

CONCLUSIONS

EVOS response and clean-up activities resulted in intensive activity in what is normally relatively undisturbed harbor seal habitat. Pre- and postspill count data for oiled and unoiled sites in PWS enabled us to estimate directly the number of seals missing in these areas following the EVOS. Calculations indicated that 135 seals were missing at the seven oiled sites that were surveyed, and that a minimum total of 302 seals were missing in PWS as a whole. We conclude that it is likely that these missing seals died as a result of the EVOS. There is no evidence that seals were displaced from oiled sites by the spill and associated activities. Other components of the harbor seal damage assessment study showed that seals became coated with oil, oil was incorporated into some of their tissues, they behaved abnormally, and there was pathological damage in their brains (see Chapters 12, 17, and 19). Our calculations of mortality are likely an underestimate of the total impact of the EVOS on harbor seals since only seals counted on haulout sites were included in the mortality estimate and the calculations do not include seals in the Gulf of Alaska, which were also impacted by the EVOS (Frost and Lowry 1993).

Available data indicate that pup production and survival were also affected by the EVOS. In oiled areas dead pups were found both before and during the normal pupping period. Aerial counts indicated that 26% fewer pups were produced at oiled sites in 1989 than would have been expected without the oil spill.

The trend in harbor seal numbers in PWS since the EVOS is unclear, with molting counts indicating relative stability but pupping counts showing a steady decline. Nonetheless, harbor seal numbers in PWS and adjacent parts of the Gulf of Alaska are greatly reduced compared to historical levels. Other factors must

≻◇ ≻◇ ≻◇

also have contributed to the overall decline in this region. The harbor seal population is not likely to recover until limiting factors are identified and mitigated. This study has demonstrated the importance of historical data in evaluating the effects of human-caused perturbations, such as oil spills, on the environment. Without historical data on distribution and abundance it would not have been possible to measure the impacts of the EVOS on harbor seals. We strongly recommend that population monitoring studies for key species be implemented or continued in areas where significant industrial activities are likely to occur, so that if an accident happens it will be possible to compare the changes that result with historical patterns.

ACKNOWLEDGMENTS

We thank S. Rainey and L. Lobe who flew the survey aircraft for their careful and conscientious support. J. Lewis, K. Wynne, and J. Harvey collected some of the survey data used in this paper. We thank E. Becker for his assistance with preliminary statistical analysis of survey data. Special thanks to K. Pitcher who began the program for monitoring the number of harbor seals in Prince William Sound, and to T. Loughlin for allowing us to use unpublished data collected by the National Marine Mammal Laboratory. The draft manuscript was improved by helpful comments by G. Antonelis, T. Loughlin, and an anonymous reviewer. Financial support for some of the aerial surveys referred to in this paper was provided by the ADF&G, the U.S. Marine Mammal Commission, and the National Marine Fisheries Service. This study was conducted as part of the Natural Resource Damage Assessment Study funded by the *Exxon Valdez* Oil Spill Trustee Council.

REFERENCES

Agresti, A. 1990. Categorical data analysis. John Wiley and Sons, Inc., New York, New York. 558 p.

Allen, S. G. 1980. Notes on births and deaths of harbor seal pups at Double Point, California. Murrelet 61:41-43.

Bigg, M. A. 1969. The harbour seal in British Columbia. Fisheries Research Board of Canada Bulletin 172. 33 p.

Boulva, J. 1971. Observations on a colony of whelping harbour seals, *Phoca vitulina concolor*, on Sable Island, Nova Scotia. Journal of the Fisheries Research Board of Canada 28:755–759.

Calambokidis, J., B. L. Taylor, S. D. Carter, G. H. Steiger, P. K. Dawson, and L. D. Antrim. 1987. Distribution and haulout behavior of harbor seals in Glacier Bay, Alaska. Canadian Journal of Zoology 65:1391–1396.

Calkins, D. G., and K. W. Pitcher. 1984. Pinniped investigations in southern Alaska: 1983-84. Unpublished Report, Alaska Department of Fish and Game, Anchorage, Alaska. 16 p.

Calkins, D. G, K. W. Pitcher, and K. Schneider. 1975. Distribution and abundance of marine mammals in the Gulf of Alaska. Unpublished Report, Alaska Department of Fish and Game, Anchorage, Alaska. 67 p.

Efron, B., and R. Tibshirani. 1986. Bootstrap methods for standard errors, confidence intervals, and other measures of statistical tendency. Statistical Science 1:54–75.

Frost, K. J. 1991. Assessment of injury to harbor seals in Prince William Sound, Alaska, and adjacent areas. 1991 Status Report, Marine Mammal Study Number 5, State-Federal Natural Resource Damage Assessment. 41 p.

Frost, K. J., and L. F. Lowry. 1993. Assessment of injury to harbor seals in Prince William Sound, Alaska, and adjacent areas following the *Exxon Valdez* oil spill. Final Report, Marine Mammal Study Number 5, State-Federal Natural Resource Damage Assessment. 95 p.

Hoover-Miller, A. 1989. Impact assessment of the T/V *Exxon Valdez* oil spill on harbor seals in the Kenai Fjords National Park, 1989. Unpublished Report, Kenai Fjords National Park, Seward, Alaska. 21 p.

Neter, J., and W. Wasserman. 1974. Applied linear statistical models. Irwin Inc., Homewood, Illinois. 842 p.

Pitcher, K. W. 1986. Harbor seal trend count surveys in southern Alaska, 1984. Unpublished Report, Alaska Department of Fish and Game, Anchorage, Alaska. 10 p.

Pitcher, K. W. 1989. Harbor seal trend count surveys in southern Alaska, 1988. Final Report Contract MM4465852-1 submitted to U.S. Marine Mammal Commission, Washington, D.C. 15 p.

Pitcher, K. W. 1990. Major decline in number of harbor seals, *Phoca vitulina richardsi*, on Tugidak Island, Gulf of Alaska. Marine Mammal Science 6:121–134.

Pitcher, K. W. 1991. Harbor seal trend counts on Tugidak Island. Final Report ACT Number T 75133261, submitted to U.S. Marine Mammal Commission, Washington, D.C. 5 p.

Pitcher, K. W., and D. G. Calkins. 1979. Biology of the harbor seal, *Phoca vitulina richardsi*, in the Gulf of Alaska. U.S. Department of Commerce, Environmental Assessment of the Alaskan Continental Shelf, Final Reports of Principal Investigators 19(1983):231–310.

Pitcher, K. W., and D. C. McAllister. 1981. Movements and haulout behavior of radio-tagged harbor seals, *Phoca vitulina*. Canadian Field-Naturalist 95:292–297.

Pitcher, K. W., and J. Vania. 1973. Distribution and abundance of sea otters, sea lions, and harbor seals in Prince William Sound. Unpublished Report, Alaska Department of Fish and Game, Anchorage, Alaska. 18 p.

Rosenberger, J. L., and M. Gasko. 1983. Comparing location estimators: trimmed means, medians, and trimean. Pages 297–338 *in*, D. C. Hoaglin, F. Mosteller, and J. W. Tukey (eds.), Understanding Robust and Exploratory Data Analysis. John Wiley and Sons, Inc., New York, New York.

Sease, J. L. 1992. Status review: harbor seals (*Phoca vitulina*) in Alaska. U.S. Department of Commerce, NOAA, National Marine Fisheries Service, Alaska Fisheries Science Center, Processed Report 92-15. 74 p.

Steiger, G. H., J. Calambokidis, J. C. Cubbage, D. E. Skilling, A. W. Smith, and D. H. Gribble. 1989. Mortality of harbor seal pups at different sites in the inland waters of Washington. Journal of Wildlife Disease 25:319–328.

Stratton, L., and E. B. Chisum. 1986. Resource use patterns in Chenega, Western Prince William Sound: Chenega in the 1960s and Chenega Bay 1984–1986. Alaska Department of Fish and Game, Anchorage, Alaska, Division of Subsistence Technical Paper 139. 161 p.

Stratton, L. 1990. Resource harvest and use in Tatitlek, Alaska. Alaska Department of Fish and Game, Anchorage, Alaska, Division of Subsistence Technical Paper 181. 163 p.

Thompson, P. M., and D. Miller. 1992. Phocine distemper virus outbreak in the Moray Firth common seal population: an estimate of mortality. Science of the Total Environment 115:52–65.

Winer, B. J. 1971. Statistical principles in experimental design. 2nd edition. McGraw-Hill, New York, New York. 907 p.

⋙ ⋙ ⋙

Appendix 6-A. Number of counts (n), mean, and maximum (max) number of harbor seals counted during aerial surveys in Prince William Sound, August-September 1983-1992. Data for 1983, 1984, and 1988 are from Pitcher (1986, 1989, and unpublished).

Site	1983			1984			1988			1989			1990			1991			1992		
	n	mean	max	n	mean	max	n	mean	max	n	mean	max	n	mean	max	n	mean	max	n	mean	max
1	6	14	47	8	46	90	8	13	31	8	0	0	8	<1	2	9	1	4	10	<1	1
2	6	12	52	8	27	49	9	12	38	5	20	54	8	5	13	10	13	28	10	24	41
3	6	50	86	8	45	66	9	42	65	8	33	50	7	21	37	10	27	38	10	31	42
4	6	86	149	8	150	239	9	80	129	7	43	66	8	69	104	10	80	125	10	41	76
5	6	12	49	8	31	54	9	4	16	7	7	13	8	1	4	10	14	21	9	8	20
6	6	77	170	8	98	133	7	42	74	8	33	53	8	22	43	8	17	26	8	12	17
7	6	22	39	8	12	16	9	2	9	8	2	4	8	4	13	9	5	11	9	<1	1
8	5	22	37	8	40	54	9	12	20	8	7	13	8	10	17	9	10	16	9	4	8
9	6	15	73	8	23	43	9	20	32	8	24	32	8	23	33	8	23	29	9	13	17
10	5	30	67	8	28	35	9	18	32	8	23	27	8	15	23	8	10	15	9	7	9
11	6	22	39	8	12	20	9	5	14	8	3	10	8	3	10	9	<1	2	9	<1	1
12	6	80	114	8	83	109	8	39	56	8	35	60	8	36	50	8	39	61	9	45	61
13	6	91	171	8	79	127	8	31	60	7	22	40	8	29	43	10	25	28	9	33	41
14	5	153	240	8	99	162	8	78	98	6	41	52	7	30	40	9	33	42	9	44	53
15	6	116	216	8	115	166	8	70	85	7	36	59	6	39	50	7	63	78	8	52	71
16	6	259	398	8	227	435	6	154	219	4	83	103	7	115	151	9	106	169	8	65	108
17	6	23	58	8	62	105	8	42	66	7	18	32	8	23	47	8	25	40	9	37	49
18	6	143	327	8	283	501	7	83	195	1	116	116	2	41	45	8	105	235	8	78	119
19	5	80	199	8	60	128	5	51	95	3	32	47	5	28	46	8	15	34	8	56	71
20	5	41	68	8	73	143	7	69	98	5	61	78	5	104	131	8	109	152	9	62	83
21	6	29	65	8	35	75	8	46	76	6	44	63	8	49	59	8	47	57	9	42	54
22	6	41	58	8	47	76	8	32	46	7	37	48	8	36	49	9	28	34	9	10	22

(Appendix Continues)

Appendix 6-A. Continued

Site	1983			1984			1988			1989			1990			1991			1992		
	n	mean	max	n	mean	max	n	mean	max	n	mean	max	n	mean	max	n	mean	max	n	mean	max
23	6	38	61	8	37	53	8	11	24	8	11	19	8	11	18	9	21	28	9	24	30
24	6	108	117	8	72	112	8	67	86	8	59	87	8	43	58	9	56	81	9	57	67
25	6	50	86	8	14	31	8	36	91	9	19	71	8	23	61	8	51	104	10	25	54

Appendix 6-B. Number of counts (n), mean, and maximum (max) number of harbor seals and harbor seal pups counted during aerial surveys in Prince William Sound, June 1989-1992. Data for 1992 are from National Marine Mammal Laboratory (unpublished).

Site	1989			1990			1991			1992		
	n	non-pups/pups mean	max	n	non-pups/pups mean	max	n	non-pups/pups mean	max	n	non-pups/pups mean	max
1	9	0/ 0	0/ 0	9	2/ 0	4/ 0	9	0/ 0	1/ 0	4	3/ 1	8/ 1
2	9	3/ 0	19/ 1	9	11/ 0	18/ 0	10	3/ 0	11/ 1	4	2/ 0	6/ 0
3	9	5/ 0	13/ 1	9	5/ 1	9/ 1	10	1/ 0	4/ 1	4	9/ 0	10/ 0
4	7	68/16	88/25	8	55/21	69/33	9	23/11	46/15	4	25/ 9	32/17
5	8	9/ 2	24/ 4	9	2/ 1	3/ 1	9	7/ 2	12/ 4	4	2/ 0	3/ 1
6	9	17/ 5	29/ 9	7	10/ 4	17/ 9	8	11/ 4	17/ 6	3	16/ 4	23/ 6
7	9	4/ 3	11/10	7	0/ 0	1/ 1	8	4/ 1	8/ 2	3	3/ 0	8/ 1
8	8	10/ 2	17/ 4	7	3/ 1	6/ 1	8	3/ 1	7/ 2	3	1/ 1	1/ 1
9	8	11/ 3	18/ 5	7	8/ 1	10/ 2	8	3/ 0	8/ 0	3	5/ 0	6/ 1
10	8	3/ 0	6/ 1	7	1/ 0	3/ 0	8	3/ 0	7/ 1	3	1/ 0	1/ 0
11	9	3/ 0	8/ 1	7	5/ 1	8/ 3	7	1/ 0	1/ 1	3	1/ 0	3/ 0
12	8	29/ 9	34/13	8	43/15	54/18	9	40/14	52/17	4	40/13	50/16
13	9	11/ 3	36/ 9	8	19/ 6	25/11	9	14/ 7	19/ 8	4	13/ 5	17/ 6
14	8	18/ 7	28/13	8	18/ 5	24/11	9	24/ 5	32/ 7	4	15/ 4	20/ 5
15	9	46/14	68/23	8	47/20	54/23	9	71/29	87/39	4	46/22	54/30
16	6	151/31	199/56	8	137/36	158/43	9	143/45	177/54	4	84/36	104/45
17	9	22/ 8	32/11	8	28/16	33/22	8	25/11	36/15	4	50/13	61/19
18	8	91/12	152/20	7	73/ 3	96/ 5	9	61/ 3	94/ 5	4	69/ 8	78/19
19	8	88/15	118/30	7	68/ 6	100/ 9	8	45/ 6	62/ 9	4	36/ 7	50/10
20	9	75/19	104/23	7	95/24	110/30	9	66/18	94/28	4	38/14	62/24

(Appendix continues)

Appendix 6-B. Continued

Site	1989			1990			1991			1992		
		non-pups/pups			non-pups/pups			non-pups/pups			non-pups/pups	
	n	mean	max	n	mean	max	n	mean	max	n	mean	max
21	9	20/ 4	32/ 9	7	28/ 0	37/ 0	9	13/ 0	24/ 0	4	6/ 1	18/ 3
22	9	15/ 4	32/ 8	6	23/ 1	28/ 2	9	16/ 1	20/ 2	4	13/ 1	16/ 2
23	9	25/ 8	32/11	6	21/ 6	28/ 7	9	19/ 5	27/ 8	4	10/ 3	14/ 6
24	9	29/ 6	54/10	8	25/ 3	42/ 5	8	24/ 3	39/ 4	4	30/ 6	38/ 8
25	9	0/ 0	1/ 0	8	1/ 1	3/ 2	10	1/ 1	5/ 1	4	0/ 0	1/ 0

Chapter 7

Impacts on Steller Sea Lions

Donald G. Calkins, Earl Becker, Terry R. Spraker, and Thomas R. Loughlin

INTRODUCTION

Steller sea lions (*Eumetopias jubatus*) inhabit coastal and offshore areas in the north Pacific Ocean from southern California to Japan (Loughlin et al. 1984). They are the largest otariid and are regularly found in coastal waters and on the beaches recently affected by the *Exxon Valdez* oil spill (EVOS). Steller sea lions breed and pup at rookeries during late May to July. Rookeries are considered haulout sites if they are used at other times of the year. Generally, most haulout sites are used seasonally or throughout the year and few pups are born at these sites. Females usually produce a single pup and within 14 days *post partum* enter into estrous and breed. Fetal implantation is delayed until approximately mid-October (Pitcher and Calkins 1981). Pup dependency can be long, up to 3 years in some cases, but most pups wean by the end of their first year (Pitcher and Calkins 1981). Reproductive failures in the form of resorption in early pregnancy or late-term abortion occur at a high rate in the Gulf of Alaska (Calkins and Pitcher 1982; Calkins and Goodwin 1988).

No rookeries occur within Prince William Sound (PWS), but three haulout sites which are used on a seasonal basis and two year-round haulout sites are found there. One rookery occurs at the entrance to PWS (Seal Rocks) and three other rookeries occur along the Kenai Peninsula (Outer and Sugarloaf Islands) and in the Kodiak Island area (Marmot Island) which were in the general area affected by the EVOS. Approximately 25 additional haulout sites in the Kodiak Island and Alaska Peninsula areas were in the path of the spilled oil (Calkins and Pitcher 1982).

It is not practical to determine the total number of sea lions in a given area at a given time because an unknown number will be at sea and will not be counted. Therefore, sea lion estimates are based on counts of animals on land at rookeries and haulout sites. Counts are generally lower in winter than in summer (Calkins and Pitcher 1982).

Steller sea lion population declines have been reported throughout most of the species' range (Loughlin et al. 1992), including southern and central California, the Gulf of Alaska, the Aleutian Islands, the central Bering Sea, and Russia (National Marine Fisheries Service 1992). Populations in Oregon, Washington, British Columbia, and southeastern Alaska appear stable. Precipitous declines in the Gulf of Alaska and Aleutian Islands were detected during a 1989 range-wide survey (Loughlin et al. 1992) resulting in the National Marine Fisheries Service (NMFS) listing the species as threatened under the U.S. Endangered Species Act and depleted under the Marine Mammal Protection Act in November 1990.

Sea lion population declines first occurred in the eastern Aleutian Islands in the mid-1970s (Braham et al. 1980), spread into the Kodiak Island area in the early 1980s, into the central and western Aleutian Islands area in the mid-1980s, and into the eastern Gulf of Alaska in the late 1980s and early 1990s (Loughlin et al. 1984; Merrick et al. 1987; Byrd 1989; Loughlin et al. 1992). Counts of adult and juvenile sea lions from the Kenai Peninsula to Kiska Island in the Aleutian Islands were used as the basis for listing the species as threatened. Counts in this area declined from 89,364 adult and juvenile sea lions in 1960 to 21,737 individuals in 1991, a decline of 76% (Merrick et al. 1992). No specific cause for the decline has been identified.

Our study began in June and July 1989 with aerial surveys of rookeries and haulout sites in the area affected by the EVOS. Surveys were also conducted in the remainder of the species' range that year (Loughlin et al. 1992). Beaches were searched and animals were collected and sampled for toxicological and histological impacts from the spilled oil. Recognizing the severe decline in sea lion abundance over much of the affected area, we attempted to measure changes beyond that which could be attributed to the ongoing decline.

We used two approaches to measure effects of the spill on sea lions. The first was designed to detect effects at the population level and included assessment of sea lion numbers and their distribution. The objective was to determine if sea lion abundance on rookeries and haulout sites had changed significantly as a result of the spill.

The second approach was designed to detect effects at the individual level. Sixteen sea lions were collected and 12 were found dead during response and clean-up efforts (Zimmerman et al., Chapter 2). When practical, tissue samples were taken from these animals and tested for toxicological effects through histological and hydrocarbon analyses. The objective was to determine if sea lions were accumulating hydrocarbon contaminants in their tissues and if tissue damage had resulted. We suspected that the toxic effects of hydrocarbons could have accelerated the already high rate of premature pupping in sea lions noted by Pitcher and Calkins (1981) and Calkins and Goodwin (1988) for the Gulf of Alaska.

METHODS

Study Area

Surveys were conducted at all rookeries and haulout sites from Cape St. Elias through the north Gulf of Alaska to Chowiet Island (Fig. 7-1). Although both sea lions and oil contamination from the spill were found offshore, it was not practical to conduct our study beyond 10 km from shore.

Distribution and Abundance

Aerial surveys were conducted from 15 June to 15 July 1989, the period when sea lion counts normally are highest (Withrow 1982). Also, this is the time period for which we had the best comparable information. Note that for PWS the highest annual abundance of sea lions on shore has been recorded in March and April (Calkins and Pitcher 1982).

Aerial surveys were timed to maximize the number of animals available for counting on land (Withrow 1982; Merrick et al. 1987). Surveys were conducted between 1000 and 1600 local time, when possible. In some cases, the timing of a survey was altered due to weather or time constraints. Survey aircraft were flown at approximately 90 knots, at an altitude of between 150 to 300 m, in a vertical plane over each survey site. Replicate photographs of sea lions were taken with a hand-held, 35-mm, auto-focus, auto-exposure, SLR camera equipped with a mo-tor-drive and a 70- to 210-mm zoom lens using high-speed (ASA 400) color transparency film. Survey time and date and environmental conditions were recorded for each site. Estimates of the number of sea lions observed were made for each survey location. Later the photographic slides were projected on a white paper screen and each adult or juvenile sea lion within the survey was counted. Some of the counts reported here were also reported by Loughlin et al. (1992).

Pups were counted from land at all rookeries from Seal Rocks to Chowiet Island in 1989 (Fig. 7-1). In 1990, all rookeries were counted except Marmot Island. Counts were timed to be completed between 25 June and 4 July, the period when most pups were born and before they were old enough to readily enter the water. Biologists landed at each site by helicopter or boat. Generally, one or two biologists walked through the rookery causing the adults to enter the water while the pups remained on land. Another biologist followed the first, counting each pup indi-vidually. In some areas a second count was conducted, but because additional disturbance may have undesirable effects, a second count was conducted only where necessary. If a second count was conducted, the two counts were averaged.

In 1989, all rookeries and haulout sites were visited and searched for aborted fetuses. Premature pupping ratios were measured by stationing two observers each at Cape St. Elias and at Chirikof Island between 5 April and 16 May. Premature pupping had been documented at that time of year at Cape St. Elias (Calkins and

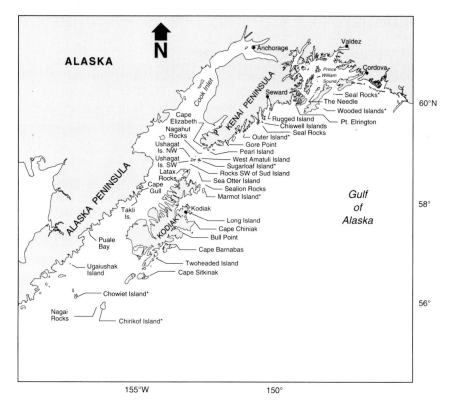

Figure 7-1. Steller sea lion study area with all haulout sites and rookeries (*) from Prince William Sound to Chowiet Island.

Pitcher 1982). In our study, observations were made for 25 days at Chirikof Island and for 34 days at Cape St. Elias. Each haulout site was searched daily using spotting scopes and binoculars. The presence of all dead, prematurely born pups was noted. Visual cues such as bird activity or activity of adult female sea lions, as well as periodic visual scans, were used to determine the presence of dead pups. Once a dead pup was located, it was retrieved, taken from the haulout site, and necropsied. Tissue samples were preserved for histological analyses according to accepted protocols (Appendix I, this volume). Adults, juveniles, and premature births were counted daily at each location.

Analyses of adult/juvenile and pup count data were accomplished by fitting simple linear regression models (Neter and Wasserman 1974) to counts conducted prior to 1990. The regression residuals (the difference between predicted and actual counts) were analyzed to ensure an adequate model fit, and that the assumptions of normality and constant variance were met (Cook 1977). The haulout site data set was limited to only three data points because of the absence of consistent counts. As a result, it was difficult to assess model fit and the assumptions of normality and constant variance for the haulout site counts. The models were used to generate a predicted value and variance for a 1990 count without an oil spill effect. A one-sided *t* test (Neter and Wasserman 1974) was used to test if the observed counts were lower than the 1990 counts predicted by the model. If they were, then an oil spill effect is inferred. Power calculations were based on the one-sided *t* statistic, and were used to assess our ability to detect meaningful biological differences.

The adult/juvenile rookery data set used in the regression analysis was constructed by combining counts from Sugarloaf, Marmot and Chirikof Islands for the years 1976, 1978, 1979, 1985, and 1989. These rookeries were close to the spill and the most likely to be affected. Outer Island and Seal Rocks, two rookeries which were closer to the spill than the three we used, were not used in the analysis due to missing count data. The 1989 data point was included with the previous counts and treated as a nonoiled year because it seemed likely that any possible oil spill effect in 1989 would have been slight to nonexistent. Observations following the spill suggested that large numbers of sea lions were not killed in 1989. We speculated that the most likely effect would be an increase in premature birth rates which should have been detected in the 1990 counts. Also, we did not have the ability to measure premature births in 1989 because the spill occurred shortly before normal parturition.

We used the counts from 17 locations for the years 1956–1957, 1976, and 1989 for the haulout-site data set for our linear regression analysis. These 17 areas include all of the haulout sites that had at least three counts in the same years. We treated 1989 as a nonoiled year in the haulout site counts for the same reasons as the rookery counts.

Pup counts were analyzed in a similar manner to adult/juvenile counts. Previous pup counts were not adequate to analyze for an effect in 1989, so we summed and analyzed counts through 1989 to predict the count expected in 1990. The combination of pup counts from Outer, Sugarloaf, Chirikof, and Chowiet Islands for 1978, 1979, 1984, 1986, and 1989 was used as the data set for construction of the linear regression model.

To determine if premature pupping ratios were different between the two haulout sites, a Poisson model (Agresti 1990) was fitted to the number of premature pups. The model adjusted for differences in the number of adults using the natural log of

⋈   ⋈   ⋈

adults counted as an offset parameter and used oil treatment as an explanatory variable.

Toxicology and Histology

An extensive effort was mounted to search the beaches of PWS following the oil spill to locate and collect carcasses of animals possibly killed by toxic effects of the spill. Twelve sea lions were recovered during this effort. Detailed necropsies were performed on these animals according to established protocols. In addition, 16 animals were collected under the Natural Resources Damage Assessment (NRDA) study conducted on Steller sea lions in 1989 following the spill. Animals were collected from haulout sites in the vicinity of PWS, along the Kenai coast, and in the Barren Islands (southern tip of the Kenai Peninsula). Necropsies were begun as soon after death as possible with tissues taken and preserved according to protocol. A complete gross and histological study was done on each animal. Tissues were taken from all major organs and systems and preserved in 10% buffered formalin for histological analysis. Three complete sets of tissues were frozen in chemically clean containers for toxicological analysis. Tissues for toxicological analysis were also obtained from one sea lion taken during the subsistence harvest in Kachemak Bay in 1990.

Premature pups were collected at Cape St. Elias and Chirikof Island during April 1990. Complete gross and histological analyses were performed on the collected tissues, but toxicological analyses were not.

Analysis of bile was performed by the Environmental Conservation Division (ECD), Northwest Fisheries Science Center, NMFS, Seattle, Washington, using high-pressure liquid chromatography with fluorescence detection. Metabolites of naphthalene (D) and phenanthrene (PHN) were measured as indicators of hydrocarbon contamination (Krahn et al. 1984). Concentrations of PHN were detected at 260/380 nm and D at 290/335 nm (excitation/emission). The PHN wavelength pair was used to estimate total concentration of 3-ring aromatic compounds, and the D wavelength pair was used to estimate 2-ring aromatic compounds. Fluorescent Aromatic Compounds (FACs) values were reported in units of ng PHN (D) equivalents/gram bile. Krahn et al. (1992) have shown a high correlation between PHN and D equivalents and the summed concentrations of metabolites of phenanthrene and naphthalene/dibenzothiophene, respectively, using gas chromatography and mass spectroscopy. In the current study, results of bile analysis were compared by area with a square root transformation and one-way analysis of variance (Neter and Wasserman 1974). A Tukey Studentized Range test (Stoline 1981) was used to test all possible group differences while maintaining an overall α of 0.05.

Other tissue samples were analyzed by either the Geochemical and Environmental Research Group (GERG), Texas A&M University, College Station, Texas, or the ECD. Analysis was performed using capillary column gas chromatography

⋖⋗ ⋖⋗ ⋖⋗

with mass spectrometry (GC/MS). Because of the large number of tissues collected in other NRDA oil spill studies, only a representative number of all tissue samples were analyzed. Thus, all tissue samples collected in our study were not analyzed. Although similar analytical methods were used (Krahn et al. 1988; Sloan et al. 1993), analyses performed by GERG were done under NRDA Technical Study No. 1 and were subject to different quality assurance and reporting requirements than the ECD analyses which were conducted as part of an ADF&G/NOAA study of subsistence foods. Aromatic compounds (ACs) consisted of the aromatic compounds naphthalene, dibenzothiophene, phenanthrene, fluorenes, and chrysenes as both parent compounds and the alkylated series. The ECD analysis consisted of and was reported in terms of the sums of lower molecular weight 2,3-ring aromatic compounds (LACs), the sums of higher molecular weight 4,7-ring aromatic compounds (HACs), and total aromatic compounds (ACs). GERG analysis was similar but in addition they analyzed and reported on aliphatic hydrocarbons n-C_{10} through n-C_{34}, pristane, phytane, and unresolved complex mixture (UCM). This allowed calculation of a pristane/phytane ratio and a carbon preference index (Boem et al. 1987; Farrington and Tripp 1977).

Tissues for histological analysis were examined by one of us (TRS) at the Colorado State Diagnostic Laboratory, University of Colorado, Fort Collins. The tissues were imbedded in paraffin and sectioned at 5–6 μ. Tissue sections were stained with eosin and hematoxylin and mounted on slides for microscopic examination. Approximately 65 to 110 slides were prepared for each collected sea lion, including multiple slides from the following major organ systems: nervous, integument, cardiovascular, respiratory, lymphohematopoietic, digestive, urogenital, endocrine, musculoskeletal, and special senses. Because of possible hydrocarbon contaminant-related pathology found in harbor seals (Spraker et al., Chapter 17), particular attention was given to the nervous system. Tissues from sea lions found dead were in varying states of autolysis when taken and many were unsuitable for microscopic examination.

RESULTS

Distribution and Abundance

During observations of sea lions near or in oil following the EVOS, it became apparent that oil did not persist on sea lions as it did on harbor seals (Lowry et al., Chapter 12). Oil did not persist on the rookeries and haulout sites either, probably due to their steep slopes and high surf activity. However, some oil fouling was noted on Seal Rocks and Sugarloaf Island in April 1989. Insignificant amounts of oil were seen at each site during pup counts in late June 1989, but none were seen in 1990.

∞ ∞ ∞

Adult and juvenile sea lion numbers from Cape St. Elias to Chowiet Island totaled 18,135 in 1989 (Table 7-1) which was substantially lower than the comparable 1976 count of 46,204 (Calkins and Pitcher 1982). Combined rookery counts totaled 6076 sea lions in 1989 (Table 7-2). The linear regression of the sample year (1976–1989) on the combined rookery counts resulted in a model (R^2=0.99) with an estimated y-intercept of 1,825,666.8 (SE=120,622.3) and an estimated year effect of -914.7 (SE=60.9). Examination of residuals suggested the model fit well and assumptions of normality and constant variance were met.

The actual count of 4243 was not significantly lower than the regression model predicted count of 5502.8 (SE=891.8) adult and juvenile sea lions in 1990 (t=-1.41, df=3, P≅0.126) at α=0.05. The power of the test (0.21) was low. In order to obtain a power of 85% or 95%, we would have had to count as few as 2532 and 1305 sea lions, respectively.

Pups counted at rookeries totaled 4195 in 1989 (Table 7-3). Pup numbers at these sites have fallen from 13,145 in 1979. The linear regression of year (1979–1989) on the combined pup counts resulted in a model (R^2=0.99) with an estimated y-intercept of 1,773,776.4 (SE=99,162.8) and an estimated year effect of -889.6 (SE=50). Examination of the residual diagnostics revealed that the model fit fairly well and that assumptions of normality and constant variance did not appear to have been violated. The 1990 count of 2952 pups was not significantly lower than the model predicted count of 3577.5 (SE=490.9, t=-1.27, df=2, P≅0.165) at α=0.05. Power of this test was also low at 0.12. We would have had to count 1624 or 711 sea lion pups, respectively, to obtain a power value of 85% or 95%.

Haulout site counts totaled 4782 in 1989 (Table 7-4). Comparison of the 1989 count to the 1976 count of 9277 illustrates the severity of the decline; it also suggests that the haulout sites declined at a similar rate and that redistribution from the rookeries to haulout sites did not occur. The linear regression of these data resulted in a model (R^2=0.99) with an estimated y-intercept of 812,914.0 (SE=2105.4) and an estimated year effect of -406.4 (SE=10.7). The model appeared to fit well. The average count was 4890.5 sea lions on haulout sites in 1990. This value was not significantly lower than the model predicted count of 4079.7 (t=0.99, df=1, P≅0.748) at α=0.05. Power of the test was extremely low at 0.05. A 1990 count of zero sea lions on the haulout sites would have resulted in a power value of 0.20.

One premature birth was observed at Chirikof Island and 11 were observed at Cape St. Elias in 1990 (Table 7-5). The Poisson model rejected the hypothesis of equivalent premature pupping ratios [Deviance (G^2) = 5.01, P=0.025, χ^2=4.01, P=0.045]. This model estimated that if premature pupping ratios were equivalent at the two sites, we should have counted 4.33 premature pups at Chirikof Island and 7.67 premature pups at Cape St. Elias.

Table 7-1. Counts of adult and juvenile Steller sea lions at all rookeries and haulout sites from Cape St. Elias to Chowiet Island, Alaska. (— = no data)

Location	1956-57	1973	1976	1985	1989	1990[a]
Cape St. Elias	—	1548	1628	—	1883	948
Middleton Island	—	—	2901	—	1	—
Wooded Islands	2500	1261	878	—	1333	1232
Seal Rocks, PWS	183	1733	1709	—	2159	1471
Glacier Island	—	0	0	—	0	0
Perry Island	—	0	0	—	0	0
Point Eleanor	—	0	0	—	0	0
The Needle	130	563	537	—	668	926
Point Elrington	250	236	725	—	487	382
Rugged Island	—	—	0	—	190	25
Chiswell	2012	—	1106	—	456	408[b]
Seal Rocks, Kenai	250	—	320	—	65	0-52[b]
Outer Island	2989	—	3847	—	1127	589-732
Gore Point	200	—	535	—	76	63-125
Perl Island	737	—	33	—	159	97-125
Elizabeth Island	129	—	124	—	180	85-15
Nagahut Rocks	—	—	344	—	43	28-73
Sugarloaf Island	11,963	—	5226	—	2467	1319-1513
Rks. SW Sud	—	—	670	2991	93	0-111
SW Ushagat	834	—	902	—	245	441-276
NW Ushagat	—	—	106	—	4	0-14
W Amatuli Island	1576	—	57	—	0	0-0
Latax Rocks	3334	—	1164	—	354	519
Sea Otter Island	—	—	541	—	450	164
Sea Lion Rocks	343	—	432	—	46	93

(Table continues)

Table 7-1. Continued

Location	1956-57	1973	1976	1985	1989	1990[a]
Marmot Island	4157	—	9862	4983	2331	1766
Long Island	75	—	0	—	30	93
Cape Chiniak	772	—	365	—	0	95
Gull Point	—	—	145	—	0	91
Cape Barnabas	1598	—	364	—	0	1
Twoheaded Island	2810	—	1615	—	479	268
Cape Sitkinak	343	—	120	—	204	234
Chirikof Island	1742	—	5199	2346	1278	1061
Nagai Rocks	—	—	657	—	233	196
Cape Ikokik	—	—	0	—	0	0
Cape Ugat	—	—	0	—	0	0
Cape Gull	—	—	207	—	0	0
Takli Island	—	—	1877	—	0	0
Puale Bay	—	—	1877	—	0	0
Ugaiushak Island	—	—	125	—	138	55
Sutwik Island	—	—	6	—	210	153
Chowiet Island	6323	—	—	2059	737	897
Total	45,250	5341	46,204	10,320	18,135	14,277[c]

[a] 1990 counts were conducted jointly with NMFS. Data are from Merrick et al. (1991).
[b] Some areas were counted twice - first count by NMFS on 12 June, second count by ADF&G on 13 June.
[c] Where two counts were available, the mean was used to calculate total numbers.

128

Table 7-2. Counts of adult and juvenile Steller sea lions at rookeries in the oil spill area. Only rookeries with counts conducted in the years shown are presented.

Rookery	Year					
	1976	1978	1979	1985	1989	1990
Sugarloaf Island	5226	4810	4374	2991	1467	1416
Marmot Island	9862	8506	6381	4983	2331	1766
Chirikof Island	2391	3699	5199	2346	1278	1061
Total	17,479	17,015	15,954	10,320	6076	4243

Table 7-3. Counts of Steller sea lion pups at rookeries within the oil spill area. Only those rookeries with counts for all years are shown.

Rookery	Year					
	1978	1979	1984	1986	1989	1990
Outer Island	431	888	1034	993	557	363
Sugarloaf Island	5021	5123	3114	3072	2109	1638
Chirikof Island	1573	1649	1913	1476	709	607
Chowiet Island	4670	5485	3207	1731	820	344
Total	11,695	13,145	9268	7272	4195	2952

129

Table 7-4. Counts of adult and juvenile Steller sea lions at haulout sites from Wooded Islands to Cape Sitkinak.

Haulout	Year			
	1956-57	1976	1989	1990[a]
Wooded Islands	2500	878	1333	1232
The Needed	130	537	668	926
Point Elrington	250	725	487	382
Chiswell Island	2012	1106	456	408[b]
Seal Rocks, Kenai	250	320	65	0-52
Gore Point	200	535	76	63-125
Perl Island	737	33	159	97-125
Elizabeth Island	129	124	180	85-15
SW Ushagat	834	902	245	441-276
W Amatuli Island	1576	57	0	0-0
Latax Rocks	3334	1164	354	51
Sea Lion Rocks	343	432	46	93
Long Island	75	0	30	93
Cape Chiniak	772	365	0	95
Cape Barnabas	1598	364	0	1
Twoheaded Island	2810	1615	479	268
Cape Sitkinak	343	120	204	234
Total	17,893	9277	4782	4702-5079

[a] Counts were conducted jointly between ADF&G and NMFS.
[b] Some areas were counted twice, the first count was by NMFS on 12 June, the second by AFD&G on 13 June.

Toxicology

Fifteen of the 16 sea lions collected under NRDA studies (Table 7-6) were tested for FACs (D and PHN) in the bile (Table 7-7). No sample was obtained from the sixteenth animal. Bile was also collected from three animals found dead but only one sample was analyzed. Bile from one sea lion (ST-SL-1; Table 7-7) taken in the subsistence harvest in Kachemak Bay was also tested.

PHN values ranged from 350 ng equivalents/g bile to 29,000 and averaged 5800. D values ranged from 4200 to 120,000 ng equivalents/g bile and averaged 35,000. Total FACs ranged from 4600 to 149,000 averaging 63,000. The mean PHN value for four sea lions collected in PWS in October 1989 (14,900 ng equivalents/g) was significantly higher than the mean for sea lions collected in PWS in June and July 1989 (3900 ng equivalents/g) or those collected in June, July, and October 1989 outside of PWS (2000 ng equivalents/g). The mean PHN values from all of the bile samples collected outside PWS in either June/July or October were relatively

                     ⋙   ⋙   ⋙

Table 7-5. Counts of adult (AD), juvenile (JUV) and premature pup (PREM) Steller sea lions at Chirikof Island (CHIR) and Cape St. Elias (CSE), April and May 1990. (--- = no data)

1990	AD & JUV CHIR	PREM PUPS CHIR	AD & JUV CSE	PREM PUPS CSE
5 April	400	0	---	---
7 April	681	0	---	---
8 April	796	0	---	---
9 April	631	0	---	---
10 April	797	0	---	---
11 April	669	0	---	---
12 April	765	1	---	---
13 April	665	0	732	1
14 April	736	0	637	1
15 April	748	0	609	1
17 April	690	0	300	0
18 April	670	0	297	0
19 April	799	0	838	0
20 April	859	0	814	1
21 April	759	0	724	0
22 April	628	0	908	0
23 April	919	0	1246	1
24 April	701	0	1081	0
25 April	939	0	1359	0
26 April	757	0	1344	1
27 April	804	0	1064	0
28 April	773	0	1105	0
29 April	630	0	913	0
30 April	---	---	952	0
1 May	---	---	924	1
2 May	---	---	703	1
4 May	---	---	1740	0
4 May	---	---	1139	0
5 May	---	---	1232	1
8 May	---	---	943	1
9 May	---	---	1650	0
10 May	---	---	1500	0
11 May	---	---	1107	0
12 May	---	---	933	0
13 May	---	---	670	1
14 May	---	---	638	0
15 May	---	---	921	0
16 May	---	---	744	0

Table 7-6. Steller sea lions examined for histological or toxicological analyses during 1989.

Specimen number	Date found dead (F) or killed (K)	Location	Sex	Age
AF-SL-1	F 12 May	Nuka Island	M	Adult
BD-SL-1	F 6 June	Swikshak Bay	F	Juvenile
GA-SL-1	F 23 June	Lower Cook Inlet	F	Adult
MH-SL-1	F 2 May	Kachemak Bay	F	Juvenile
MH-SL-4	F 9 June	Naked Island	F	Adult
MH-SL-5	F 9 June	Naked Island	F	Fetus of MH-SL-4
TS-SL-1	F 27 April	Sea Lion Rocks	F	Juvenile
TS-SL-2	F 13 June	Eleanor Island	M	Juvenie
TS-SL-3	F 19 June	Axel Island	M	Juvenile
TS-SL-579	K 29 June	NW Ushagat Island	F	Juvenile
TS-SL-580	K 29 June	NW Ushagat Island	F	Juvenile
TS-SL-581	K 29 June	NW Ushagat Island	M	Juvenile
TS-SL-583	K 3 July	The Needle	F	Juvenile
TS-SL-584	K 3 July	The Needle	M	Juvenile
TS-SL-585	K 3 July	The Needle	F	Juvenile
TS-SL-586	K 3 July	The Needle	F	Juvenile
TS-SL-587	K 4 July	Chiswell Islands	M	Adult
TS-SL-588	K 26 October	Sea Otter Island	F	Adult
TS-SL-589	K 28 October	Nagahut Rock	F	Adult
TS-SL-590	K 29 October	The Needle	F	Juvenile
TS-SL-591	K 29 October	The Needle	F	Adult
TS-SL-592	K 29 October	The Needle	F	Juvenile
TS-SL-593	K 30 October	The Needle	F	Juvenile
TS-SL-594	K 3 November	Chiswell Island	M	Adult

low and were not significantly different from those collected in PWS in June/July. Bile samples from sea lions taken inside PWS during June/July, with one exception, were also moderately low. The single exception was an immature female taken at the Needle on 3 July 1989 (TS-SL-586).

Selected tissues were analyzed for polycyclic aromatic hydrocarbons (PAHs) from 8 sea lions found dead, 1 sea lion taken in the subsistence harvest, and 13 sea lions that were collected (Table 7-8). These analyses were performed by ECD on liver, muscle, kidney, and blubber on sea lions TS-SL-579 through TS-SL-587. All other tissue analyses were conducted by GERG. In general, tissues analyzed by GERG were reported to have consistently higher levels of PAHs than those analyzed by ECD (in no case did both groups analyze the same tissue from the same animal). The difference in reported values between the two laboratories was probably due to the differences in detection levels and reporting requirements. Differences were also seen in harbor seal tissues analyzed by the two groups (Frost et al., Chapter 19).

Table 7-7. Levels of fluorescent aromatic hydrocarbon metabolites of phenanthrene (PHN) and napthalene (NPH) found in the bile of Steller sea lions in 1989. Units are in ng PHN (NPH) equivalents/g bile.

Specimen Number	PHN	NPH
ST-SL-1	4600	38,000
TS-SL-579	1200	18,000
TS-SL-580	2600	14,000
TS-SL-581	2100	45,000
TS-SL-582	890	19,000
TS-SL-583	3100	29,000
TS-SL-584	3000	31,000
TS-SL-585	3700	19,000
TS-SL-586	9000	36,000
TS-SL-587	350	4200
TS-SL-588	730	16,000
TS-SL-590	15,000	54,000
TS-SL-591	6600	32,000
TS-SL-592	29,000	120,000
TS-SL-593	9900	41,000
TS-SL-594	3000	20,000
TS-SL-1	3600	58,000

The highest total PAH level reported in this study was 607 parts per billion (ppb) from the liver of an adult male sea lion found dead on Nuka Island on the Kenai Peninsula on 12 May 1989 (AF-SL-1; Table 7-8). Six percent of the tissues tested had PAH levels exceeding 200 ppb. The highest PAH levels were seen in the sea lions found dead. Fewer than 6% of the tissues tested from collected animals had PAH levels exceeding 100 ppb. PAH levels detected in all tissues from collected sea lions were from low to not detectable, regardless of location, time of collection, or laboratory conducting the analysis.

Phytane levels were generally very low in sea lion tissues (Table 7-9). The highest phytane ratio was from the liver sample of AF-SL-1. Brain tissue from some of the sea lions collected in PWS had the lowest pristane/phytane ratios (Table 7-9). The unresolved complex mixture was low for all tissues analyzed. The carbon preference index was 1 for the liver sample from TS-SL-3 and ranged higher for all other samples.

Histology

Numerous lesions were identified in adult and juvenile sea lions. The most common gross lesions were caused by parasites in the nasal cavity, stomach, and intestine. Multiple histological lesions were seen in all organ systems. No gross

⋙ ⋙ ⋙

Table 7-8. Polycyclic aromatic hydrocarbons (PAH) found in Steller sea lion tissues collected in 1989. LAC=lower molecular weight aromatic contaminants; HAC=higher molecular weight aromatic contaminant; nd=non detected; PAH=total polycyclic aromatic hydrocarbons. All measurements in part per billion ng/g wet weight.

Specimen number	Liver			Muscle			Kidney			Blubber			Brain		
	LAC	HAC	PAH	LAC	HAC	PAH	LAC	HAC	PAH	LAC	HAC	PAH	LAC	HAC	PAH
TS-SL-2	67.8	39.5	107.3												
TS-SL-3	120.2	89.5	209.6												
MH-SL-1	96.4	49.1	145.5							206	160	366			
MH-SL-4	82.4	38.5	120.9												
MH-SL-5	72.9	105.6	182.5							160.1	105.2	265.3			
AF-SL-1	251.4	355.1	606.5												
BC-SL-1	46.4	38.2	84.76												
GA-SL-1	75.3	28.3	103.6												
TS-SL-579	2	0.9	2.0	nd	nd	nd	nd	nd	nd	1	nd	1			
TS-SL-580	nd	nd	nd	nd	nd	nd	nd	nd	nd	nd	0.9	0.9			
TS-SL-581	5	2	7	nd	nd	nd	0.1	nd	0.1	nd	0.3	0.3			
TS-SL-582	1	1	2	nd	nd	nd	nd	nd	nd	nd	0.6	0.3	1		
TS-SL-583	3	2	5	0.3	nd	0.3	nd	0.3	0.3	1	nd	nd		nd	nd
TS-SL-584	nd	0.2	0.2	0.6	nd	0.6	nd	nd	nd	nd	nd	nd			
TS-SL-585	1	0.3	1.3	1	nd	1	nd	nd	nd	nd	0.2	0.2			
TS-SL-586	0.9	0.5	1.5	0.9	nd	0.9	nd	nd	nd	nd	0.3	0.3	40.4	33.1	73.5
TS-SL-587	nd	nd	nd	0.7	nd	0.7	nd	nd	nd	nd	0.6	0.6			
TS-SL-590	17.8	11.9	29.7							64.4	51.8	116.2	29.6	37.8	67.4
TS-SL-591	29.4	53.7	83.1							53.3	39.5	92.8	35.4	24.6	60
TS-SL-592	20.3	13.4	33.7							67	53.2	120.8	75.8	51.3	107.1
TS-SL-593	18.5	13.6	32.1							48	33.7	81.2	46.4	29.4	75.8
ST-SL-1	1	nd	1	2	0.2	2.2	4	nd	4	4	0.4	4			

Table 7-9. Pristane (PR), phytane (PH), pristane/phytane ratio, unresolved complex mixture (UCM) and carbon preference index (cpi) (Farrington and Tripp 1977) for Steller sea lion tissues in 1989. (— = no data)

Specimen number	Liver					Blubber					Brain				
	PR	PH	ratio	UCM	cpi	PR	PH	ratio	UCM	cpi	PR	PH	ratio	UCM	cpi
GH-SL-1	1538.6	0	—	13	80	138,009	0	—	30	938					
MH-SL-4	1908	0	—	21	43										
MH-SL-5	0	0	—	18	45.6	127	0	—	46	182.8					
AF-SL-1	327,890	267	1228	5	1.2										
TS-SL-3	3428	0	—	0	1										
MH-SL-1	28,011	0	—	0	15										
TS-SL-2	848	0	—	0	42										
BC-SL-1	761	0	—	0	4										
TS-SL-586											126	24	5.3	0	3.4
TS-SL-582											37	13	2.9	1.1	170.6
TS-SL-590	29,928	46	651	6.2	1.8	47,203	29	1628	0	73.9	296	15	20	2.5	121.6
TS-SL-591	5877	27	218	12.7	2	106,563	40	2664	127	114.8	323	0	—	77.4	4.1
TS-SL-592	157	0	—	21.3	1.8	105,161	81	1298	81	31.7	45	18	2.5	46.5	143
TS-SL-593	118,679	37	505	30.4	4.3	71,681	43	1667	67.2	219.6	60	22	2.7	76.2	148.8

or histological lesions were found in premature pups. None of the lesions found were linked to exposure to hydrocarbon contamination.

DISCUSSION

Historically, sea lion abundance has been determined by aerial surveys and ground counts. (Calkins and Pitcher 1982; Loughlin et al. 1984; Merrick et al. 1987). We attempted to take advantage of the historical data base in our investigation of the effects of the EVOS on sea lions. Aerial surveys of sea lions traditionally cover broad geographical areas and are expensive, time consuming, and dangerous. Consequently, surveys have not been conducted on a regular basis and the data are incomplete. Generally, the types of data collected on sea lions in the past have been used to show trends in the population but are far less appropriate for separating and identifying sources of impact in terms of population reductions. The power calculations in the linear regression analyses from our study were consistently low because of these complications.

In the analysis of adult and juvenile counts at rookeries, 1260 (23%) fewer sea lions were counted than the model predicted. However, the adult/juvenile counts at haulout sites totaled 811 (20%) more sea lions than the model predicted. It is possible that some redistribution or even reduction in numbers of adults and juveniles did occur beyond that which was expected from the ongoing sea lion decline. This was not detected in terms of the statistical analysis because of the limitations in the data from prior counts. These limitations are illustrated by the consistently low power of the statistical tests. Power of the analysis of adult/juvenile counts at haulout sites was particularly low. It may have been difficult to statistically confirm an oil spill effect unless all of the sea lions were missing from the haulout sites. Counts at haulout sites viewed alone cannot be used for more than an assessment of redistribution.

The actual number of pups counted at rookeries was 11% lower in 1990 than the model predicted. This analysis also suffered from the same limitations as the adult/juvenile count analyses. The difference between the actual and predicted pup counts was not as large as the difference in the adult/juvenile counts at rookeries, but the power of the pup analysis was lower.

Premature pupping ratios were different at the two areas, with the highest occurring nearest the spill. However, this by itself cannot be considered an oil spill effect because overall pup abundance was not shown to have been significantly affected by the spill. Lack of historical data on Chirikof Island prevented us from separating site differences from an oil spill effect.

Sea lions that were near or contacted oil could have become contaminated with hydrocarbons internally through inhalation, contact and absorption through the skin, or ingestion either directly or by consuming contaminated prey (Engelhardt

≫ ≫ ≫

et al. 1977; Engelhardt 1987). There is little doubt that sea lions were exposed to oil inside PWS and in the northern Gulf of Alaska. Sea lions were sighted swimming in or near oil slicks, oil was seen near numerous haulout sites, and oil fouled the rookeries at Seal Rocks and Sugarloaf Island. One of the collected sea lions (TS-SL-582) was sighted at The Needle in PWS on 18 April 1989, then collected at the same location on 3 July 1989. This animal likely remained in that vicinity and should have contacted oil. However levels of contaminants measured in the bile or other tissues from this animal were very low.

Mammals have the ability to metabolize hydrocarbon contaminants to some degree (Addison et al. 1986), however, the amount of hydrocarbon contaminants that can be tolerated by marine mammals is unknown. All of the sea lions collected in PWS in October had high enough levels of metabolites of aromatic hydrocarbons in the bile to confirm exposure and active metabolism at the tissue level. These sea lions had a mean level of PHN FACs of 14,900 ng equivalents/g. Harbor seals (*Phoca vitulina*) sampled in PWS during 1989 had a mean phenanthrene level of 43,200 ng/g (Frost et al., Chapter 19), 2.9 times higher than sea lions. The highest levels of PHN FACs measured in harbor seals were from a heavily oiled female (110,000 ng/g) and a nursing pup of another heavily oiled female (215,000 ng/g). These harbor seals had PHN levels 2.8 and 7.4 times the highest PHN level measured in sea lions. Note, however, that it is difficult to make interspecies comparisons of PHN equivalents. For example, Dolly Varden and halibut injected with the same relative amount of *Exxon Valdez* oil had different proportions of AC metabolites (Krahn et al. 1992).

CONCLUSIONS

None of the data presented and analyzed provided conclusive evidence of an effect of the *Exxon Valdez* oil spill on Steller sea lions. Predicted values generated by regression models were not significantly different from actual counts. Data collected on premature pupping showed significantly higher premature pupping ratios at a haulout site nearer the oil spill compared to a haulout site farther away. However, an oil spill effect cannot be shown based on this evidence alone. This investigation was designed to detect large changes in premature pupping ratios which should have been coupled with a reduction in pup production, but numbers of premature pups were relatively small which complicated the analysis.

The data collected and analyzed in this study showed that the previously observed decline (Loughlin et al. 1984; Merrick et al. 1987; Byrd 1989; Loughlin et al. 1992) continued throughout the period. The rate of decline was as predicted by the regression models. Data from previous counts allowed only very low power statistical analysis and no significant oil spill effect was detected.

⋈ ⋈ ⋈

When all tissues from a single individual were compared, toxicant levels were not consistently high enough to confirm contamination. However, contamination cannot be determined by looking for aromatic compounds in tissues only, because the hepatobiliary is very efficient and eliminates the metabolized AC compounds via bile. As a result, ACs are found only at relatively low levels in other tissues. Our studies showed that some sea lions were exposed to oil and that they were metabolizing and excreting metabolites of aromatic hydrocarbons into the bile. No lesions related to hydrocarbon contamination were found during histological examinations. No evidence indicated damage caused to sea lions from toxic effects of oil.

ACKNOWLEDGMENTS

Although too numerous to mention by individual, we are grateful to the many ADF&G and NMFS employees who contributed to this study. M. Krahn and D. Brown provided useful comments on our reporting of the chemical analysis. Two anonymous reviewers provided useful comments on an earlier version of the chapter. This work was funded by the NMFS, ADF&G, and as an NRDA study through the *Exxon Valdez* Oil Spill Trustee Council.

REFERENCES

Addison, R. F., P. F. Brodie, A. Edwards, and M. C. Sadler. 1986. Mixed function oxidase activity in the harbour seal (*Phoca vitulina*) from Sable Is., N.S. Comparative Biochemical Physiology 85C(1):121–142.

Agresti, A. 1990. Categorical data analysis. John Wiley and Sons, New York, New York. 558 p.

Boem, P. D., M. S. Steinhauer, D. R. Green, B. Fowler, B. Humphrey, D. L. Feist, and W. J. Cretney. 1987. Comparative fate of chemically dispersed and beached crude oil in subtidal sediments of the Arctic nearshore. Arctic 40 (supplement 1):133–148.

Braham, H. W., R. D. Everitt, and D. J. Rugh. 1980. Northern sea lion decline in the eastern Aleutian Islands. Journal of Wildlife Management 44:25–33.

Byrd, G. V. 1989. Observations of northern sea lions at Ugamak Island, Buldir and Agatu Islands, Alaska in 1989. Unpublished report, U.S. Fish and Wildlife Service. Alaska Maritime National Wildlife Refuge, P.O. Box 5251, NSA Adak, FPO Seattle, Washington 98791.

Calkins D. G., and E. Goodwin. 1988. Investigation of the declining sea lion population of the Gulf of Alaska. National Marine Mammal Laboratory contract NA-ABH-00029. Alaska Department of Fish and Game, 333 Raspberry Road, Anchorage, Alaska 99502. 76 p.

Calkins, D. G., and K. W. Pitcher. 1982. Population assessment, ecology, and trophic relationships of Steller sea lions in the Gulf of Alaska. Environmental Assessment of the Alaskan Continental Shelf, Final Reports 19:445–546.

Cook, R. D. 1977. Detection of influential observations in linear regression. Technometrics, 19:15–18.

〜〜 〜〜 〜〜

Engelhardt, F. R. 1987. Assessment of the vulnerability of marine mammals to oil pollution. Pages 101-115, *in* J. Kiuper and W. J. Van Den Brink (eds.), Fate and effects of oil in marine ecosystems. Martinus Publishing, Boston, Massachusetts.

Engelhardt, F. R., J. R. Geraci, and T. J. Smith. 1977. Uptake and clearance of petroleum hydrocarbons in the ringed seal, *Phoca hispida.* Journal of the Fisheries Research Board Canada 34:1143–1147.

Farrington, J. W., and B. W. Tripp. 1977. Hydrocarbons in western north Atlantic surface sediment. Geochemica et cosmochemica Acta 41:1627–1641.

Krahn, M. M., M. S. Myers, D. G. Burrows, and D. C. Malins. 1984. Determination of metabolites of xenobiotics in the bile of fish from polluted waterways. Xenobiotica 14:633–646.

Krahn, M. M., L. K. Moore, R. G. Bogar, C. A. Wigren, S. L. Chan, and D. W. Brown. 1988. A rapid high-pressure liquid chromatographic method for isolating organic contaminants from tissue and sediment extracts. Journal of Chromatography 437:161–175.

Krahn, M. M., D. G. Burrows, G. M. Ylitalo, D. W. Browwn, C. A. Wigren, T. K. Collier, S. -L. Chan, and U. Varanasi. 1992 Mass spectrometric analysis for aromatic compounds in bile of fish sampled after the *Exxon Valdez* oil spill. Environmental Science and Technology 26:116–126.

Loughlin T. R., A. S. Perlov, and V. A. Vladimirov. 1992. Range-wide survey and estimation of total number of Steller sea lions in 1989. Marine Mammal Science. 8:220–239.

Loughlin, T. R., D. J. Rugh, and C. H. Fiscus. 1984. Northern sea lion distribution and abundance: 1956-80. Journal of Wildlife Management 48:729–740.

Merrick, R. L., T. R. Loughlin, and D. G. Calkins. 1987. Decline in abundance of the northern sea lion, *Eumetopias jubatus,* in Alaska, 1956-86. Fishery Bulletin, U.S. 85:351–365.

Merrick, R. L., L. M. Ferm, R. D. Everitt, R. Ream, and L. A. Lessard. 1991. Aerial and ship-based surveys of northern sea lions (*Eumetopias jubatus*) in the Gulf of Alaska and Aleutian Islands during June and July 1990. U.S. Department of Commerce NOAA Technical Memorandum NMFS F/NWC-196. 34 p.

Merrick, R. L., D. G. Calkins, and D. C. McAllister. 1992. Aerial and ship-based surveys of Steller sea lions (*Eumetopias jubatus*) in southeast Alaska, the Gulf of Alaska, and Aleutian Islands during June and July 1991. U.S. Department of Commerce NOAA Technical Memorandum NMFS-AFSC-1. 41 p.

National Marine Fisheries Service. 1992. Recovery Plan for the Steller sea lion (*Eumetopias jubatus*). Prepared by the Steller Sea Lion Recovery Team for the National Marine Fisheries Service, Silver Springs, Maryland. 92 p.

Neter, J., and W. Wasserman. 1974. Applied linear statistical models: regression, analysis of variance,and experimental designs. Richard D. Irwin, Inc., Homewood, Illinois. 842 p.

Pitcher K. W., and D. G. Calkins. 1981. Reproductive biology of Steller sea lions in the Gulf of Alaska. Journal of Mammalogy 62:599–605.

Sloan, C. A., N. G. Adams, R. W. Pearce, D. W. Brown, and S. L. Chan. 1993. Northwest Fisheries Science Center organic analytical procedures. *In* G. G. Lauenstein and A. Y. Cantillo (eds.), Benthic surveillance and mussel watch projects analytical protocols 1984-1992. U.S. Department of Commerce, NOAA Technical Memorandum NOS ORCA 71..

Stoline, M. R. 1981. The status of multiple comparisons: simultaneous estimation of all pairwise comparisons in one-way ANOVA designs. The American Statistician 35:134-141.

Withrow, D. E. 1982. Using aerial surveys, ground truth methodology, and haul-out behavior to census Steller sea lions, *Eumetopias jubatus.* Master of Science thesis, University of Washington, Seattle, Washington. 102 p.

✂ ✂ ✂

Chapter 8

Status of Killer Whales in Prince William Sound, 1985–1992

Craig O. Matkin, Graeme M. Ellis, Marilyn E. Dahlheim, and Judy Zeh

INTRODUCTION

Two forms of killer whales exist sympatrically in British Columbia and Puget Sound (Bigg et al. 1987; Morton 1990), and in Prince William Sound (PWS) (Leatherwood et al. 1990; Saulitis 1993). The two forms, termed resident and transient, have never been observed in association, nor have individuals from one form emigrated to the other. Resident and transient killer whales in British Columbia have been separated into two distinct maternally based assemblages by mtDNA analysis (Stevens et al. 1989; Hoelzel and Dover 1991). Differences in the vocal repertoire of the two forms have been reported (Ford and Hubbard-Morton 1990). Behavioral differences between forms (i.e., dive times, call types, prey species) have been described in British Columbia (Morton 1990) and are supported by observations in PWS (Saulitis 1993). The social structure of transient groups is not well understood but appears to differ substantially from that of resident whales. Interchange of individuals for periods of years has been reported for transient groups but not for resident pods (Bigg et al. 1987).

Photographs of individual killer whales in PWS have been collected since the late 1970s. Members of some PWS resident pods and some transient killer whales were photographed as early as 1977 and 1978, respectively, indicating consistent use of PWS by individual killer whales for at least 14 years (von Ziegesar et al. 1986). Killer whales were first assigned to pods in 1984 based on these photographs and known associations between killer whales (Leatherwood et al. 1984). By 1987, 221 individual whales had been photographed in PWS and vital rates estimated for some resident pods (Leatherwood et al. 1990; Matkin et al. 1986; Matkin et al. 1987).

The social structure of resident pods in PWS appears similar to that reported for resident pods in British Columbia (Leatherwood et al. 1990). In both areas, a resident pod is the largest group of whales that travel together over 50% of the time (Bigg et al. 1990). They can be separated by their unique vocal repertoires (Ford 1991). Resident pods are composed of maternal groups which include a female and her offspring of either sex. Maternal groups demonstrate long-term stability; individuals in these groups are added only by birth and are lost only by death (Bigg 1982; Bigg et al. 1990; Leatherwood et al. 1990). Formation of new pods may occur when maternal groups split off from the existing pod. Demographic rates can be estimated by matching photographs of pod members from year to year. In the fall of 1988, prior to the *Exxon Valdez* oil spill (EVOS), demographic data were available for six pods totaling 122 whales of the nine major resident pods that use PWS (Matkin and Saulitis 1994; Matkin, unpublished data).

Studies in British Columbia (Bigg 1982; Bigg et al. 1990; Olesiuk et al. 1990) and Puget Sound (Balcomb et al. 1982) indicate that resident killer whales are long lived with low reproductive and mortality rates. Males may live as long as 50–60 years and females may live as long as 80 years, based on analysis of mortality rates (Olesiuk et al. 1990). Reproductive maturity for killer whales is reported at 12–16 years for both sexes in British Columbia and Puget Sound. Mortality rates did not exceed 5% in any of the 17 years of observation in these areas (Olesiuk et al. 1990)

This chapter describes the decline of the resident AB pod in PWS from 36 whales to 23 whales in the period 1989-1990. The assumed losses due to mortality are viewed in the context of demographics of British Columbia and Washington State resident pods and other PWS resident pods prior to and following the EVOS. Trends in the resighting of transient AT1 group whales are presented which suggest changes within that population. Also described are the distribution and movements of PWS killer whales in the summer following the EVOS. Some observations on the reaction of killer whales to sheen oil on the water are included.

METHODS

Research Platforms

Three field camps were used in 1989 (Fig. 8.1). Camp 1 operated from 27 May to 31 August on Squire Island in southwestern PWS. Camp 2 operated in Double Bay, Hinchinbrook Island, in the southeastern portion of PWS and operated from 7 July to 4 September. Camp 3 operated in northwestern PWS on the south end of Perry Island from 6 June to 3 July and moved 14 miles south to Point Nowell from 3 July to 31 August. A 5.5-m cabin skiff with an 88-hp outboard was used at Camp 1, a 6.4-m double-ended cabin skiff with an 88-hp outboard was used at Camp 2, and a 4.7-m inflatable boat with a 40-hp outboard was used at Camp 3. Only Camp 1 was used in 1990, 1991, and 1992. From Camp 1, the 6.4-m double-ended cabin skiff was used from 15 May to 1 September 1990, 5 June to 1 September 1991, and

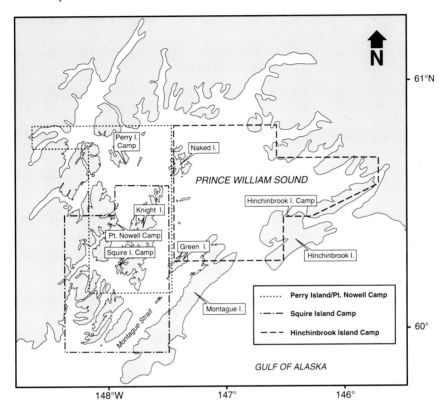

Figure 8-1. Map depicts the study area in Prince William Sound and location of three study camps at Point Nowell, Squire Island, and Hinchinbrook Island. The approximate area covered by researchers at each camp is shown.

29 May to 30 August 1992. In addition, a 7.9-m live-aboard vessel powered by a 165-hp diesel inboard/outboard was used from 21 May to 30 August 1990, from 8 July to 10 September 1991, and from 27 July to 6 September 1992. This vessel centered its range of activity in upper Montague Strait but made biweekly searches into northwestern PWS and Hinchinbrook Entrance (southeastern PWS) in all 3 years.

In all years, the 12.8-m vessel F.V. *Lucky Star* and a 6.4-m inboard/outboard-powered skiff were used as a mobile research platform and to supply food and fuel to the camps and vessels. This vessel operated for approximately 30 days each year (1989–1992), including the period from late August to mid-September. Studies after the EVOS employed photographic techniques similar to those used to collect photographs in 1984–1988. Field work during 1984–1988 occurred during the same months as in the postspill studies and employed similar vessels, although the yearly effort was less than in 1989–1992.

Data Collection

In all years, research crews attempted to maximize the number of contacts with each pod to ensure sufficient photographs of each individual within the pod. Searches for whales were not random but were based on current and historical sighting information. An encounter was defined as the successful detection, approach, and taking of identification photographs. Notification of whales from other vessels (generally by VHF radio) were termed "reports." Although reports were used to select areas to be searched, all identifications used in our analysis were made from photographs taken during encounters.

In 1989, whales were located visually during regular searches of the region adjacent to the field camps. In prior and subsequent years, searches centered in areas that had produced the most encounters with killer whales in the past. In all years, whales were found by listening for killer whale calls with a directional hydrophone or by responding to other vessels reporting sightings of whales.

A vessel log and chart of the vessel track were kept for each day of operation. The elapsed time and distance traveled were recorded and the vessel track was plotted. A record was made of time and location of all whale sightings, and weather and sea state were noted at regular intervals. Specifics of each encounter with killer whales were recorded, including date, time, duration, and location of the encounter. Rolls of film exposed and the estimated number of whales photographed also were recorded. A chart of the whales' trackline during the encounter was completed, and the distance traveled by the vessel with the whales calculated. The presence or absence of oil on the water was recorded. General behavior of the whales (i.e., feeding, resting, traveling, socializing, and milling) was recorded.

Photographs for identification of individual killer whales were taken of the port side of each whale showing details of the dorsal fin and white saddle patch. Photographs were taken at no less than 1/1000 second using Ilford HP5, a high-speed black and white film, exposed at 1600 ASA. A Nikon 8008 autofocus camera with internal motor-drive and a 300-mm f4.5 autofocus lens was used. When whales were encountered, researchers systematically moved from one subgroup (or individual) to the next keeping track of the whales photographed. If possible, individual whales were photographed several times during each encounter to ensure an adequate identification photograph. Whales were followed until all whales were photographed or until weather and/or darkness made photography impractical.

Data Analysis

All photographic negatives were examined under a Wild M5 stereomicroscope at 9.6 power. Identifiable individuals in each frame were recorded; uncertain

∞   ∞   ∞

identifications were not included in the analysis. Unusual wounds or other injuries to the killer whales were noted.

The alphanumeric code used to label each individual was based on the system of Leatherwood et al. (1984) and Heise et al. (1992). The first character in the code is "A" to designate Alaska, followed by a letter (A–Z) indicating the individual's pod. Individuals within the pod receive sequential numbers. For example, AB3 is the third whale designated in the AB pod. New calves were identified and designated in their pod with the next available number. Both published and unpublished photographs of individual whales were used in our analysis (Heise et al. 1992).

Except for one instance, new calves were already present when fieldwork began so their birth date could not be determined. We followed the method of Olesiuk et al. (1990) and used a January date for the birth of all calves in our calculations. Thus, birth rates could not be measured, and recruitment rates represent the survival of calves to about 0.5 years of age.

The determination of mothers of new calves was based on the consistent close association of calves with a female. Although young calves may travel with other individuals, a majority of time is spent with the mother as demonstrated by association analysis of identification photographs from repeated encounters (Bigg et al. 1990). The white saddle patch of calves generally does not develop for several years, but other scars and marks, including the shape of the white eye patch, are used to reliably reidentify calves.

If a whale from a resident pod was not photographed swimming alongside other members of its maternal group during repeated encounters over the course of the summer season, it was considered missing. If it was again missing during repeated encounters in the following summer season, it was considered dead.

No individual resident whale known to be consistently missing during repeated encounters with its pod and maternal group over the course of a summer season has ever returned to its pod or appeared in another pod in all the years of research in Canada and the United States (Bigg et al. 1990; Leatherwood et al. 1990). Subgroups of resident pods may travel separately from the pod for a season or longer; however, this has not been observed for individuals. In a few instances, missing whales have been found dead on beaches, but strandings of killer whales are infrequent events and most missing whales are never found. During 1975–1987, only six killer whales were found on beaches throughout the entire Gulf of Alaska (Zimmerman 1991). One explanation for the lack of recorded dead killer whales comes from the observations of early Soviet researchers where killer whales shot for specimens sank (Zenkovich 1938).

Immigration and emigration may occur between groups of transient whales. In British Columbia, infrequently sighted transients missing from their original groups for periods ranging from several months to several years or more have been

><> ><> ><>

resighted swimming with other groups of transient whales (Ellis, unpublished data). For this reason, transient whales missing from a particular group for several years cannot necessarily be considered dead. However, members of the AT1 group of transients have been resighted consistently in PWS since 1983 and records of these resightings are presented and discussed.

Finite annual mortality rates (MR) and reproductive rates (RR) for resident pods were calculated as follows:

NM = number of whales missing from a pod in a given year,
NP = number of whales present in a pod at the end of previous year,
NR = number of calves recruited to 0.5 years in a pod in a given year, then
MR = NM/NP and RR = NR/NP.

If the year a mortality or recruitment occurred could not be determined, it was split between the possible years. A mean-weighted mortality and reproductive rate for all pods for all years was determined by pooling the data for all pods for all years.

The sex and age group of missing whales were determined, when possible, using data collected before their disappearance. In some cases, sex had been determined by viewing the ventral side of the whale. Reproductive females were identified by the presence of offspring since 1983. Adult female whales that had not calved since 1983 and were not accompanied by a juvenile(s) were considered as postreproductive. Exact ages of whales could be determined only for whales born since 1983. Juveniles born before 1984 were given approximate ages by comparing the relative size of the whale and development of its saddle patch and dorsal fin as shown in photographs. At about 15 years of age, males are readily identified because their dorsal fin is taller and less falcate than females. At sexual maturity, male fin height will exceed the width by 1.4 times (Olesiuk et al. 1990). The fin continues to grow until physical maturity (about 21 years of age).

Ideally, the analysis of mortality rates for the resident pods should be based on an age- and sex-structured population dynamics model. However, complete age and sex data were not available and mortality data were not complete because not all pods were observed in all years. In addition, only a few years of baseline data were available prior to the EVOS. Thus, only two statistical models were used to assess the significance of post-EVOS mortality.

The first model focused on the maximum annual mortality observed for each pod over the 8-year period of available data. An expected maximum annual mortality during an 8-year period was estimated from data given by Olesiuk et al. (1990). Probabilities of maximum observed mortalities in PWS resident pods were computed from binomial probability distributions with the probability parameter representing the expected maximum annual mortality.

✕◇　✕◇　✕◇

The second model used logistic regression to assess whether the observed mortality rates could be explained by pod and year. If m is the expected mortality rate, the logistic regression model assumes that the logit of m, that is, the natural logarithm of $m/(1-m)$, can be expressed as a linear combination of the predictors. Under this assumption, m will always be estimated to lie between 0 and 1 (i.e., 0% and 100%), as in fact it must. The regression coefficients in the linear combination are estimated by maximum likelihood (McCullagh and Nelder 1983), assuming that each whale in a particular pod that is alive during a particular year has the same probability of dying, and that what happens to one whale is independent of what happens to others. The estimation technique takes into account that mortality rates can be estimated more precisely for large pods than for small.

Differences in age and sex structure between pods and years that might affect their mortality rates have been taken into account by including pod and year effects in the model. Pod dynamics may lead to departures from the random distribution that is assumed in this model, resulting in a clustered distribution of deaths. To accommodate this possibility, the logistic regression model allows for the estimation of a larger variance than implied by the assumptions if the data indicate that this is appropriate.

Because of missing data for some pods for some years, and because killer whale mortality rates are so low that no deaths were observed in one of the pods, the best estimate for m in some cases was $m=0$. The logit of 0 is negative infinity. To mitigate these problems, the smallest pods were grouped together and 2-year, instead of 1-year, time periods were considered in the analysis.

RESULTS

Movements

Areas surveyed were similar in 1989, 1990, 1991, and 1992, however, less time was spent in eastern PWS after 1989. In 1991 and 1992 there was reduced effort in June and early July which resulted in fewer survey days and fewer miles surveyed than in 1989 and 1990 (Table 8-1). Survey effort during the peak field season (mid-July to mid-September) was comparable in all 4 years.

Survey areas were similar in 1984–1988; however, the greatest effort (kilometers surveyed) during these years was recorded in 1984 and was considerably reduced in later years (see Dahlheim and Matkin, Chapter 9).

For all years the area of greatest survey effort was southwestern PWS, although other areas of the Sound, particularly the northwestern portion and parts of eastern PWS (Perry Island through Hinchinbrook Entrance), were regularly surveyed. Effort was spread most evenly among the eastern, northwestern, and southwestern

≈ ≈ ≈

Table 8-1. Summary of effort expended in searching for and photographing killer whales during 1989-1992.

	Year			
	1989	1990	1991	1992
Dates	23 May - 15 September	15 May - 8 September	16 May - 13 September	29 May - 6 September
Survey days	260	247	159	113
Weather days	44	34	20	16
Km surveyed	15,397	17,045	11,744	7680
Whale encounters	89	80	54	69
Km with whales	1662	2112	1432	1582
Search km/encounter	172	212	218	112
Reports from vessels	153	119	54	47
Reports/survey day	0.58	0.48	0.33	0.42
Film frames exposed	6600	9300	5800	6200

Table 8-2. Rates of killer whale encounters and the number of pods encountered by nautical mile (nm) listed by region of Prince William Sound for 1989.

Region		Residents		Transients
	Total nm	Encounters/ 100 nm	Pods encountered/ 100 nm[a]	Encounters/ 100 nm
Southwest	4740	0.69	1.03	0.38
Northwest	2226	0.33	0.45	0.34
Southeast	2657	0.45	0.54	0.27

[a] For some encounters more than one resident pod was present.

regions in 1989 when separate field camps were established following the EVOS. Encounter rates are given by region only for this year.

Encounter rates with resident whales were higher in southwestern than in northwestern or eastern PWS (Table 8-2). The number of pods encountered per nm was also higher in the southwest than in other regions. Encounter rates for transient groups were similar in all areas.

In 1989, all major resident pods photographed in eastern PWS were also photographed in western PWS (pods AB, AI, AK, AE, AN10). However, the AJ pod was photographed only in the southwest and the AD pod was encountered only once in the northwest. All transient groups were sighted in both eastern and western PWS, except for the AT60 group which was photographed once in eastern PWS.

All of southwestern and part of northwestern PWS were oiled following the EVOS while most of eastern PWS was not. Some pods were photographed in an unoiled area and then photographed in an oiled area a few days later. For example, on 27 June 1989, members of the AK pod were photographed near Fairmont Island in the unoiled portion of northwestern PWS and 23 hours later they were photographed at Chenega Point in the oiled southwestern region 67 km distant. On 27 July, members of the AB pod were photographed near Johnstone Point in unoiled eastern PWS and subsequently were encountered on 30 July in oiled upper Knight Island Passage in the northwestern PWS, 80 km distant. Killer whales were observed swimming through oil on five occasions (Table 8-3) and did not appear to avoid oil slicks.

Table 8-3. Observations of killer whales in oil with a description of location and oil type.

Date	Location	Pod(s)	Oil type
31 March 1989	Knight Island Pass	AB, AJ	heavy sheen
2 April 1989	Knight Island Pass	AK	sheen
5 August 1989	5 mi SW Bligh Reef	AT	sheen
2 September 1989	0.5 mi W Point Grace	AB, AI	moderate sheen
3 September 1989	0.5 mi W Point Helen	AB, AN, AI	sheen

Vital Rates

Resident pods

Although 13 or more resident pods have been photographed in PWS since 1983, all but 7 (AB, AI, AE, AK, AJ, AN10, AN20) are seen rarely and could not be included in our analysis. Prior to 1989, the AN pod was considered a single pod. However, the AN10 and AN20 subpods were photographed together in only 2 of the 69 total encounters with these groups from 1988 through 1992. By definition (Bigg et al. 1990), the AN10 and AN20 subpods are now considered separate pods.

Births and deaths were recorded for each of the seven pods that have been repeatedly encountered since 1988 (Tables 8-4 and 8-5), and mortality rates have been calculated (Table 8-6).

In 1984, the AB pod contained 35 individuals. Between 1984 and 1988 there were eight mortalities and nine calves recruited to at least 0.5 years of age within the pod. In September 1988, the last time the AB pod was photographed prior to the oil spill, it contained 36 whales. Only 29 whales were present when the pod was photographed on 31 March 1989, 7 days after the EVOS. An additional 6 whales were missing between September 1989 and June 1990, totaling 13 missing from March 1989 to June 1990. The corresponding MR for the AB pod is 19.4% in 1988–1989 and 20.7% in 1989–1990. There was one additional mortality in the AB pod in 1990–1991 (4.3%) and none in 1991–1992. There were no new calves born into the pod in 1988–1989 or 1989–1990; one calf was born in 1991 and two calves were born in 1992. The pod totaled 25 whales in 1992.

The significance of the high post-EVOS mortality in the AB pod was assessed by comparing the maximum annual mortality for the AB pod during the 8-year period of observation with an expected maximum computed from data on 19 resident pods in British Columbia and Washington State. Olesiuk et al. (1990) gave complete data on births and deaths in these pods over the 8-year period starting in 1979 when the population numbered 221, and ending in 1986 when the population was 252. Mortality during this period should represent natural mortality since the

∞   ∞   ∞

Table 8-4. Identification numbers of killer whales missing from Prince William Sound pods from 1985 to 1992.

Pod	84/85	85/86	86/87	87/88	88/89	89/90	90/91	91/92
AB [35/25]	9 15 34	1 7 12	28	6	13 18 21 23 30 31 37	8 19 20 36 42 44	29	
AI [6/6]								
AK [7/10]		5						
AE [13/14]	8	4	a	7	12			
AJ	a	23						9
AN10 [12/16]			6		2			
AN20 [23/*]					a		a	a

[a] Entire pod not encountered in that year.
[ ] Number of whales in pod 1984/1992.
AN10 and AN20 were considered a single pod prior to 1989.

period postdates both the passage of protective legislation and the live-capture fishery for killer whales that operated between 1962 and 1977.

From Olesiuk et al. (1990), an estimated expected maximum mortality for a pod during an 8-year period was calculated by considering the year of maximum mortality for each pod. This estimate was 0.078, calculated from a total of 17 deaths among 217 whales. The observed number of deaths per year in all pods described by Olesiuk et al. (1990) and all PWS pods except AB pod was consistent with a binomial probability distribution in which the probability of death was 0.078. The probability of the observed number of deaths or more in each of these pods was

Table 8-5. Identification numbers of whales recruited to approximately 6 months of age in Prince William Sound resident killer whale pods during 1985-1992.

Pod	Year							
	84/85	85/86	86/87	87/88	88/89	89/90	90/91	91/92
AB [35/25]		36(23) 37(6)	38(31) 39(25)	40(14) 41(8) 42(32) 43(17) 44(22)			45(16)	46(25) 47(32)
AI [6/6]								
AK [7/10]	8(6)	9(2)				10(2)	11(6)	
AE [13/14]	13(11)		a	15(10)	16(2) 17(5)	18(11)		
AJ [25/31]		a			26(22) 27(20) 28(24)	29(8)	30(3)	31(24) 32(22) 33(13)
AN10 [12/16]			38(10)		40(35)	41(8)	45(35)	46(10) 47(11)
AN20 [23/ᵃ]		36(23)	37(17)	39(20)	a	42(26) 43(29) 44(31)	a	a

a Entire pod not encountered in that year.
[ ] Number of whales in pod 1984/1992.
( ) Identification number of mother following number of new calf.
AN10 and AN20 were considered a single pod (AN) prior to 1989.

greater than 0.15. However, the probability was only 0.023 that six or more deaths would occur in a pod of 29 whales, as occurred in the AB pod in 1989–1990.

The other six frequently photographed PWS resident pods (AI, AK, AE, AJ, AN10, AN20) totaled 86 whales in 1984. There were three mortalities and 14 calves recruited to 0.5 years between September 1988 and September 1992 in these six pods. These pods contained approximately 105 whales in 1992. None had declined since the EVOS. Data for the AN20 pod in 1990/1991 and 1991/1992 was incomplete so these years were excluded from this tally.

Table 8-6. Annual mortality rates (%) for Prince William Sound killer whale pods 1985-1992.

Pod				Year				
	84/85	85/86	86/87	87/88	88/89	89/90	90/91	91/92
AB	8.6	9.4	3.2	3.1	19.4	20.7	4.3	0.0
AI	0.0	0.0	0.0	0.0	0.0	0.0	0.0	0.0
AK	0.0	12.5	0.0	0.0	0.0	0.0	0.0	0.0
AE	7.7	7.7	0.0	8.3	8.3	0.0	0.0	0.0
AJ	0.0	4.0	0.0	0.0	0.0	0.0	0.0	3.4
AN10+ AN20	0.0	0.0	2.8	0.0	2.6	0.0	NA	NA

Mortality rate for all pods, all years = 3.4%.

Table 8-7. Mortalities in Prince William Sound resident pods following the *Exxon Valdez* oil spill (1989-1991) with information on sex and age class.

Pod	Whale	Year died	Sex	Age class
AB	13	1989	F?	juvenile (about 8 years)
	18	1989	?	juvenile (about 8 years)
	21	1989	F	adult
	23	1989	F	adult (mother of AB36)
	30	1989	?	juvenile (about 13 years)
	31	1989	F	adult (mother of AB38)
	37	1989	?	juvenile born in 1986
	8	1990	F	adult (mother of AB41)
	19	1990	M	maturing (about 15 years)
	20	1990	F?	juvenile (about 14 years)
	36	1990	?	juvenile born in 1986
	42	1990	?	juvenile born in 1988
	44	1990	?	juvenile born in 1988
	29	1991	M	recently mature (about 18 years)
AE	12	1989	F?	adult
AN	2	1989	M	adult

Table 8-7 summarizes the sex and approximate age of whales missing from three of the seven regularly photographed PWS resident pods since 1988. Of these missing whales, AE12 was considered an adult whale in 1984, probably a female, although it had not produced a calf. Male AN2 was sexually mature when first photographed in 1983. All 14 whales missing from the AB pod were juveniles (8

Table 8-8. Mortality rates for reproductive females in resident pods calculated for Prince William Sound and Canada.

	Number whales	Number whale years	Number dead	Annual rate (%)
AB pod (89-91)	10	27	3	11.1
Other pods in PWS (89-91)	21	54	0	0
Canada[a] (74-87)	73	623	3	0.48

[a] From Olesiuk et al. 1990.
Whale years = summation of the number of years each whale was observed.

Table 8-9. Annual mortality rates for juvenile killer whaler calculated for Prince William Sound and Canada.

	Number whales	Number whale years	Number dead	Annual rate (%)
AB pod (89-91)	18	42	9	21.4
Other pods in PWS (89-91)	37	95	0	0
Canada[a] (74-87)	195	1336.8	24.2	1.8

[a] From Olesiuk et al. 1990.
Whale years = summation of the number of years each whale was observed.

total) or females that had produced a calf since 1984 (Table 8-7), except for 2 maturing males (AB19, AB29) and an adult female (AB21).

The MRs for reproductive females (Table 8-8) and juveniles (Table 8-9) for resident killer whale pods in PWS were compared to rates for British Columbia killer whales (Olesiuk et al. 1990). Females were considered reproductive if they had produced a calf since first observed in 1983 or 1984. The age of one maturing male (AB19) was estimated at <15.5 years, which was considered a juvenile in our calculations. The MR for the AB pod was over 20 times higher for reproductive females and 10 times higher for juveniles than for British Columbia pods.

These analyses provide evidence that post-EVOS mortality in the AB pod exceeded natural mortality. However, they assume similar rates occur in populations of resident whales in PWS and British Columbia and Washington. For this reason, PWS data alone were used to investigate mortality in a logistic regression model. In this model, the logit of the expected annual mortality rate (m) for a given pod during a given 2-year period was assumed to be the sum of the value expected for the AB pod in 1988–1990, the difference between that pod and the AB pod, and the difference between that 2-year period and 1988–1990.

The pod effects allow for differences in age and sex composition between the pods over the 1984–1992 time period, as well as for effects of the EVOS or other sources of mortality related to human activities. Because the pod experiencing the highest mortality is contrasted with the others, the significance of the pod effects is not easily assessed. They are included in the model in order to obtain estimates of year effects that are adjusted for differences between pods.

Since year effects are averaged over pods, significant effects due to age and sex composition would not be expected. Thus, year effects indicate whether overall mortality, adjusted for differences between pods, was higher during the post-EVOS period than at other times.

The expected annual mortality rates computed from this model and the observed rates are given in Table 8-10. The model demonstrated acceptable fit with no evidence of larger variance than the assumed binomial variance. The final deviance of 13.09 falls well within one standard deviation of the value expected under the model (expected value=12, SD=4.9, χ^2, df=12).

The highest observed and expected annual mortality rates were for the AB pod in 1988–1990, with lower values for other pods and years. The model coefficients divided by their standard errors are approximately normal with mean=0 and variance=1 when the true differences that they represent are zero. Standardized coefficients are given in Table 8-11, along with P values for the year effects. The P values indicate whether differences in mortality rates between 1988 and 1990 and the other 2-year periods depart significantly from zero. P values for pod effects could not be given. Because the pod with the highest mortality rate was contrasted with the others, a negative rather than a zero mean is expected for these differences.

Table 8-11 indicates that mortality was significantly higher in 1988–1990, following the EVOS, than in 1986–1988 or 1990–1992. It was higher, but not significantly higher, than in 1984–1986, when whales were being shot as a consequence of their interactions with the sablefish (*Anoplopoma fimbria*) longline fishery.

Transient whales

Members of the AT1 group have been the most frequently encountered transient whales in PWS every year since 1984 (Matkin, unpublished data). Unlike other transient whales that have been photographed, members of this group appear to spend much of their time in PWS.

Many of the individuals in the AT1 group were not observed or photographed in 1990, 1991, and 1992. These include AT5, AT7, AT8, and the whales photographed with AT6 near the *Exxon Valdez* shortly after the spill. Also not photographed were AT15, AT16, AT19, AT20, AT21, and AT22 (Table 8-12). Of note

✂ ✂ ✂

Table 8-10. Expected annual mortality rates (m) computed from a logistic regression model which includes effects of year and pod, and observed rates (obs) by pods and by years. Rates are expressed as percentages.

	Years							
Pods	1984-1986		1986-1988		1988-1990		1990-1992	
	m	obs	m	obs	m	obs	m	obs
AB	10.8	9.0	4.6	3.2	16.2	20.0	2.9	2.2
AI + AK	1.1	3.7	0.4	0.0	1.8	0.9	0.3	0.0
AE	4.9	7.7	2.0	4.2	7.6	4.0	1.3	0.9
AJ	1.2	2.0	0.5	0.0	1.9	0.0	0.3	1.7
AN	0.9	0.0	0.4	1.4	1.4	1.3	0.2	0.0

Table 8-11. Standardized coefficients for the differences in mortality rates between AB pod and other pods and between the 1988-1990 post-EVOS period and other 2-year periods, from the logistic regression model. Significance of the year effects is indicated by P values.

Pods	Coef/SE	Two-year periods	Coef/SE	P values
AI + AK	-2.32	1984-1986	-1.09	0.178
AE	-1.52	1986-1988	-2.39	0.017
AJ	-3.12	1990-1992	-2.44	0.015
AN	-3.51			

Table 8-12. Summary of identifications of AT1 group for years 1989-1992.

Year	Whale identification number																					
	1	2	3	4	5	6	7	8	9	10	11	12	13	14	15	16	17	18	19	20	21	22
1989	x			x	x	x	x	x	x	x		x	x		x	x		x		x	x	x
1990	x	x	x	x		x			x	x	x	x	x	x			x	x				
1991	x	x	x	x		x			x	x		x	x	x			x	x				
1992	x	x	x	x		x			x	x			x	x			x	x				

x = Whale present in photographs.

is the disappearance of AT5, AT7, and AT8 who have frequently associated with AT6 in the past. AT6 was repeatedly photographed in 1990, 1991, and 1992 without these whales. Because immigration and emigration between transient groups does occur, the missing whales cannot be considered mortalities, except for AT19 which was identified dead on a beach in eastern PWS in 1990. There has been no recruitment into the AT1 group since 1984.

Strandings

Three killer whale carcasses were found in 1990 and one in 1992 on beaches in PWS. Three had marine mammal parts in their stomachs suggesting they were transient whales (one of these was identified as AT19). Transient whales have been the only killer whales observed feeding on marine mammals in PWS. One whale had a halibut hook in its stomach suggesting that the whale had interacted with the longline fishery or consumed a halibut with a hook in its mouth. There have been no other dead killer whales observed or reported within PWS since systematic killer whale photo-identification work began in 1983. Zimmerman (1991) reported no stranded killer whales in PWS from 1975 to 1987.

Fin collapse

The dorsal fins of two adult males (AB2,AB3) have folded since 1988. The collapse began in 1989, accelerated in 1990, and was complete by 1991. The dorsal fin of both whales is now flattened against the body (see Heise et al. 1992 for photographs). One of these whales (AB3) had a bullet wound in the dorsal fin in 1985. There was no external deterioration of the fin in either whale. This affliction was observed in PWS in one other whale (AN1) whose fin declined in 1985 and 1986.

DISCUSSION

The higher rate of encounter with killer whales in southwestern PWS compared to eastern or northwestern PWS may be due to several factors. The narrow passages in southwestern PWS may facilitate locating whales. Montague Strait appears to be a primary route of transit for whales entering and leaving PWS. Also, southwestern PWS is the primary point of entrance for returning salmon during summer months. Killer whales are often observed feeding on salmon in this area. The higher rate of pod encounters in the southwestern area is due to multipod aggregations that are typically found in that area in late summer and fall. Distribution of resident killer whales may change in the late fall and winter but data are lacking for these time periods.

>< >< ><

The lack of difference in rates of sighting transient killer whales between regions suggests that they may be found with equal regularity in all three regions. Since most encounters were with the AT1 group, this reflects the distribution of members of this group.

Observations indicate that killer whales did not avoid using oiled sections of southwestern PWS. It is likely that nearly all resident whales in PWS eventually swam through heavily oiled sections of southwestern PWS. In addition, observations of whales in the presence of oil sheens indicated that they did not attempt to avoid oil. However, killer whales were not observed in the presence of heavier slicks of crude oil.

Because maternal groups are stable, resident pods can be viewed as discrete units within a larger population. The significance of annual MR of a pod may be exaggerated by small pod size, hence some pods were combined in our statistical analysis. However, the MR observed in AB pod in 1988–1989 (19.4%) and 1989–1990 (20.7%) has no precedent in resident pods of comparable size (20+ whales) studied in Puget Sound, British Columbia, or southeastern Alaska. During the mid-1980s, interactions with the sablefish longline fishery were severe and many bullet wounds were observed on members of the AB pod (Matkin et al. 1986; Matkin et al. 1987; Dahlheim 1988; Leatherwood et al. 1990); however, the MR for the AB pod did not exceed 9.4%.

Comparison of the expected maximum mortality rate for British Columbia/Washington State with the year of maximum mortality for each pod in this area and in PWS was consistent with a binomial distribution for all pods except the AB pod. However, there are factors that might influence this binomial model. The live capture and removal of whales from the British Columbia/Washington population prior to the study in that region may have had some effect on the population structure and mortality rate. Although demographics and social structure appear similar for the two areas, it cannot be assumed that natural mortality rates are the same for British Columbia/Washington and PWS. The binomial distribution also treats each mortality as an independent event and does not take into account social factors and age and sex composition that might influence mortality rates. Despite these limitations, it is noteworthy that only the AB pod maximum mortality rates did not fit this model.

It is extremely unlikely, and without precedent, that the whales missing from the AB pod left the pod to join another pod or form a separate pod. Although larger pods may tend to split on a temporary or permanent basis, whales do not leave their maternal groups. The pod will split along the lines of the maternal groups. Such was the case with the AN pod, now considered by definition (Bigg et al. 1990) as the two pods, AN10 and AN20. The AN pod split about 1988, but one of the new pods was not observed in 1989. The whales in the AN20 pod could not be considered missing since there was no evidence that individuals were missing from their maternal groups. The AN20 pod was encountered and photographed in 1990.

Killer whale juveniles and reproductive females in PWS resident pods and in other areas (Olesiuk et al. 1990) have low mortality rates; AB pod is the exception. However, individuals lost from the AB pod following the EVOS were primarily juveniles and reproductive females. In 1988–1990, three reproductive females in the AB pod were not observed and were presumed dead, leaving calves which were ≤3 years of age. Again, this has no precedent in pods of comparable size in any areas where resident killer whales have been studied.

The simultaneous collapse of the dorsal fins of two adult males (AB2, AB3) within a pod is unusual. AB3 was a recently matured male and the collapse of his fin would not have been related to age. Although he suffered a bullet wound in 1985, there had been no external degradation of the fin since that time. Fin collapse preceded death in one male killer whale (G1) in British Columbia and has been observed in captive killer whales.

It is not clear why six resident pods, other than AB, have increased from 86 whales in 1984 to over 100 whales in 1992. The few mortalities within these pods were suspected to be older whales because they had adult confirmation when first photographed in 1983 or 1984, they had not developed a male dorsal fin, and those that were known to be females had not calved since 1984 or earlier. The structure of these pods has not changed appreciably since 1988 other than by the recruitment of calves. There has been no observed immigration or emigration among resident pods that use PWS or between PWS pods and pods regularly photographed in other areas (e.g., British Columbia, southeastern Alaska, or Puget Sound).

A logistic regression model indicated a significantly higher mortality rate of killer whales in PWS in 1988–1990 than in 1986–1988 or 1990–1992. This rate was not significantly higher than in 1984–1986, though, when bullet wounds were recorded on whales after interactions with the sablefish longline fishery. However, fishery interactions were not apparent factors in mortalities recorded during 1988–1990 (Dahlheim and Matkin, Chapter 9). These results argue that a severe and atypical event(s) occurred in 1988–1990 that caused these mortalities.

Prediction of short- and long-term effects of the AB pod mortalities and the potential for recovery are difficult. In 1991 there were few encounters with the pod and in most of these only part of the pod was present. The loss of three reproductive females may reduce the potential speed of recovery. However, in 1992 AB pod was repeatedly photographed traveling as a cohesive group. One calf was recruited in 1991 and two more were added in 1992. Barring additional perturbations, recovery of the pod to pre-1989 numbers now seems a possibility.

Only one of the nine missing whales from the AT1 group is a confirmed mortality. There is increasing doubt, though, that the other AT1 whales are still alive based on our resighting records from previous years. Unlike other transients

⋈ ⋈ ⋈

photographed in PWS, members of the AT1 group (22 whales in 1988) have been resighted on a regular basis since 1983.

There has been no systematic approach to measuring the stranding rate of killer whales in PWS. Zimmerman (1991) reported no stranded killer whales in PWS from 1985 to 1987. The stranding of four killer whales since 1989 (three in 1990) is noteworthy. The discovery of these carcasses may be related to increased observer effort following the EVOS. However, in years prior to EVOS, the shorelines of PWS were regularly flown during aerial surveys of herring spawning areas and salmon streams and by other private small plane traffic.

CONCLUSIONS

All seven regularly encountered, resident killer whale pods used the oiled areas of PWS in the year following the EVOS. Killer whales did not avoid oiled areas and did not avoid surfacing in oil slicks. Thus, there was a potential for whales to contact oil or consume oiled prey in the spill area. There were substantial losses of individuals within the AB pod only in the 2 years following the EVOS (1989–1990) with the death of 13 whales, primarily juveniles and reproductive females. Although there was a previous history of the AB pod interacting with the PWS sablefish fishery and a significant rise in mortality rates at that time (1985–1986), there was no evidence of fishery-related mortalities in 1988–1990. Analyses demonstrated significantly higher overall mortality rates for PWS resident killer whales in 1988–1990 than for any period except when whales were being shot during sablefish fishery interactions. In addition, it is possible that the nine whales missing for 3 years from the transient AT1 group also are dead, although it is not certain because of the variability of associations between transient whales. The factors potentially responsible for the loss of the 14 AB pod whales are discussed in Dahlheim and Matkin (Chapter 9).

ACKNOWLEDGMENTS

Funding for work from 1989–1991 was provided by the National Marine Mammal Laboratory, Seattle, with monies provided as part of the damage assessment program of the *Exxon Valdez* Oil Spill Trustee Council. Work in 1984 was supported by Hubbs SeaWorld Research Institute and fieldwork in 1986 was supported by the Alaska Sea Grant Program and the National Marine Mammal Laboratory. Data for killer whale pods in PWS in 1985, 1987, 1988, 1992, and in March 1989 following the EVOS was provided by the North Gulf Oceanic Society. The Society was funded by private donations, the Alaska State Legislature, and the Alaska Sea Grant program.

✠   ✠   ✠

The project would not have been possible without the participation of L. Barrett-Lennard, K. Heise, E. Saulitis, O. von Ziegesar, and K. Balcomb-Bartok who led efforts in the field. Field assistance or other substantial contributions were made by R. Angliss, K. Englund, F. Felleman, M. Freeman, B. Goodwin, M. Hare, M. James, L. Larsen, E. Miller, L. Saville, C. Schneider, R. Blancato, S. Sikema, and E. Weintraub. A. York provided advice on analysis of data. E. Miles of Miles Photo Lab provided all photographic development and printing services.

REFERENCES

Balcomb, K. C., J. R. Boran, and S. L. Heimlich. 1982. Killer whales in Greater Puget Sound. Reports of the International Whaling Commission 32:681–686

Bigg, M. A. 1982. An assessment of killer whale (*Orcinus orca*) stocks off Vancouver Island, British Columbia. Reports of the International Whaling Commission 32:655–666.

Bigg, M. A., G. M. Ellis, J. K. B. Ford, and K. C. Balcomb III. 1987. Killer whales: A study of their identification, genealogy and natural history in British Columbia and Washington State. Phantom Press, Nanaimo, British Columbia, 79 p.

Bigg, M. A., P. F. Olesiuk, G. M. Ellis, J. K. B. Ford, and K. C. Balcomb III. 1990. Social organization and genealogy of resident killer whales (*Orcinus orca*) in the coastal waters of British Columbia and Washington State. Pages 383-405, *in* P. S. Hammond, S. A. Mizroch, and G. P. Donovan (eds), Individual recognition of cetaceans: use of photo-identification and other techniques to estimate population parameters. Report of the International Whaling Commission Special Issue 12.

Dahlheim, M. E. 1988. Killer whale (*Orcinus orca*) depredation on longline catches of sablefish (*Anoplopoma fimbria*) in Alaskan waters. Northwest and Alaska Fisheries Center, National Marine Fisheries Service, NWAFC Processed Report, 88-14. 31 p.

Ford, J. K. B. 1991. Vocal traditions among resident killer whales (*Orcinus orca*) in coastal waters of British Columbia. Canadian Journal of Zoology 69:1454–1483.

Ford, J. K. B., and A. B. Hubbard-Morton. 1990. Vocal behavior and dialects of transient killer whales in coastal waters of British Columbia, California, and southeastern Alaska. Abstract submitted to the Third International Orca Symposium, Victoria, B.C., Canada.

Heise, K., G. Ellis, and C. Matkin. 1992. A catalogue of Prince William Sound killer whales, 1991. ISBN No. 0-9633467-3-3, North Gulf Oceanic Society, Homer, Alaska. 51 p.

Hoelzel, A. R., and G. A. Dover. 1991. Genetic differentiation between sympatric killer whale populations. Heredity 66:191–195

Leatherwood, S., K. C. Balcomb III, C. O. Matkin, and G. Ellis. 1984. Killer whales (*Orcinus orca*) in southern Alaska. Hubbs Sea World Research Institute Technical Report No.84-175. 54 p.

Leatherwood, S., C. O. Matkin, J. D. Hall, and G. M. Ellis. 1990. Killer whales, *Orcinus orca*, photo-identified in Prince William Sound, Alaska, 1976 through 1987. Canadian Field-Naturalist 104:362–371.

McCullagh, P., and J. A. Nelder. 1983. Generalized linear models. Chapman and Hall, London. 261 p.

Matkin, C. O., G. M. Ellis, O. von Ziegesar, and R. Steiner. 1986. Killer whales and longline fisheries in Prince William Sound, Alaska, 1986. Unpublished report for National Marine Mammal Laboratory, NMFS, 7600 Sand Point Way, Seattle, WA 98115.

Matkin, C. O., R. Steiner, and G. M. Ellis. 1987. Photoidentification and deterrent experiments applied to killer whales in Prince William Sound, Alaska, 1986. Unpublished report to the University of Alaska, Sea Grant Marine Advisory Program, Cordova, Alaska.

162   *Matkin et al.*

Matkin, C. O., and E. Saulitis. 1994. Killer whale (*Orcinus orca*): Biology and management in Alaska. U.S. Marine Mammal Commission, Washington, D.C.

Morton, A. B. 1990. A quantitative comparison of the behavior of resident and transient forms of killer whale off the central British Columbia coast. Pages 245-248, *in* P. S. Hammond, S. A. Mizroch, and G. P. Donovan (eds), Individual recognition of cetaceans: use of photo-identification and other techniques to estimate population parameters. Report of the International Whaling Commission Special Issue 12.

Olesiuk, P. F., M. A. Bigg, and G. M. Ellis. 1990. Life history and population dynamics of resident killer whales (*Orcinus orca*) in the coastal waters of British Columbia and Washington State. Pages 209–243, *in* P. S. Hammond, S. A. Mizroch, and G. P. Donovan (eds), Individual recognition of cetaceans: use of photo-identification and other techniques to estimate population parameters. Report of the International Whaling Commission Special Issue 12.

Saulitis, E. 1993. The behavior and vocalizations of the AT group of transient killer whales in Prince William Sound, Alaska. M.S. Thesis, Institute of Marine Science, University of Alaska, Fairbanks, Alaska.

Stevens, T. A., D. A. Duffield, E. D. Asper, K. G. Hewlett, A. Bolz, L. J. Gage, and G. D. Bossart. 1989. Preliminary findings of restriction fragment differences in mitochondrial DNA among killer whales (*Orcinus orca*). Canadian Journal of Zoology 67:2592–2595.

von Ziegesar, O., G. Ellis, C. O. Matkin, and B. Goodwin. 1986. Sightings of identifiable killer whales in Prince William Sound, Alaska 1977-1983. Cetus:9–13.

Zenkovich, B. A. 1938. On the Kosatka or whale killer (*Grampus orca*) Priroda 4:109–112 (Translated by L.G. Robbins)

Zimmerman, S. 1991. A history of marine mammal stranding networks in Alaska, with notes on the distribution of the most commonly stranded cetacean species, 1975–1987. Pages 43–53 *in* J. Reynolds and D. Odell (eds.) Marine mammal strandings in the United States: Proceedings of the Second Marine Mammal Stranding Workshop, 3-5 December 1987, Miami, Florida. NOAA Technical Report NMFS 98.

Chapter 9

Assessment of Injuries to Prince William Sound Killer Whales

Marilyn E. Dahlheim and Craig O. Matkin

INTRODUCTION

The loss of 14 killer whales (*Orcinus orca*) from the resident AB pod in Prince William Sound (PWS) over a 3-year period (Matkin et al., Chapter 8) is unprecedented for North Pacific killer whales. As top-level predators, killer whales occur in relatively low densities throughout their range. Two forms, residents and transients, have been described from the Pacific Northwest (Bigg et al. 1990). The two forms appear to differ in several aspects of morphology, ecology, and behavior; however, few studies exist which quantify the characters that separate them (Baird and Stacey 1988; Morton 1990).

Demographic rates of killer whale pods occurring in the inland waterways of Puget Sound, Washington, and British Columbia have been calculated through photo-identification studies. The resident forms are characterized by having low birth and death rates (<2.2% per year; Olesiuk et al. 1990). Annual birth rates for resident pods are based on the number of new calves observed within a pod in a given year divided by the total number of whales present at the end of the previous year. Mortality rates are based on the number of known individuals absent within a pod for at least 1 year divided by the number of whales present at the end of the previous year. If a pod is not encountered each year, mortality rates are averaged over the entire period the pod was not seen. Stranding information also provides information on mortality and, in some cases, individual identification.

Resident pods are organized into a series of progressively smaller units (subpods) and maternal groups. Pods are thus composed of one or more subpods that frequently travel together; subpods are known to leave the main pod for weeks or months. The subpod consists of one or more maternal groups that always travel together. The maternal group, the smallest traveling unit of resident whales, contains a mother and her offspring. The social structure of resident pods is

Marine Mammals and the *Exxon Valdez* **163**

considered so stable that an animal not seen for more than 1 year is considered dead (Bigg et al. 1990). Conversely, the social organization of transient killer whale pods is not well understood and the dynamic nature of these pods makes determination of mortality rates difficult.

Killer whale photo-identification studies in PWS were initiated in 1984 (Leatherwood et al. 1990) and continued through 1988 (Matkin et al., Chapter 8). These studies provided baseline information on PWS killer whale pod structure and abundance. Resident and transient pods were reported to exist in PWS based on similar patterns of social behavior observed in Washington and British Columbia.

We used these earlier studies to explain the loss of 14 whales in Prince William Sound (from the AB pod) over the 3-year period 1989–1991. We considered whether the loss of 14 whales could be explained by: (1) errors in photo-identification; (2) bias in survey coverage; (3) movement of whales out of the survey area; (4) natural mortality; (5) fisheries interactions; or (6) the *Exxon Valdez* oil spill (EVOS).

PHOTO-IDENTIFICATION

The number of whales present in PWS from May to September in 1989 to 1991 was obtained through examination of the photographic database of individual animals. Presence or absence of pod members was determined by comparing photographs taken during our 3-year study to photographs from 1984 to 1988 (Matkin et al., Chapter 8). The results showed the absence of 14 whales in the AB pod (7 whales missing in 1989, 6 missing in 1990, and 1 missing in 1991). Continuing investigations conducted in PWS during 1992 and 1993 also confirmed the absence of these whales.

To determine whether or not errors were made during the initial identification process, four subsequent photographic comparisons were made by independent examiners. The results confirmed the initial conclusion.

BIAS IN SURVEY COVERAGE

Another possible bias that could have resulted in the 14 whales not being seen and photographed was the amount of effort to establish presence or absence of individuals in the pod. This included the effort in survey days to locate whales and the number of times each pod was encountered. The total effort (number of km surveyed) conducted during 1989 to 1991 while searching for PWS killer whales exceeded that reported for previous years (Table 9-1).

Since individual whales or subpods may temporarily move away from the main pod or may be missed by chance, it was important that each pod was encountered

⤛⤜ ⤛⤜ ⤛⤜

Table 9-1. Kilometers and days surveyed each year for Prince William Sound killer whales, 1984 to 1991. (— = no data)

Year	Effort	
	(km)	(days)
1984	10,297	—
1985	2495	—
1988	2785	—
1989	17,822	260
1990	19,729	247
1991	13,594	159

several times (Bigg et al. 1987). An adequate number of encounters are also needed with each pod since the chance of "capturing" each individual is not equal. For example, an animal with conspicuous features may be photographed more often than an individual with no obvious markings. In addition, the behavior of an individual whale may make it easier to capture (i.e., one whale may be more prone to come close to the boat).

Encounters with pods other than the AB pod were also a consideration since the AB pod members could temporarily be traveling with other groups. Between 1986 and 1991, numerous encounters (mean=16.9, range 8-37; Table 9-2) occurred each year with PWS killer whales. The amount of effort searching for whales and the number of encounters with PWS killer whales during 1989, 1990, and 1991 was sufficient for locating and identifying the presence of individual animals, thus eliminating the potential bias in survey coverage.

MOVEMENT OF WHALES OUT OF THE SURVEY AREA

We considered the possibility that individual whales moved out of PWS and were not available to be photographed from 1989 to 1991. Two resident pods described in southeastern Alaska (AF and AG pods) are known to move between southeastern Alaska and PWS. Killer whales are also known to move between Kodiak Island and PWS. In southeastern Alaska, three separate research teams collectively spent 230 days between early June and late September 1989 searching for PWS killer whales. Sixty-three whales were photographed, but the missing PWS whales were not seen. From 1989 to 1991, no effort was expended near Kodiak Island and the waters adjacent to PWS to locate the missing whales.

⋙ ⋙ ⋙

Table 9-2. Number of encounters (the approach and collection of photographic data) with the AB killer whale pod by year in Prince William Sound.

Year	Number of encounters
1984	37
1985	12
1986	15
1987	12
1988	7
1989	16
1990	28
1991	8

The possibility that the missing PWS whales moved out of the area is not supported by our current understanding of the social structure and behavior of resident killer whales. Examination of the 20-year killer whale database from British Columbia and Washington suggests that no resident killer whale missing for more than 1 year has ever returned to its pod or appeared in another pod (Bigg et al. 1990).

The 14 missing whales represent individuals from four of seven different subpods comprising the AB pod. Although subpods may occasionally travel away from the main pod, individual whales from different subpods have never been documented to leave the pod. Two of the missing females left behind juveniles that were 2- to 3-year-old animals. Based on historical life history information presented by Bigg et al. (1990), it is likely that the missing resident whales have not moved to other areas but are dead. However, an environmental perturbation as severe as the EVOS and its direct impact on cetaceans has never been investigated. It is possible that such a major catastrophe had unpredicted impacts on killer whales.

NATURAL MORTALITY

Given that the natural mortality rate for killer whales is 2.2% per year or less, it is unlikely that natural mortality would account for the loss of 14 whales over a 3-year period. The 1989 mortality rate for the AB pod, including the loss of seven whales, was 19.4% (Matkin et al., Chapter 8). With the six additional whales missing in 1990, the mortality rate was 20.7%. In 1991, one more whale was missing from the AB pod (1991 mortality rate of 4.3%). The 1989–1990 rates are higher than would be expected from natural causes. Of the 14 missing whales, 8 were juveniles, and 6 were adults. Two were males, four were females (of which three were reproductively active), and the sex of the remaining eight whales was not determined. It is unlikely that natural mortality would account for more than one to three animals, particularly in the age categories given above (i.e., six adults).

FISHERIES INTERACTIONS

The apparent mortality of the 14 missing whales is complicated by the past history of the AB pod's interactions with the PWS sablefish longline fishery (Dahlheim 1988). Numerous methods have been employed by fishermen to either trick the whales or discourage them from stealing fish off the longlines (Dahlheim 1988). In 1985, we received reports of fishing crews shooting at killer whales to frighten whales away from their gear. Subsequent photographic data collected on the AB pod in 1985 and 1986 suggested the presence of bullet wounds on 10 whales; 5 more whales had possible bullet wounds. Five of the 10 whales with certain wounds have not been seen since and are assumed dead.

In 1985, three whales were reported missing; one had evidence of a bullet wound. In 1986, three additional whales were gone; of these only one had been seen with a bullet wound. In 1987 and 1988, the AB pod lost two more individuals (neither whale had been noted with bullet wounds). Of the five whales with possible bullet wounds, four are missing. Two losses occurred in 1986, one in 1987, and one in 1989. The loss of at least some of the whales between 1985 and 1988 was attributed to shooting (although never confirmed). These whales have not been seen since they were first classified as missing. Of the 14 missing whales between 1989 and 1991, at least three and possibly four (all missing during the 1989 season) had earlier indications of bullet wounds.

It is possible that some proportion of the 14 whales missing after the EVOS could have been shot during or before the 1989–1991 period. However, this is unlikely because: (1) sablefish longline fishing was closed between the last time all whales were accounted for (September 1988) and the time when the first seven whales were missing (March 1989); (2) there were no reports of shootings; and (3) no new bullet wounds have been observed on individuals of the AB pod since 1986. If a significant amount of shooting occurred, the remaining animals in the pod would probably show bullet wounds. It has been suggested recently that other pods may also interact with longline fisheries (e.g., AI pod, which associates frequently with the AB pod, and possibly AN pod). These two pods have not had comparable losses, suggesting that the mortalities noted in the AB pod may not be fisheries related. However, since the AB pod appears in the fishing area more frequently than other pods, there is a higher probability that this pod may simply be involved in more fishery interactions.

Underwater explosives have also been used by PWS fishermen to reduce or eliminate killer whale depredation on longline-caught sablefish (Dahlheim 1988).

∞ ∞ ∞

It is possible that whales may have been either severely injured or killed when explosive charges were detonated.

EXXON VALDEZ OIL SPILL

The last explanation examined for the missing whales was the effect of the oil spill. Six killer whale pods (AB, AJ, AI, AK, AT, and AN) were observed swimming directly through oil (light sheen) but only the AB pod suffered losses. However, since the AB pod frequently inhabits the area around Knight and Naked Islands, it is probable that this pod, and not others, would be the one most likely impacted by oil. It is possible that the AB pod was in the Naked Island area when fresh oil was blown down in that area on 27 March 1989. The AB pod is known to frequent the Naked Island area in early spring, presumably to feed on herring; the pod also associates with the sablefish fishery that typically opens 1 April. Although killer whale pods are seen tightly grouped when resting and socializing, when feeding or traveling they may be spread out across distances of a mile or greater. Thus, within a pod, some whales may have had direct contact with oil resulting in death or causing them to be more susceptible to contamination.

The mechanism/pathway of death, if related to oil, is equivocal. The loss of the first seven animals from the AB pod could have been through direct contact with oil, such as from inhalation of toxic volatile gases or ingestion. However, controversy still exists about whether the level of volatiles released from the spilled oil was at high enough concentrations to kill whales. The loss of the six additional whales during the 1990 season (the loss would have occurred between 4 September 1989 and 1 June 1990) is more difficult to explain from oil effects, but might have been associated with residual effects or other indirect effects (e.g., eating contaminated prey).

None of the missing whales were found stranded, although killer whales typically sink upon death (Zenkovich 1938). Four killer whale carcasses were found during 1989–1991, of which only one whale could be identified and it was not from the AB pod. Stranding rates of killer whales in Alaska are low with only six whales reported stranded for the entire Gulf of Alaska (none of which occurred in PWS) from 1975 to 1987 (Zimmerman 1991). The increase in strandings (four whales in a 3-year period which coincided with the oil spill) in PWS may simply be a result of increased observer effort after the EVOS. Blubber samples and scrapings from the stomach linings of the stranded whales were analyzed for hydrocarbons. There was no indication of oil contamination in these tissues and cause of death could not be determined. Caution, however, must be used when interpreting these results since the animals had been dead for some time when found and decomposition decreases the utility of tissue samples for hydrocarbon analysis.

✂ ✂ ✂

RECOVERY OF THE AB POD

Between 1985 and 1988, the AB pod experienced a high mortality rate; eight AB pod whales were missing during the 4-year period (Matkin et al., Chapter 8). In 1988, five calves were born into this pod resulting in 36 members in 1989. Studies conducted in PWS during 1992 and 1993 report that no additional whales were missing from the AB pod. Two new calves were observed in 1992 and one new calf observed in 1993. In September 1993, the AB pod had 26 members (Dahlheim et al., 1993). Recovery of the AB pod to prespill levels (36 animals), if it occurs, could take 10 to 15 years given the current age and sex structure of the population (Table 9-3). The overall number of killer whales in PWS is similar to

Table 9-3. Age and sex structure of the AB pod in Prince William Sound during 1993. Full grown males were defined as 20 years of age and full grown females as 12 years of age (Bigg et al. 1987). The age and sex structure of the AB pod was derived from data in Ellis (1987), Matkin et al. (Chapter 8), and Dahlheim et al. (1993).

Specimen Number	Sex	Age	Year born (mother)	Comments
AB2	M	>29		Full grown in 1984
AB3	M	>29		Full grown in 1984
AB4	M	>27		Almost full grown in 1984
AB5	M	>29		Full grown in 1984
AB10	F	>21		Full grown in 1984
AB11	M	>15		Not yet full grown in 1993
AB14	F	>21		Full grown in 1984
AB16	F	>21		Full grown in 1984
AB17	F	>19		Full grown in 1986
AB22	F	>21		Full grown in 1984
AB24	M	~20		Not yet full grown in 1986
AB25	F	>19		Full grown in 1986
AB26	F	>13		Young juvenile in 1984
AB27	?	>10		Young juvenile in 1984
AB32	F	>21		Full grown in 1984
AB33	?	>10		Yearling in 1984
AB35	M	~13		Not yet full grown in 1993
AB38	?	6		Mom AB31 (dead)
AB39	F	6	1987 (AB25)	
AB40	?	5	1988 (AB14)	
AB41	?	5		Mom AB8 (dead)
AB43	?	5	1988 (AB17)	
AB45	?	2	1991 (AB16)	
AB46	?	1	1992 (AB25)	
AB47	?	1	1992 (AB32)	
AB48	?	newborn	1993 (AB26)	

><> ><> ><>

that reported prior to 1989 which may enhance recovery of the injured pod. Recovery could also be enhanced by protecting foraging habitats, minimizing fishery interactions, reducing or redirecting other human-use impacts, and promoting public education.

CONCLUSIONS

The cause(s) for 14 killer whales missing from the AB pod is unknown. We are confident that: (1) whales have not been misidentified; (2) adequate effort was made in PWS to locate the missing animals; and (3) the number of times the pod was encountered was sufficient to evaluate the presence or absence of an individual whale. It is unlikely that the whales left their pod and moved elsewhere. Therefore, we assume that some of the whales may have died of natural causes and the remainder are dead from either, or a combination of, a result of interactions with fisheries or the EVOS. The highest mortality rates ever reported for North Pacific resident killer whales occurred in 1989 and 1990, coinciding with the EVOS. There is a spatial and temporal correlation between the loss of the 14 whales and the EVOS, but there is no clear cause-and-effect relationship.

ACKNOWLEDGMENTS

We extend our appreciation to the numerous field crews who worked long hours collecting the photographic data on killer whales. G. Ellis, J. Waite, D. Ellifrit, M. Merklein, and D. Cassano completed individual identifications. We thank H. Braham, T. Loughlin, J. Hall, J. Stern, and B. Würsig for reviewing the manuscript.

REFERENCES

Baird, R. W., and P. J. Stacey. 1988. Variation in saddle patch pigmentation in populations of killer whales (*Orcinus orca*) from British Columbia, Alaska, and Washington State. Canadian Journal of Zoology 66:2582–2585.

Bigg, M. A., G. E. Ellis, J. K. B. Ford, and K. C. Balcomb III. 1987. Killer whales: A study of their identification, genealogy, and natural history in British Columbia and Washington State. Phantom Press, Nanaimo, British Columbia. 79 p.

Bigg, M.A., P. F. Olesiuk, G. M. Ellis, J. K. B. Ford, and K. C. Balcomb. 1990. Social organization and genealogy of resident killer whales (*Orcinus orca*) in the coastal waters of British Columbia and Washington State. Pages 383–405, *in* P. S. Hammond, S. A. Mizroch, and G. P. Donovan (eds), Individual recognition of cetaceans: use of photo-identification and other techniques to estimate population parameters. Report of the International Whaling Commission Special Issue 12.

Dahlheim, M. E. 1988. Killer whale (*Orcinus orca*) depredation on longline catches of sablefish (*Anoplopoma fimbria*) in Alaskan waters. U.S. Department of Commerce, NOAA, NWAFC Processed Report 88-14. 31 p.

Dahlheim, M. E., D. Bain, and J. Waite. 1993. Recovery monitoring of Prince William Sound killer whales injured by the *Exxon Valdez* oil spill using photo-identification techniques. Unpublished Report. (Available through the National Marine Mammal Laboratory, 7600 Sand Point Way NE, Seattle, Washington.) 12 p.

Ellis, G. 1987. Killer whales of Prince William Sound and Southeast Alaska — A catalogue of individuals photoidentified, 1976–1986. Sea World Research Institute, Technical Report No. 87-200. 76 p.

Leatherwood, S., C. O. Matkin, J. D. Hall, and G. M. Ellis. 1990. Killer whales, *Orcinus orca*, photo-identified in Prince William Sound, Alaska, 1976 through 1987. Canadian Field-Naturalist 104:362–371.

Morton, A. B. 1990. A quantitative comparison of behavior in resident and transient killer whales of the central British Columbia coast. Pages 245–248, *in* P. S. Hammond, S. A. Mizroch, and G. P. Donovan (eds), Individual recognition of cetaceans: use of photo-identification and other techniques to estimate population parameters. Report of the International Whaling Commission Special Issue 12.

Olesiuk, P. F., M. A. Bigg, and G. M. Ellis. 1990. Life history and population dynamics of resident killer whales (*Orcinus orca*) in the coastal waters of British Columbia and Washington State. Pages 109–243, *in* P. S. Hammond, S. A. Mizroch, and G. P. Donovan (eds), Individual recognition of cetaceans: use of photo-identification and other techniques to estimate population parameters. Report of the International Whaling Commission Special Issue 12.

Zenkovich, B. A. 1938. On the Kosatka or whale killer (*Grampus orca*). Prioda 4:109–112. (In Russian, translated by L. G. Robbins).

Zimmerman, S. 1991. A history of marine mammal stranding networks in Alaska, with notes on the distribution of the most commonly stranded cetacean species, 1975 to 1987. Pages 43–53, *in* J. Reynolds and D. Odell (eds.) Marine mammal strandings in the United States: Proceedings of the 2nd Marine Mammal Stranding Workshop -5 December 1987, Miami, Florida. NOAA Technical Report NMFS 98.

Chapter 10

Impacts on Humpback Whales in Prince William Sound

Olga von Ziegesar, Elizabeth Miller, and Marilyn E. Dahlheim

INTRODUCTION

About 10,000 humpback whales (*Megaptera novaeangliae*) occur world-wide (Johnson and Wolman 1984) of which at least 1400 are in the North Pacific Ocean (Wada 1980; Johnson and Wolman 1984; Baker et al. 1986). In the North Pacific, humpback whales spend winter principally in waters off Hawaii, Mexico, and Japan. During summer they range widely across the North Pacific from California (Calambokidis et al. 1990), to southeastern Alaska (Baker et al. 1985, 1992), and to Prince William Sound (von Ziegesar 1984, 1992). Summer aggregations have also been documented off Kodiak Island, the Aleutian Islands, and the Bering Sea (Berzin and Rovnin 1966; Wada 1980; Rice and Wolman 1982; Leatherwood et al. 1983; Morris et al. 1983).

Population studies on humpback whales in Prince William Sound (PWS) have been in progress since 1977. These studies include line transects from aircraft and ships (Hall 1979; Matkin and Matkin 1981), radiotelemetry (Watkins et al. 1981), and photo-identification using pigmentation patterns on the ventral side of the fluke, and the shape of the trailing edge of the fluke (Hall 1981; von Ziegesar 1984, 1992).

Photo-identification was first described as a reliable method of recognizing individual humpback whales over time in the North Atlantic Ocean (Katona et al. 1979; Katona and Kraus 1979; Katona and Whitehead 1981). Photographing humpback whale flukes has become a reliable tool for individual whale recognition, determining reproductive rates, and assessing movement patterns. Approximately 60–100 humpback whales feed in PWS between June and September each year, with many of the same individuals returning from year to year (von Ziegesar 1992). Recent observations suggest that some whales may even overwinter in PWS (Charles Monnett, Enhydra Research, Homer, Alaska, personal communication).

Although animals from different winter feeding areas intermix at breeding locations, they have a high degree of fidelity to feeding areas. Prior to 1989, only six
individual whales were known to cross over between PWS and southeastern Alaska
(Darling and McSweeney 1985; Baker et al. 1986; McSweeney, unpublished data;
Straley and Baker, unpublished data). One whale from PWS has been matched to
a sighting in the Gulf of Alaska near Seward (von Ziegesar, unpublished data). Site
fidelity with respect to feeding grounds appears to be passed genetically through
maternal lines (Martin et al. 1984; Baker et al. 1986; Clapham and Mayo 1990;
Baker et al. 1992).

The potential impacts to humpback whales in PWS caused by the *Exxon Valdez*
oil spill (EVOS) may have included displacement from their normal feeding areas
in PWS, reduction in prey, or possible physiological impacts resulting in reproductive failure or mortality. We used photo-identification techniques to test the
hypotheses that humpback whale abundance and distribution in PWS had not
changed as a result of the EVOS, that female calving rates and seasonal residency
of female-calf pairs had not changed, and that humpback whale mortality rates had
not changed since the EVOS. We used data collected during studies in 1988-1990
in our analysis.

METHODS

Field Procedures

Similar methods were used in 1988, 1989, and 1990. However, the overall effort
expended and the size of the study area examined during 1988 was smaller. In
1988, one camp (Camp no. 1) was used in southwestern PWS (Fig. 10-1); in 1989
three field camps (Camp nos. 1–3) were used in eastern and southwestern PWS. A
small outboard-motor boat was used at each camp to locate and photograph whales.
In 1990, only Camp no. 1 was used and two boats (Whale 1 and Whale 3) operated
from this station. Camp nos. 2 and 3 were replaced in 1990 with an 8.5-m
bowpicker-style fishing vessel (Whale 2) that was used as a mobile camp and
research platform. A 14-m fishing vessel (*Lucky Star*) was used intermittently
throughout 1989 and 1990 as a mobile research platform. A concurrent killer whale
survey was conducted in 1988–1990 with all vessels collecting data on both
humpback and killer whales (Matkin et al., Chapter 8). In 1990, Whale 3 was
dedicated to humpback whale investigations.

Whales were visually located during regular searches from the camps or by
examining areas that had been historically used by humpback whales (Hall 1979,
1981; von Ziegesar 1984, 1988). In addition, researchers also responded to radio

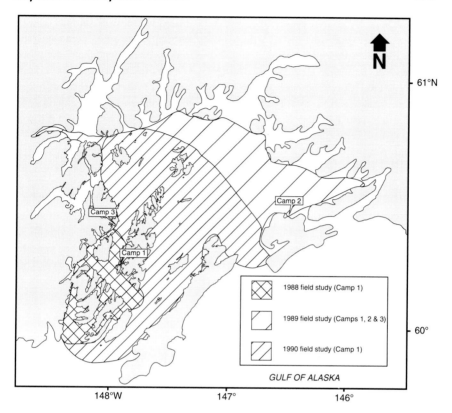

Figure 10-1. General range of surveys for humpback whales during 1988-1990 and the locations of field camps in Prince William Sound.

calls from pleasure, commercial, and government vessels reporting sightings of whales. The successful detection, approach, and taking of identification photographs was defined as an "encounter," even if there were small groups or pairs of whales moving synchronously within feeding aggregations.

When humpback whales were located, each animal was slowly approached from behind to a distance of 25 to 50 m. When the animal dived, the ventral side of its tail was photographed. Each 35-mm camera was equipped with a motor-drive and a 300-mm lens set at 1/1000-second shutter speed or the highest speed possible. Black and white Ilford HP5 film (ASA 400) was used and developed at ASA 1600. At least two photographs of each whale were taken before resuming surveys. Whales that did not raise their flukes upon diving were followed for at least five dive sequences. When their flukes were not raised, the dorsal fin was photographed. Individuals identified by dorsal fin photographs were not included in the final tally unless the dorsal fin photograph was matched with one taken later when the fluke of the whale was also photographed. Dorsal fin and fluke photographs were collected for females and calves because calves often do not raise their flukes. Because whale group size in PWS is small (2–3 animals) and calves stay close to their mothers, we assumed a calf was the offspring of the adult whale even if it was photographed only once.

Data were recorded on individual forms for each humpback whale encounter. The form included date, time, location, count, other species interactions, film frames, behavioral observations, and sketches of the fluke and dorsal fin. Other information such as weather and sea conditions, time spent searching for whales, and oil-spill-related information, such as an assessment of the amount of oil clean-up activities, oil or oily debris in the water, and incidents of whale harassment (whales changing their behavior as a result of the approach of a vessel or aircraft) were recorded in a daily log book. Charts of the research vessel routes were kept and a log of sightings reported by other vessels or aircraft was compiled.

Data Analysis

Individual humpback whales were identified using both the photographic negatives and contact sheets. Individuals were labeled according to an alphanumeric system developed by von Ziegesar and Matkin (1985) and matched to humpback whales photographed in PWS since 1977 (von Ziegesar 1992). Photographs taken under poor light conditions or from a bad angle were often enlarged and reexamined. Questionable identifications and new whales were confirmed by a second experienced researcher.

A rate-of-discovery plot was constructed for identifications made each season. The cumulative number of "new" (previously unidentified) whales was plotted against the number of identifications arrayed in the order in which they were obtained (Darling and Morowitz 1986). If every whale identified was new (i.e., there were no resightings), the slope of the curve was steep (rising at 45°). However, as individuals were identified the rate of discovery of new whales decreased and the slope declined toward zero when all whales were known or identified. An analysis of variance was used to compare monthly abundance for each year.

A subset of the whales (core whales) that were observed in PWS during all 3 years was compared to all other documented whales for differences in patterns of occurrence and length of stay (seasonal residency). The seasonal residency of humpback whales in PWS during the study was inferred from the interval between the first and last sighting of individual whales each summer and compared using a *t* test.

⋈ ⋈ ⋈

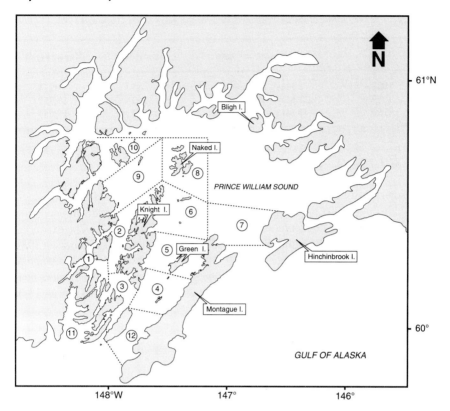

Figure 10-2. Map of Prince William Sound showing the 12 areas used for identifying humpback whale sightings and usage.

To identify distribution patterns of humpback whales, the study area was divided into relatively equal-size areas (12) along natural boundaries (Fig. 10-2). But because these areas were not searched equally, statistical analysis of distribution was not attempted.

Annual crude calving rates (CCR) were calculated as a percentage of the total population using,

$$CCR = \frac{NR}{NP}$$

where NR=the number of new calves documented in a given year and NP=the number of whales present in that year, including calves.

Table 10-1. Survey effort in Prince William Sound for the years 1988, 1989, and 1990.

Year	Beginning/end date	Survey days	Km surveyed
1988	8 June - 8 September	45	3213
1989	23 May - 15 September	260	15,397
1990	15 May - 8 September	247	17,045

RESULTS

Research effort in 1988 was less (45 days, 3213 km) than in 1989 (260 days, 15,397 km) and 1990 (247 days, 17,045 km) (Table 10-1), but in 1988 the encounter rate was highest (1.71 encounters/day) of the 3 years (Table 10-2). The expanded effort in 1989 and 1990 resulted in 254 whales identified in 119 encounters in 1989 and 474 whales in 201 encounters in 1990, compared to 134 whales in 77 encounters in 1988 (Table 10-2). Most whales were photographed and identified on more than one occasion during each year. The number of individuals identified was 35 in 1988, 59 in 1989, and 66 in 1990 (Table 10-2).

Between June and September 1989, concurrent with PWS research, photo-identification studies were conducted in southeastern Alaska to determine if PWS whales were displaced there. Humpback whales were encountered 2448 times and 500 individual whales were identified. One whale (X54) was seen in both PWS (20 July) and southeastern Alaska (29 August) in 1989.

Photo-identification

In the years following the EVOS (1989–1990), the number of whales identified leveled off by the end of the season, while the number of whales photographed continued to increase (Fig. 10-3). The rate of discovery for these years suggests that most or all the whales in the study area were identified at the end of the season. The number of whales identified differed significantly by month (June to August) in 1990 ($F=30.4$, df=2, 198, $P=0.0001$) but not in 1988 ($F=0.986$, df=2, 71, $P=0.38$) or 1989 ($F=0.58$, df=2, 104, $P=0.85$). The monthly difference in 1990 was largely the result of a large aggregation seen in August.

The number of new whales was highest in 1989 (37%). Seventeen whales (core whales) were seen in all 3 years (Fig. 10-4). Core whales and all other whales that were resighted in the area were analyzed to determine if patterns of occurrence were different for the two groups. Core whales were encountered more frequently than others ($t=3.41$, $P<0.002$). From 1988 to 1990 the 17 core whales averaged 7.43 encounters ($n=51$, SE=0.40, range=1–30). The mean number of encounters with other individual humpbacks was 4.09 ($n=107$, SE=0.98, range=1–18).

⋈   ⋈   ⋈

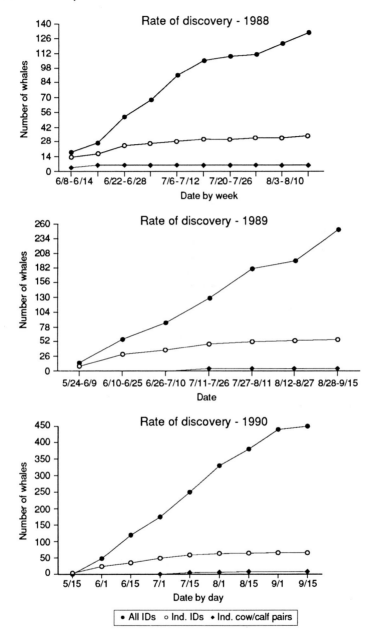

Figure 10-3. Rate of discovery of humpback whales in Prince William Sound in 1988, 1989, and 1990.

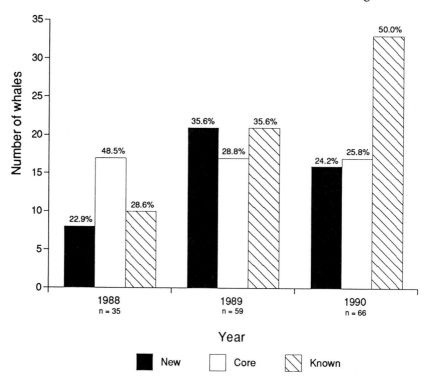

Figure 10-4. A comparison of the total number of humpback whales identified in Prince William Sound (PWS) during 1988-1990. The percentage of the total for each year is shown on the top of the bar. New whales=whales identified in the year indicated; core whales=whales sighted in PWS 1988-1990; known whales=whales sighted previously in PWS before 1988.

Over the 3-year study, core whales averaged 49.1 days (n=51, SE=5.69, range=1–113) in PWS and other whales averaged 27.35 days (n=107, SE=3.09, range=1–111). The seasonal residency for the core whales was significantly longer than for other whales (t=3.82, P<0.001).

Seasonal residency time of females with calves averaged 22.25 days (n=18, SE=5.02, range=1–69,). Residency time of females with calves after the spill was not significantly different from 1988 before the spill (F=0.02, df=2, 16, P=0.98). Residence time of females with calves was not significantly different from the residency time of other humpback whales in PWS.

Distribution

During all 3 years, humpback whales were seen primarily in Knight Island Passage (Areas 2 and 3), the south end of Chenega Island, and in the entrances of Icy and Whale Bays (Area 1) (Figs. 10-5, 10-6, and 10-7). In 1989, whale

≫◇ ≫◇ ≫◇

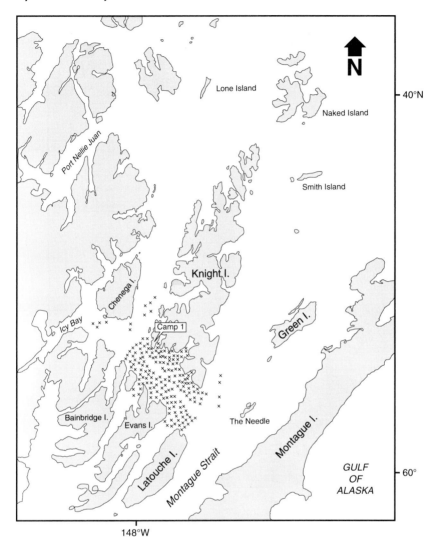

☒ = one sighting of Humpback whale in 1988.

Figure 10-5. Distribution of humpback whale sightings during 1988 (von Ziegesar 1988).

182   *von Ziegesar et al.*

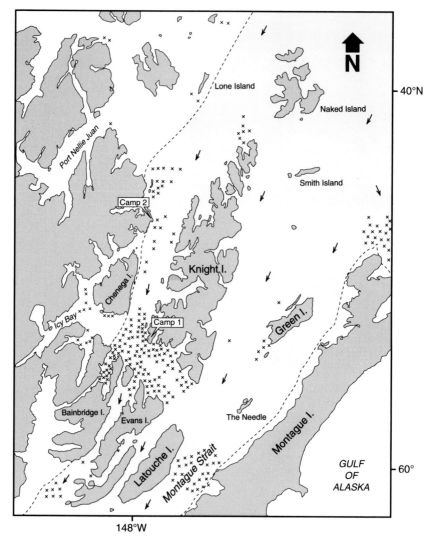

☒ = one sighting of Humpback whale in 1989

⟋ = Path/direction of oil

Figure 10-6. Distribution of humpback whale sightings during 1989 with the path of oil from the *Exxon Valdez* shown. Even though most of the oil had drifted out of PWS before the peak in humpback whale abundance, the whales may have been exposed to residual oil during the time they occurred in southwestern PWS.

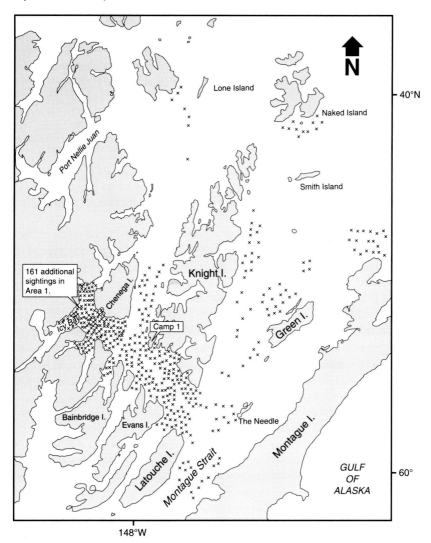

= one sighting of Humpback whale in 1990.

Figure 10-7. Distribution of humpback whale sightings during 1990.

Table 10-2. Number of individually identified humpback whales in Prince William Sound (1988-1990).

Year	Encounters	Encounters/day	Number of identifications
1988	77	1.71	35
1989	119	0.45	59
1990	201	0.81	66

aggregations were seen in Montague Strait (Areas 4 and 12) and in Hinchinbrook Entrance in late August and September. In 1988, survey effort was concentrated in lower Knight Island Passage resulting in a bias in the distribution data obtained for that year.

Humpback whales were not seen feeding in water with spilled oil on the surface, however they were seen feeding in water which comprised the primary path of the oil as it drifted through southwestern PWS and out into the Gulf of Alaska. Figure 10-6 shows the path of the oil overlaid on the location of humpback whale sightings in 1989. Even though most of the oil had drifted out of PWS before the peak in humpback whale abundance, the whales may have been exposed to residual oil during the time they occurred in southwestern PWS. Oil clean-up activity throughout 1989 and 1990 provided a constant source of oil and residue from the beaches into common humpback whale areas. In addition, there was an increase in aerial and vessel activity, underwater noise, and water pollution in the form of runoff from steam-cleaning processes and from vessel fuel and bilge spills.

During the period of increased clean-up activity in 1989 (25 June to 24 July), there were fewer whales in lower Knight Island Passage compared to other years (21 whales in 1989 compared to 85 in 1988 and 68 in 1990) (Table 10-3). In addition to the increased human activity, 113,550 liters of diesel fuel spilled from an anchored clean-up vessel in Mummy Bay on 7 July 1989 (Area 3). The resulting slick was observed until 18 July in an area of lower Knight Island Passage that was used by the whales.

In 1989 Hinchinbrook Entrance was surveyed regularly and 19 whales were identified, including two of four female-calf pairs. This area had not been surveyed regularly in previous years and ten new whales were documented there. Nine (47.3%) of the 19 Hinchinbrook whales (including one female with a calf) were not documented farther inside PWS that year. Four (44%) were animals that had been documented inside PWS in past years. In 1990, Hinchinbrook Entrance was

Table 10-3. A comparison of the total number of humpback whales encountered in lower Knight Island Passage (Area 3) in years 1988, 1989, 1990. Clean-up activities intensified in 1989 between 25 June and 24 July.

	25 May - 24 June	25 June - 24 July	25 July - 24 August	25 August - 15 September
1988	49	85	10	2
1989	48	21	25	10
1990	66	68	12	1

surveyed less frequently. One of the seven individuals (14.3%) documented there did not venture further into PWS.

During August 1990, aggregations of 10 to 18 humpback whales were consistently found at the south end of Chenega Island and in the entrance of Icy Bay. Though this area has been surveyed annually, humpback whale aggregations had not been seen there since 1980 (von Ziegesar 1984).

The distribution of sightings of core whales was compared to those of all other whales, since core whales might have a greater tendency to use historical feeding areas. No difference was detected, suggesting that humpback whale movements within PWS were not habitual but probably determined by prey or other factors.

Natality

Four females with calves were encountered and identified in 1989 and eight female-calf pairs were identified in 1990. Seven of the 1990 females were encountered in PWS in 1989 suggesting that they were pregnant and feeding there after the spill. The mean CCR for 1980 through 1988 was 9.47% (SE=1.2, range 3.6–14.6%) (von Ziegesar 1992). The CCR for 1989 was 6.3% (below the mean) and in 1990 it was 10.8% (above the mean) (Fig. 10-8).

Mortality

There were no humpback whale deaths or strandings observed or reported during 1988–1990 in PWS. Humpback whales were never observed swimming directly in oil.

≫ ≫ ≫

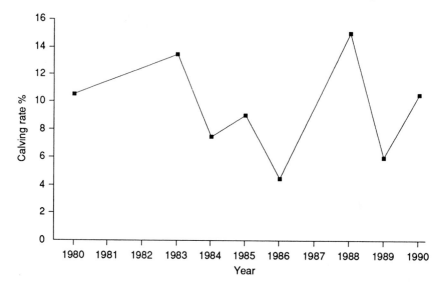

Figure 10-8. Line graph depicting calving rates of humpback whales in PWS 1980-1990.

DISCUSSION

Fifty-nine humpback whales were identified in PWS during 1989 and 66 in 1990. We believe that the study area was comprehensively surveyed because encounter rates were not significantly different throughout the season and the rates of discoveries of new whales leveled off. A significant decline in the number of humpback whales using PWS was not detected when comparing minimum counts from 1989–1990 to counts prior to the spill. In fact, more humpback whales were encountered in 1989–1990 than previously reported for PWS. The increase in the number of individual whales identified during 1989 and 1990 is likely related to the increased level of search effort following the EVOS.

Although the absolute number of humpback whales identified in 1989 and 1990 was higher than in 1988, the number of encounters per day or per kilometer was lower (Table 10-2). However, there were problems when comparing the 1988 data to 1989 and 1990 data. The 1988 study was conducted primarily in the southwest section of PWS which is known for its high concentration of humpback whales. In contrast, the geographic coverage of the 1989 and 1990 studies was more extensive. The lower number of humpback whales found per day and per kilometer in 1989 and 1990 may have been due to the greater area covered that included areas that had low humpback whale usage. Humpback whale distribution and movements within PWS varied considerably among years. Whale movements are probably correlated with abundance and distribution of prey. Displacement of whales due to the EVOS or to clean-up activity was difficult to detect or to demonstrate. Some changes in humpback whale distribution occurred during the 1989 season but the changes appear to have been temporary. Even with increased survey coverage following EVOS, the greatest distribution of whales was still found in the areas surrounding lower Knight Island Passage (Areas 1, 2, and 3).

Nine out of 19 whales encountered in Hinchinbrook Entrance in 1989 did not venture farther into PWS. Four of these were known PWS whales including one female with a calf. In 1990, seven whales were documented at Hinchinbrook Entrance; all but one entered PWS. Since Hinchinbrook Entrance was out of the main path of the spilled oil it is possible that whales in this area were avoiding heavily oiled areas further in the Sound. It is also possible that there was an abundance of prey at Hinchinbrook Entrance in 1989 and some whales from other areas in the Gulf of Alaska gathered there to feed and then returned to the outer coast.

The whale (X54) that was seen in both PWS and southeastern Alaska in 1989 was seen once in PWS (at Hinchinbrook Entrance). This whale was also documented in southeastern Alaska in 1982 and 1985 (Jan Straley, Sitka, Alaska, personal communication) so its movements in 1989 are not necessarily a displacement associated with the EVOS.

Results from studies in the North Atlantic (Clapham and Mayo 1990) and southeastern Alaska (Baker et al. 1992) report CCR in the same range that we calculated for PWS. High variability in annual CCR is not unusual for baleen whales (Mizroch and York 1984) making it difficult to determine the effect of EVOS on humpback whale reproductive rates. The number of calves observed in 1989 was below average with four identified female-calf pairs. Interestingly, seven of the eight calves documented in PWS in 1990 were born to females that were feeding in PWS while pregnant in 1989.

Because humpback whales do not remain in groups, and PWS is not a "closed" area (all whales are not seen there all years), it is difficult to determine mortality rates for humpback whales. No dead humpbacks were observed or reported in PWS between 1988 and 1990, and no whales encountered in PWS during 1989 or 1990 were reported dead in other areas of the species's range.

In 1989, humpback whales fed in areas that had been heavily oiled, but none were observed feeding in oil, nor did they seem to prefer areas that had not been oiled. During the months after the spill, some animals may have consumed contaminated prey or ingested small amounts of residual oil while feeding at the surface. Irritation to the eyes or skin or inhalation of toxic fumes (Hensen 1985; Geraci and St. Aubin 1985) was also possible. Core whales may have had more exposure to oil since they occur in the area for longer periods of time. However,

≫ ≫ ≫

residency times for other whales (averaging over a month) may have been enough for them to be affected as well.

Given that this endangered species exhibits site fidelity and that the PWS humpback whale population is small and somewhat distinct from other populations in Alaska, this population could have been vulnerable to a major disturbance such as the EVOS. The peak occurrence of whales within the area occurred after much of the spilled oil had moved out of PWS.

CONCLUSIONS

The largest number of humpback whales reported in PWS occurred in 1989–1990 coinciding with the highest survey effort associated with the EVOS. However, the number of whales encountered per kilometer surveyed was lower in 1989-1990 than in 1988. This may be the result of surveying a larger portion of PWS in 1989 and 1990, including areas where humpback whales occurred in low density. The reduction in the number of whales seen in lower Knight Island Passage in 1989 was temporary and may have been related to increased vessel and aircraft traffic associated directly with oil contamination, with clean-up activities, or it may have been caused by natural fluctuations in prey. Whales were seen in lower Knight Island Passage in 1990, as in prespill years. Movements of whales into PWS from Hinchinbrook Entrance were also more common in 1990, 1 year after the spill. The reproductive rate was low in 1989 but in 1990 it was above historical levels. When compared to previous years, no significant differences could be detected in CCR. There were no humpback whale deaths or strandings observed during 1989–1990 in PWS.

The results of this study do not indicate a change in abundance, calving rates, seasonal residency time of female-calf pairs, or mortality. Temporary displacement from certain areas of PWS was observed. During the 2-year study it was difficult to determine whether the EVOS had any measurable impact on the number of humpback whales occurring in PWS. Long-term physiological impacts to the whales or cumulative impacts to the ecosystem affecting humpback whale prey would not have been detected in this time period. However, because of the wide distribution of the North Pacific humpback whale and the unequal survey effort in PWS between prespill and postspill years, effects, if any, of the EVOS may never be known.

≫⊂ ≫⊂ ≫⊂

ACKNOWLEDGMENTS

The National Marine Mammal Laboratory funded and directed the 1989 and 1990 studies with funds provided by the EVOS Trustee Council, and the North Gulf Oceanic Society funded and directed prespill studies. This project could not have been completed without the aid in the collection of field data from researchers R. Angliss, L. Barrett-Lennard, M. Beak, F. Felleman, M. Freeman, M. Hare, K. Heise, C. Jurasz, L. Larsen, C. Matkin, D. Matkin, D. McSweeney, E. Saulitis, and J. Straley. H. Spence of the Homer News provided photographic services. We also thank all the skippers, crew people, biologists, hatchery workers, and pilots who have helped us by reporting whale sightings. In addition to the researchers listed above, K. Balcomb-Bartok, K. Englund, B. Goodwin, J. Hall, B. Hansen, S. Leatherwood, C. Matkin, C. Miller, K. Moore, J. Reinke, W. Schevill, and W. Watkins have all contributed data to the Prince William Sound Humpback Whale Catalogue. The combined effort of all involved helped create the database against which the status of PWS humpback whales after the spill could be compared. The manuscript was improved with comments provided by J. Calambokidis, D. Rugh, and T. Loughlin.

REFERENCES

Baker, S. C., L. M. Herman, A. Perry, W. S. Lawton, J. M. Straley, and J. H. Straley. 1985. Population characteristics and migration of summer and late-season humpback whales (*Megaptera novaeangliae*) in southeastern Alaska. Marine Mammal Science 1:304–323.

Baker, S. C., L. M. Herman, A. Perry, W. S. Lawton, J. M. Straley, A. A. Wolman, G. D. Kaufman, H. E. Winn, J. D. Hall, J. M. Reinke, and J. Ostman. 1986. Migratory movement and population structure of humpback whales (*Megaptera novaeangliae*) in the central and eastern North Pacific. Marine Ecological Progress Series 31:105–119.

Baker, S. C., A. Perry, and L. M. Herman. 1987. Reproductive histories of female humpback whales (*Megaptera novaeangliae*) in the North Pacific. Marine Ecological Progress Series 41:103–114.

Baker, S. C., J. M. Straley, and A. Perry. 1992. Population characteristics of individually identified humpback whales in Southeast Alaska: Summer and fall 1986. Fishery Bulletin, U.S. 90:429–437.

Berzin, A. A., and A. A. Rovnin. 1966. The distribution and migration of whales in the northeastern part of the Pacific Ocean and in the Bering Sea and the Sea of Chukotsk. Izvestia Tikhookeanskogo Nauchno-Issledovatel'skogo Institute Rybnogo Khozyalstva (Okeanografil 58:179–207).

Calambokidis, J., J. C. Cubbage, G. H. Steiger, K. C. Balcomb, and P. Bloedel. 1990. Population estimates of humpback whales in the Gulf of the Farallons. Pages 325–334, *in* P. S. Hammond, G. Donovan, and S. Mizroch (eds.) Individual recognition of cetaceans: Use of photo-identification and other techniques to estimate population parameters. Report of the International Whaling Commission Special Issue 12.

◇◇   ◇◇   ◇◇

Clapham, P. J., and C. A. Mayo. 1990. Reproduction of humpback whales (*Megaptera novaeangliae*) observed in the Gulf of Maine. Pages 171–175, *in* P. S. Hammond, G. Donovan, and S. Mizroch (eds.) Individual recognition of cetaceans: Use of photo-identification and other techniques to estimate population parameters. Report of the International Whaling Commission Special Issue 12.

Darling, J. D., and D. McSweeney. 1985. Observations on the migrations of North Pacific humpback whales (*Megaptera novaeangliae*). Canadian Journal of Zoology 63:308–314.

Darling, J. D., and H. Morowitz. 1986. Census of "Hawaiian" humpback whales (*Megaptera novaeangliae*) by individual identification. Canadian Journal of Zoology 64:105–111.

Geraci, J. R., and D. J. St. Aubin. 1985. Expanded studies on the effects of oil on cetaceans. Final report. Part 1. U.S. Department of the Interior, Minerals Management Service, Washington, D.C. 144 p.

Hall, J. D. 1979. A survey of cetaceans of Prince William Sound and adjacent vicinity, their numbers and seasonal movements. U. S. Department of Commerce, NOAA OCS Environmental Assessment Program, Final Report of Principal Investigators 6:631–726.

Hall, J. D. 1981. Aspects of the natural history of cetaceans in Prince William Sound, Alaska. Ph.D. dissertation, University of California, Santa Cruz, California. 149 p.

Hensen, D. J. 1985. The potential effects of oil spills and other chemical pollutants on marine mammals occurring in Alaskan waters. Unpublished report to Outer Continental Shelf Environmental Assessment Program, Department of the Interior, Minerals Management Service, Washington, D.C. 22 p.

Johnson, J. H., and A. A. Wolman. 1984. The humpback whale, *Megaptera novaeangliae*. Marine Fishery Review 46 (4):30–37.

Katona, S., B. Baxter, O. Brazier, S. Kraus, J. Perkins, and H. Whitehead. 1979. Identification of humpback whales by fluke photographs. Pages 33–44, *in* H. E. Winn and B.L. Olla (eds.), Behavior of marine animals- current perspectives in research, Vol. 3: Cetaceans. Plenum Press, New York, New York.

Katona, S., and S. Kraus. 1979. Photographic identification of individual humpback whales (*Megaptera novaeangliae*): evaluation and analysis of the technique. NTIS PB-298 740, Report No. MMC-77/17. Springfield, Virginia.

Katona, S., and H. Whitehead. 1981. Identifying humpback whales using their natural markings. Polar Record 20:439–444.

Leatherwood, S., A. E. Bowles, and R. R. Reeves. 1983. Aerial surveys for marine mammals in the southeastern Bering Sea. U.S. Department of Commerce, NOAA OCS Environmental Assessment Program. Final Report 42 (1986):147–490.

Martin, A. R., S. K. Katona, D. Mattila, D. Hembree, and T. D. Waters. 1984. Migration of humpback whales between the Caribbean and Iceland. Journal of Mammalogy 65:330–333.

Matkin, C. O., and D. R. Matkin. 1981. Marine mammal survey of southwestern Prince William Sound, 1979–1980. Unpublished report U.S. Fish and Wildlife Service, Anchorage, Alaska, Contribution Number 70181-0125-81.

Mizroch, S. A., and A. E. York. 1984. Have pregnancy rates of southern hemisphere fin whales (*Balaenoptera physalus*) increased? Report of the International Whaling Commission Special Issue 6:401–410.

Morris, B. F., M. S. Alton, and H. W. Braham. 1983. Living marine resources of the Gulf of Alaska, a resource assessment for the Gulf of Alaska/Cook Inlet proposed oil and gas Lease Sale 88. U.S. Department of Commerce, NOAA Technical Memorandum NMFS F/AKR-5. 232 p.

Nishiwaki, M. 1966. Distribution and migration of the larger cetaceans in the North Pacific as shown by whaling results. Pages 171–191, *in* K.S. Norris (ed.), Whales, dolphins and porpoises. University of California Press, Berkeley, California.

Rice, D. W. and A. A. Wolman. 1982. Whale census in the Gulf of Alaska, June to August, 1980. Report of the International Whaling Commission 32:491–498.

von Ziegesar, O. 1984. Survey of humpback whales (*Megaptera novaeangliae*) in southwestern Prince William Sound, Alaska 1980–1981–1983. Unpublished report to the Alaska Council on Science and Technology. Juneau, Alaska. 37 p.

von Ziegesar, O. 1988. Survey of humpback whales in Prince William Sound, summer 1988. The North Gulf Oceanic Society, P.O. Box 15244, Homer, AK 99603.

von Ziegesar, O. (ed.). 1992. Humpback whale fluke catalogue for Prince William Sound, Alaska. The North Gulf Oceanic Society, P.O. Box 15244, Homer, AK 99603.

von Ziegesar, O., and C. O. Matkin (eds.). 1985. Humpback whale fluke catalogue for Prince William Sound, Alaska identified by fluke photographs between the years 1977 and 1984. The North Gulf Oceanic Society, P.O. Box 15244, Homer, AK 99603.

Wada, S. 1980. Japanese whaling and whale sighting in the North Pacific 1978 season. Report of the International Whaling Commission 30:415–424.

Watkins, W. A., K. E. Moore, D. Wartzok, and J. H. Johnson. 1981. Radio tracking of finback (*Balaenoptera physalus*) and humpback whales (*Megaptera novaeangliae*) in Prince William Sound, Alaska. Deep Sea Research 78:577–588.

Chapter 11

Sea Otter Foraging Behavior and Hydrocarbon Levels in Prey

Angela M. Doroff and James L. Bodkin

INTRODUCTION

Following the *Exxon Valdez* oil spill (EVOS), Prudhoe Bay crude oil from the vessel spread on the sea surface and covered coastal shores from western Prince William Sound (PWS) to the Alaska Peninsula. In PWS alone, acute mortality of sea otters at the time of the spill was estimated to be greater than 2000 (Doroff et al. 1993; Garrott et al. 1993).

Shoreline oiling was observed on approximately 24% of the 1891 km of coastline surveyed within PWS (*Exxon Valdez* Oil Spill Damage Assessment Geoprocessing Group 1991). The effect of oil on the abundance of nearshore marine invertebrate populations is unclear, and the concentration and persistence of hydrocarbons present in tissues of most of these invertebrate species still remains unknown. What is known is that marine bivalves can accumulate petroleum hydrocarbons from both chronic and acute sources (Blumer et al. 1970; Ehrhardt 1972; Boehm and Quinn 1977). Potential long-term chronic effects of oiled intertidal and subtidal prey on the sea otter population are of concern.

Sea otters prey on a wide variety of benthic marine invertebrates (Riedman and Estes 1990) and forage in shallow coastal waters (Wild and Ames 1974), which vary widely in exposure to the open ocean, substrate type, and community composition. Sea otters have high metabolic demands relative to other marine mammals and can consume 20–25% of their body weight per day in invertebrate prey (Kenyon 1969; Costa and Kooyman 1984).

Sea otters have occupied southwestern PWS since at least the early 1950s (Lensink 1962; Garshelis et al. 1986). The sea otter population in the PWS spill region was likely near equilibrium density and limited by prey availability before the oil spill occurred (Estes et al. 1981; Garshelis et al. 1986; Johnson 1987). Sea otters in this region spent 59% of the daylight hours foraging, while otters in

Marine Mammals and the *Exxon Valdez* **193**

Copyright © 1994 by Academic Press, Inc.
All rights of reproduction in any form reserved.

recently reoccupied habitats of eastern PWS spent only 27% (Garshelis et al. 1986). Therefore, small differences in abundance of prey or net caloric availability due to heavy oiling in portions of southwestern PWS may have led to reduced carrying capacity and delayed recovery for the sea otter population in this region.

Recovery of the PWS sea otter population may be influenced by several factors. Decreased food availability caused by oil-related prey mortality or consumption of contaminated prey may be detrimental. Prey availability in western PWS may have declined due to increased mortality of invertebrates at the time of shoreline oiling, or by oil-removal activities. In addition, relative prey availability may have been decreased by sea otters avoiding invertebrate prey contaminated with petroleum hydrocarbons. However, we lack the baseline data on abundance and distribution of nearshore invertebrates necessary to estimate a reduction in prey availability. In addition, the effects of ingesting prey contaminated with petroleum hydrocarbons on sea otters are unknown.

Our objectives were to determine if sea otter foraging success and prey composition differed between oiled and nonoiled areas and to assess hydrocarbon levels in sea otter prey between oiled and nonoiled areas.

METHODS

Study Sites

The study area included sea otter foraging sites at Squirrel, Green, and Montague Islands in western PWS (Fig. 11-1). Sites were selected on the basis of two criteria: (1) degree of shoreline oiling (based on Alaska Department of Environmental Conservation shoreline oiling maps) with Squirrel, Green, and Montague Islands representing heavy (>50% of the beach area covered or penetrated with oil), moderate (10–50% of the beach area covered or penetrated with oil), and no shoreline oiling, respectively; and (2) sufficient sea otter densities to obtain foraging data (determined by sea otter survey and capture data from other spill-related studies). In general, the study area was a female-occupied area where breeding and pup rearing occurred (Estes et al. 1981; Garshelis 1983; Riedman and Estes 1990). Sea otter foraging data were collected in the study area between mid-April and July 1991 and subtidal sea otter prey were collected during August 1991.

Foraging Observations

Visual observations of foraging sea otters were made with high-resolution telescopes (Questar Corporation, New Hope, PA) and 10X40 binoculars. Foraging behavior was documented using a focal animal sampling method (Altmann 1974). A foraging otter was located and observed until a maximum of 50 identifiable prey

⋙   ⋙   ⋙

items were observed, or until visual contact with the animal was lost, or foraging ceased. When possible, data recorded for each focal animal on each dive included age (i.e., adult, juvenile, or unknown), sex, number of prey and relative prey size, dive interval (seconds), surface interval between foraging dives (seconds), and prey item to lowest identifiable taxon. Prey were classified into one of five size classes (<5 cm; ≥5 to <7 cm; ≥7 cm to <9 cm; ≥9 to <12 cm; and ≥12 cm). Size class of prey was estimated by observers based on the mean forepaw width (4.5 cm) and mean skull width (10 cm) for adult sea otters in this region (Johnson 1987; U.S. Fish and Wildlife Service, unpublished data). Adult animals were categorized as male, independent female, or female with a pup. Small (estimated at ≤18 kg), dark-headed otters were identified as juveniles. Foraging dives were classified as successful (prey item captured), unsuccessful (no prey item captured), or as producing an unknown result (observer could not determine if the dive was successful or unsuccessful). The locations of foraging sea otters were recorded on a Geographic Information System coverage map gridded with a Universal Transverse Mercator projection. Data were collected only during daylight hours and during all tidal cycles.

Scat Analysis

From 20 April to 3 May 1991, 253 sea otter scat samples were examined in the field along 8.5 km of beach within the Green Island study site (Fig. 11-1). For each scat sample encountered, the prey species (when possible) were recorded within each scat. The estimated percentage that each prey type (mussel, clam, crab, or other) contributed to the entire scat was categorized as follows: 100, 90, 75, 50, 25, 10, and 5%.

Collection and Hydrocarbon Analysis of Prey

At each study site, clam species identified as sea otter prey were collected and tissues were analyzed for hydrocarbon content. Coordinates of foraging observations were plotted for each study site. The outermost coordinate locations delineated a polygon over which a grid of 100-m^2 plots was laid. Ten 100-m^2 plots were chosen randomly within each study site, and SCUBA divers searched for prey within each plot, beginning at the boat anchor. The boat anchor location was haphazard within each of the plot boundaries. Clams were recovered using a venturi dredge (Keene Engineering, Northridge, California). Water depth averaged 8 m (range 5–12 m). Clams were brought to the surface in nylon-mesh dive bags, wrapped in chemically cleaned aluminum foil (acetone and hexane washed), and frozen whole. During prey collection, divers attempted to obtain three *Saxidomus giganteus* within each plot. However, this could not be accomplished in all plots and, where possible, three of each clam species encountered were submitted for

>≍≍≍   ≍≍≍   ≍≍≍

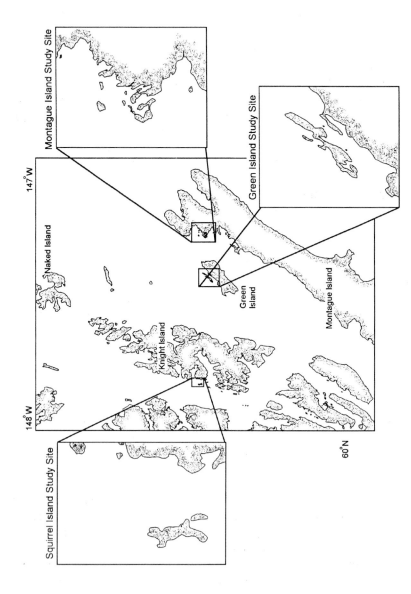

Figure 11-1. Sea otter forage study site locations in western Prince William Sound, 1991.

analysis. When more than three clams of the same species were retrieved from a single plot, three were randomly selected for hydrocarbon analyses. Clams were thawed in the laboratory and soft tissue was removed (using instruments cleaned with acetone and hexane) from the shell and placed in chemically clean jars, weighed, and refrozen. Samples were shipped to the Geochemical and Environmental Research Group (GERG), Texas A&M University, College Station, Texas, for analysis of the hydrocarbon content. The tissue extraction method used in the analysis was developed by McLeod et al. (1985) and modified by Wade et al. (1988, 1993) and Jackson et al. (in press). Laboratory methodology for the hydrocarbon analysis for this study was standardized with all Natural Resources Damage Assessment Studies by GERG (GERG standard operating procedures 8901-8905).

Data Analysis

The foraging record is defined in this chapter as the foraging data specific to a focal animal and was used as the sample unit in the analyses of foraging behavior. The sample unit in the analysis of dive and surface intervals was individual dives.

The percentage of successful dives was determined for all foraging records of adult and juvenile sea otters having ≥10 dives. Dives of unknown result were not included in this analysis. The proportion of successful dives was normalized by an arcsine square-root transformation. An analysis of variance (ANOVA) was used to test for differences in foraging success among sites and between adults and juveniles.

The number of prey items captured per dive was averaged for each foraging record by site. Dives resulting in the capture of mussels were excluded from this analysis due to the difficulty in obtaining accurate counts on a per-dive basis. Dives of unknown result were not used in this analysis. An ANOVA was used to test for differences in the number of prey retrieved per dive among sites.

Mean dive and surface intervals were tested among study sites and prey types (clams, crabs, and mussels) by a two-way ANOVA for an unbalanced sample.

Foraging records for each focal animal having ≥10 foraging dives were summarized into the proportion of dives resulting in the capture of clams, crabs, or mussels within each study site. The Kruskal-Wallace nonparametric procedure was used to test for differences in the proportion of clams, crabs, and mussels captured among sites for adult sea otters and between adults and juveniles (sample sizes were sufficient to test age differences only for the Green Island study site).

Hydrocarbon concentrations were reported by GERG in ng/g wet weight for alkanes and aromatics, and in μg/g wet weight for the unresolved complex mixture (UCM). Mean concentrations of total alkanes, total aromatics, and UCM were tested among study sites by ANOVA procedures. Prior to analysis, data were transformed by a $\log_{10}(x_i + 1)$ to normalize the distribution.

⋊⋉ ⋊⋉ ⋊⋉

RESULTS

Foraging Behavior

At Squirrel Island, 69 foraging records were observed (68 adults and 1 juvenile). Thirty-eight foraging records (29 adults and 9 juveniles) were observed at Green Island and 72 foraging records (69 adults and 3 juveniles) were observed at Montague Island.

Sea otters recovered prey items on 87–92% of their foraging dives and foraging success did not differ among sites (F=1.23, P=0.29) (Table 11-1). Mean foraging success rates were 90% (*n*=82) for adult and 92% (*n*=10) for juvenile sea otters in all study sites combined and did not differ significantly (F=0.50, P=0.48).

Mean number of prey retrieved per dive were 1.2, 1.0, and 1.3 for Squirrel, Green, and Montague Islands, respectively; differences were not detected among sites (F=2.19, P=0.11). Size class was estimated for 1867 prey items; the majority of prey items, 96% or greater, were <9 cm in all sites (Table 11-1).

Mean dive intervals varied from 43 to 88 seconds, and surface intervals varied from 37 to 48 seconds for all prey types within the study sites. Dive intervals differed significantly for dives retrieving clams (80–119 seconds), mussels (20–35 seconds), and crabs (63-82 seconds) among study sites (F=19.83, P<0.001) and among prey types (F=135.92, P<0.001), and the interaction between site and prey type also differed (F=24.16, P<0.001).

Prey Composition

Adults differed in the proportion of dives resulting in the capture of clams (χ^2=9.73, P=0.01), crabs (χ^2=7.03, P=0.03), and mussels (χ^2=7.21, P=0.03) among sites (Table 11- 2). The median proportion of dives resulting in the capture of clams was higher than that for mussels or crabs in all study sites for adults and was less (0.29) for Squirrel Island than for Green (0.75) or Montague (0.62) Islands. Sample sizes were insufficient to allow testing for differences in prey composition related to sex or reproductive status. Juvenile sea otters at the Green Island site captured mussels on a significantly higher proportion of dives than did adults (χ^2=5.73, P=0.02) (Table 11-2). Differences in the proportion of dives in which clam or crab were captured (in the Green Island area) were not detected between adult and juvenile sea otters. There again, sample sizes were insufficient to allow for testing of age group differences in the proportion of dives resulting in the capture of prey at Squirrel and Montague Island study sites.

Clams were retrieved on 34%, 61%, and 44% of the successful sea otter foraging dives at Squirrel (*n*=833), Green (*n*=759), and Montague (*n*=752) Islands, respectively (Table 11-3). *Saxidomus giganteus* was the most commonly identified clam in the sea otter diet for all study sites. Other clam species identified were *Mya* spp.,

Table 11-1. Prey type, size class, proportion of successful dives, and mean number of prey retrieved per dive estimated for sea otters (*Enhydra lutris*) at three sites in western Prince William Sound, Alaska, during April-July 1991.

Prey type	Size class (cm)	Squirrel Island	Green Island	Montague Island
Clam	< 5	63%	79%	49%
	≥ 5 < 7	28%	20%	46%
	≥ 7 < 9	8%	1%	5%
	≥ 9 <12	1%	0%	0%
	≥12	<1%	0%	0%
		(n = 296)	(n = 479)	(n = 351)
Mussel	< 5	100%	100%	100%
		(n = 142)	(n = 159)	(n = 53)
Crab	< 5	18%	21%	43%
	≥ 5 <7	43%	71%	52%
	≥ 7 <9	30%	7%	5%
	≥ 9 <12	7%	0%	0%
	≥12	2%	0%	0%
		(n = 90)	(n = 14)	(n =112)
All Prey[a]	< 5	63%	79%	49%
	≥ 5 <7	23%	17%	42%
	≥ 7 < 9	10%	4%	8%
	≥ 9 <12	3%	<1%	<1%
	≥12	1%	0%	1%
		(n = 598)	(n = 690)	(n = 579)
Mean number of prey per dive[b]		1.2	1.0	1.3
Percentage of successful dives		87%	92%	90%

[a] Includes clams, mussels, crab, and all other prey identified as to size class.
[b] Dives resulting in capture of mussels were excluded for this analysis due to the difficulty in obtaining accurate counts on a per dive basis.

⨾⨾ ⨾⨾ ⨾⨾

Table 11-2. Median proportion of dives resulting in the capture of clams, crabs, and mussels for adult and juvenile sea otters (*Enhydra lutris*) in Prince William Sound, Alaska, 1991. (— = no data)

Age class	Green Island				Squirrel Island				Montague Island			
	Clam[a]	Crab[b]	Mussel[b]	N[c]	Clam[a]	Crab[b]	Mussel[b]	N[c]	Clam[a]	Crab[b]	Mussel[b]	N[c]
Adults	0.75	0.0	0.0[d]	15 (356)	0.29	0.03	0.06	34 (754)	0.62	0.07	0.0	28 (531)
Juveniles	0.16	0.0	0.44[d]	8 (365)	—	—	—	—	0.17	0.41	0.0	2 (59)

[a] Significant differences among areas in the proportion of dives resulting in the capture of clam (P = 0.01) by adults determined by a Kruskal-Wallace test.
[b] Significant differences among areas in the proportion of dives resulting in the capture of crab (P = 0.03) and mussel (P = 0.03) by adults determined by Kruskal-Wallace tests.
[c] Number of foraging records (total number of foraging dives).
[d] Significant differences among age classes in the proportion of dives capturing mussels at Green Island (P = 0.02) determined by a Kruskal-Wallace test.

Table 11-3. Composition of sea otter (*Enhydra lutris*) prey determined by visual observation at three sites in western Prince William Sound Alaska, during April-July 1991. (--- = no data)

	Squirrel Island (%)	Green Island (%)	Montague Island (%)
Clam[a]	34	61	44
Mya spp.	2	---	3
Protothaca staminea	3	5	<1
Saxidomus giganteus	21	20	9
Tresus capax	<1	<1	<1
Unknown clams	73	75	87
Mussel[a]	17	20	7
Mytilus edulis	100	100	100
Crab[a]	11	2	14
Telmessus spp.	46	27	72
Unknown crabs	54	73	28
Other	5	4	4
Balanus spp.	3	12	---
Chlamys spp.	---	---	6
Clinocardium spp.	21	3	33
Cucumaria spp.	5	---	---
Echiurus echiurus	3	67	12
Notoacmea spp.	3	---	---
Octopus spp.	3	---	3
Pisaster ochraceus	47	12	39
Pododesmus macrochisma	---	3	---
Pycnopodia helianthoides	3	---	---
Strongylocentrotus spp.	10	---	3
Chiton (class Polyplacophora)	3	---	---
Tunicate (class Ascidiacea)	---	3	3
Unknown prey	33	12	30

[a] Adults differed in the proportion of dives retrieving clam (P = 0.01), crab (P = 0.03), and mussel (P = 0.03) among study areas.

Protothaca staminea, and *Tresus capax*. Mussels (*Mytilus edulis*), and crabs (primarily *Telmessus* spp.) each contributed 20% or less to the identified species for each study site. Other prey included: limpets (*Notoacmea* spp.), barnacles (*Balanus* spp.), cockles (*Clinocardium* spp.), scallops (*Chlamys* spp.), sea cucumbers (*Cucumaria* spp.), fat innkeepers (*Echiurus echiurus alaskensis*), octopus (*Octopus* spp.), sea stars (*Pisaster* spp.), jingles (*Pododesmus* spp.), sunflower sea stars (*Pycnopodia helianthoides*), sea urchins (*Strongylocentrotus* spp.), chitons (Class Polyplacophora), and tunicates (Class Ascidiacea). These species contributed 5% or less to otter diets at each study site (Table 11-3).

Table 11-4. Estimated percentage of prey type (mussel, clam, crab, and other small invertebrates) found in 253 scat samples examined during 20 April to 2 May 1991 in western Prince William Sound, Alaska.

Prey type	Estimated percentage							Occurrence in sample (percentage)	
	100%	90%	75%	50%	25%	10%	5%		
Mussel[a]	76	24	10	13	14	6	10	153	(60%)
Clam[b]	23	22	8	15	21	10	17	116	(46%)
Crab[c]	0	2	2	5	21	10	7	47	(19%)
Other[d]	13	4	5	8	4	6	10	50	(20%)

[a] *Mytilus edulis*
[b] *Protothaca staminea, Saxidomus giganteus, Humilaria kennerleyi, Gari californira:* includes unidentified shell fragments.
[c] Species not identified.
[d] Other is equivalent to one or more of the following species: scallop (*Chlamys* spp.), snail (*Natica* sp.), cockle (*Clinocardium* spp.), limpet (*Notoacmea scutum*), and other unidentified shell fragments.

Fifty-six percent of the 253 scat samples examined in the Green Island study site contained more than one prey species (Table 11-4). Mussels were observed in 153 of 253 (60%) sea otter scat and clams were observed in 116 of 253 (46%) scat examined. Clam species were primarily *Protothaca staminea* and *Saxidomus giganteus* with trace amounts of *Humilaria kennerleyi* and *Gari californica*. Crab and other small invertebrates were found in 19% and 20%, respectively, of scat sampled. Of scats containing a single prey type, 76 contained only mussels, 23 contained only clams, and 13 contained either scallops (*Chlamys* spp.), snails (*Natica* sp.), cockles (*Clinocardium* spp.), or limpets (*Notoacmea scutum*).

Prey Hydrocarbon Analysis

A total of 79 prey samples were collected for hydrocarbon analyses. Twenty-five prey were collected in seven plots at Squirrel Island; 33 prey in seven plots at Green Island, and 21 prey in six plots at Montague Island. *Protothaca staminea* (*n*=24), *Mya* spp. (*n*=23), and *S. giganteus* (*n*=20) were most frequently collected. Species composition and size class samples within sites are presented in Table 11-5.

Tissue samples of subtidal bivalves obtained from sites which had received heavy to moderate shoreline oiling in 1989 had no detectable differences in mean total alkane (F=2.35, P=0.10), aromatic (F=0.16, P=0.85), and UCM (F=0.56, P=0.57) concentrations from the site where no shoreline oiling occurred (Table 11-5). Mean concentrations of total alkanes and aromatics were slightly higher, however, for tissue samples collected at Green Island than those from Squirrel and Montague Islands. At all sites, *Mya arenaria* contained the highest concentration of total alkanes of all species sampled.

DISCUSSION

Although foraging success was high (90% for all observations), the majority of clams (95% of 1126) observed were small (estimated to be <7cm). Garshelis et al. (1986) reported clams captured by sea otters rarely exceeded 6 cm in the Green Island site during 1980–1981. During 1991, 79% (*n*=479) of the clams captured at Green Island were estimated to be <5 cm, 20% ranged between 5 and 7 cm, and none were estimated to be larger than 9 cm. Mean shell length for clams recovered in the dredge samples in the Green Island area ranged from 3.3 to 4.7 cm.

Dive duration and surface intervals between dives were variable for individuals but significantly different depending on the type of prey captured. Individual animals, water depth, geographic location, and food item all contribute to variation in duration of foraging dives (Estes et al. 1981; Garshelis 1983). Sea otters at Squirrel, Green, and Montague Islands foraged on the same principal species in

Table 11-5. Hydrocarbon and size class means for bivalves collected subtidally near Squirrel (oiled), Green (oiled), and Montague (non-oiled) Islands in western Prince William Sound, Alaska, summer 1991.

Sample location and species sampled	Total alkanes (ng/g)	Total aromatics (ng/g)	Unresolved complex mixture (µg/g)	Mean shell length (mm)	Mean wet meat mass (g)	N
Squirrel Island						
Humilaria kennerleyi	788.5	48.4	14.8	46	7.8	3
Mya arenaria	1752.6	79.4	0.0	41	4.4	4
Protothaca staminea	629.0	55.6	3.9	44	10.0	6
Saxidomas giganteus	900.3	51.7	6.3	51	14.6	11
Serripes groenlandicus	1225.1	57.0	6.3	56	16.2	1
Site mean ± SD	971.2 ± 712.0	56.9 ± 18.7	5.7 ± 8.3	47 ± 17.4	11.2 ± 6.1	25
Green Island						
Gari californica	1278.2	39.7	3.7	47	0.4	4
Humilaria kennerleyi	1034.2	74.7	56.1	33	2.7	1
Mya arenaria	1494.4	68.8	4.0	40	4.1	15
Protothaca staminea	790.2	54.8	0.5	41	8.0	9
Saxidomas giganteus	897.8	45.9	0.5	41	8.2	4
Site mean ± SD	1189.9 ± 1033.0	58.9 ± 18.7	4.1 ± 10.9	41 ± 6.1	6.3 ± 3.5	33
Montague Island						
Gari californica	823.3	49.2	3.8	49	5.5	1
Humilaria kennerleyi	569.1	59.7	11.6	52	13.7	2
Mya arenaria	996.4	72.5	1.2	48	7.1	4
Protothaca staminea	806.2	52.2	4.7	41	8.3	9
Saxidomas giganteus	843.0	49.0	0.3	33	4.0	5
Site mean ± SD	829.4 ± 163.9	55.9 ± 12.0	3.6 ± 6.4	42 ± 7.9	7.4 ± 4.0	21

1991, as were observed in previous years (Calkins 1978; Garshelis et al. 1986; Johnson 1987), suggesting there has been no detectable shift in prey composition over time or as a result of shoreline oiling at these study sites. Clams, mussels, and crabs were the primary prey of sea otters at all sites; however, there were differences in the proportion with which these prey were captured among sites. The difference in the proportions of prey type captured by sea otters among sites may have been influenced by the proportion of unidentified prey within each site (Table 11-3) or by variation in prey availability within each site. There was no replication of treatment types (heavy oil, moderate oil, and no oil); therefore, we have no measure of natural variation within each treatment.

Prey composition determined from scat contents also indicated mussels, clams, and crabs were important prey of sea otters. Sea otters haul out most frequently during the winter in PWS; therefore, these data primarily represent the overwinter diet near Green Island (Johnson 1987; VanBlaricom 1988). Johnson (1987) examined 3275 scat in the Green Island site during 1974–1984 and found 58%, 34%, 36%, and 16% of the scat contained clams, mussels, crabs, and other species, respectively. In our sample from the same region, we observed mussels most frequently (60%). Whether the observed differences reflect changes in prey use over time, changes in the ratio of adults and juveniles using the haulout site through time, or variation in scat content between observation periods is unknown.

Determination of sea otter prey composition through visual observation or scat analysis can yield different results; both methods have inherent biases. Prey composition based on visual observations is biased toward: (1) prey captured from nearshore areas; (2) larger prey items (greater than the paw size of the animal); and (3) prey captured during daylight hours. Prey composition based on scat analysis is biased against larger prey when no hard parts are ingested. Scat analysis also cannot reveal any potential variation in diet between adult and juvenile or male and female otters.

Adult sea otters foraged primarily on species found in the subtidal zone, whereas juveniles had a higher proportion of an intertidal species, the mussel, in their diet based on visual observation. Johnson (1987) also reported dietary differences between adult (19% mussel and 59% clam) and juvenile (63% mussel and 16% clam) sea otters at Green Island during 1974–1984. In California, Estes et al. (1981) found that juveniles commonly foraged in water ranging from 1 to 2 fathoms while adults nearly always foraged in deeper water. Mussels can easily be obtained by foraging sea otters because they occur in the intertidal zone and require little effort to capture (Estes et al. 1981; VanBlaricom 1988). Mean dive intervals for mussels were shorter than those recorded for other prey. However, mussels are less valuable calorically than other sea otter prey (Garshelis 1983).

Mean total aromatic and UCM concentrations in intertidal mussel tissue collected at our study site on Green Island during 1989 were 2566 ng/g (± 853) and

171.4 µg/g (± 58.6), respectively (Andres and Cody 1993). These values are as much as 40 times greater than the mean levels that we observed in the subtidal clam tissue samples at Green Island in 1991. Unfortunately, no intertidal mussels were collected in 1991 to assess the persistence of hydrocarbons in the mussel tissues at the Green Island site. Andres and Cody (1993) also reported hydrocarbon levels in mussel tissue of 82 ng/g (± 21) and 7.4 µg/g (± 1.7) for total aromatic and UCM concentrations, respectively, from our Montague Island study site; aromatic and UCM levels were slightly lower in the subtidal bivalve tissue collected 1991. Other sites in PWS were sampled annually (1989–1992) and, at some sites, mussel tissue and the underlying sediments consistently contained high concentrations (up to 50 parts per million) of total aromatic hydrocarbons (Babcock et al. 1993; Rounds et al. 1993).

Juvenile sea otters foraged on mussels to a greater extent than adults. However, individual adults and juveniles may specialize on only a few species, some of which occur in the intertidal region (Ralls et al. 1988; Riedman and Estes 1990). There-fore, juveniles and individual adults specializing in intertidal species could have a higher probability of encountering hydrocarbon contamination in their prey than individuals foraging in the subtidal regions.

CONCLUSIONS

Sea otter foraging success, in terms of the percentage of successful dives or mean number of prey items captured per dive, was not affected in the oiled area 2 years after the EVOS. Prey composition (primarily clam, mussel, and crab) was similar among oiled and nonoiled study sites and to prespill data from the western PWS region. Adult sea otters foraged primarily in the subtidal region, while juveniles foraged more frequently intertidally. Tissues of subtidal bivalve prey tested for hydrocarbon content did not differ regardless of the degree of shoreline oiling. Mussel tissue sampled in 1989–1992 in the intertidal regions exhibited, in site-spe-cific areas, hydrocarbon concentrations similar to crude oil (Babcock et al. 1993). Contamination of mussels and other intertidal prey species may be of concern for juvenile sea otters and for adults specializing in the use of intertidal prey.

ACKNOWLEDGMENTS

This work was conducted as part of *Exxon Valdez* Oil Spill Natural Resources Damage Assessment program. We thank K. Modla, who was a primary observer during data collection and assisted with data entry. D. Bruden, C. Doroff, and M. Fedorko also assisted with data collection. G. VanBlaricom assisted with the

✂ ✂ ✂

species identification of prey. B. Ballachey, L. Holland-Bartles, A. DeGange, D. Mulcahy, K. Oakley, M. Riedman, and two anonymous reviewers provided draft reviews of the manuscript and made many valuable suggestions. We thank M. Ronaldson for her help in preparing the manuscript.

REFERENCES

Altmann, J. 1974. Observational study of behavior: Sampling methods. Behaviour 49:227–267.

Andres, B. A., and M. M. Cody. 1993. The effects of the *Exxon Valdez* spill on black oystercatchers breeding in Prince William Sound. Bird Study No. 12, Restoration Study No. 17. Unpublished Report, U.S. Fish and Wildlife Service, 1011 East Tudor Road,Anchorage, Alaska 99503.

Babcock, M., G. Irvine, S. Rice, P. Rounds, J. Cusick, and C. C. Brodersen. 1993. Oiled mussel beds in Prince William Sound two and three years after the *Exxon Valdez* oil spill. Pages 184–185, *in Exxon Valdez* oil spill symposium, 2–5 February 1993, Anchorage, Alaska. (Available, Oil Spill Public Information Center, 645 G Street, Anchorage, Alaska 99501.)

Blumer, M., G. Souza, and J. Sass. 1970. Hydrocarbon pollution of edible shellfish by an oil spill. Marine Biology 5:195–202.

Boehm, P. D., and J. G. Quinn. 1977. The persistence of chronically accumulated hydrocarbons in the hard shell clam *Mercenaria mercenaria*. Marine Biology 44:227–233.

Calkins, D. G. 1978. Feeding behavior and major prey species of the sea otter, *Enhydra lutris*, in Montague strait, Prince William Sound, Alaska. Fishery Bulletin, U.S. 76:125-131.

Costa, D. P., and G. L. Kooyman. 1984. Contribution of specific dynamic action to heat balance and thermoregulation in the sea otter *Enhydra lutris*. Physiological Zoology 57:199–203.

Doroff, A., A. R. DeGange, C. Lensink, B. E. Ballachey, J. L. Bodkin, and D. Bruden. 1993. Recovery of sea otter carcasses following the *Exxon Valdez* oil spill. Pages 285–288, *in Exxon Valdez* oil spill symposium, 2–5 February 1993, Anchorage, Alaska. (Available, Oil Spill Public Information Center, 645 G Street, Anchorage, Alaska 99501.)

Ehrhardt, M. 1972. Petroleum hydrocarbons in oysters from Galveston Bay. Environmental Pollution 3:257–271.

Estes, J. A., R. J. Jameson, and A. M. Johnson. 1981. Food selection and some foraging tactics of sea otters. Pages 606–641, *in* J. A. Chapman and D. Pursley (eds.), Worldwide Furbearer Conference Proceedings, 3–11 August 1980. Frostburg, Maryland.

Exxon Valdez Oil Spill Damage Assessment Geoprocessing Group. 1991. The *Exxon Valdez* oil spill natural resource damage assessment and restoration: A report on oiling to environmentally sensitive shoreline. Draft. Pages 1–31, *in Exxon Valdez* Oil Spill Damage Assessment Geoprocessing Group. *Exxon Valdez* Oil Spill Technical Services No. 3, GIS Mapping and Statistical Analysis. Alaska Department of Natural Resources and U.S. Fish and Wildlife Service, Anchorage, Alaska.

Garrott, R. A., L. L. Eberhardt, and D. M. Burn. 1993. Impact of the *Exxon Valdez* oil spill on sea otter populations. Marine Mammal Science 9:343–359.

Garshelis, D. L. 1983. Ecology of sea otters in Prince William Sound, Alaska. Ph.D. thesis, University of Minnesota, Minnesota. 321 p.

Garshelis, D. L., J. A. Garshelis, and A. T. Kimker. 1986. Sea otter time budgets and prey relationships in Alaska. Journal of Wildlife Management 50:637–647.

Jackson, T. J., T. L. Wade, T. J. McDonald, D. L. Wilkinson, and J. M. Brooks. In press. Polynuclear aromatic hydrocarbon contaminants in oysters from the Gulf of Mexico (1986–1990). Environmental Pollution.

Johnson, A. M. 1987. Sea otters of Prince William Sound, Alaska. Unpublished Report, U.S. Fish and Wildlife Service, Alaska Fish and Wildlife Research Center, 1011 East Tudor Road, Anchorage, Alaska.

Kenyon, K. W. 1969. The sea otter in the eastern Pacific Ocean. North American Fauna 68. 352 p.

Lensink, C. J. 1962. The history and status of sea otters in Alaska. Ph.D. dissertation, Purdue University, West Lafayette, Indiana. 188 p.

McLeod, W. D., D. W. Brown, A. J. Friedman, D. G. Burrow, O. Mayes, R. W. Pearce, C. A. Wigren, and R. G. Bogar. 1985. Standard analytical procedures of the NOAA National Analytical Facility 1985–1986. Extractable Toxic Compounds. 2nd Edition. U.S. Department of Commerce, NOAA Technical Memorandum NMFS F/NWC-92. 121 p.

Ralls, K., B. Hatfield, and D. B. Siniff. 1988. Feeding patterns of California sea otter. Pages 84–105, *in* D. B. Siniff and K. Ralls (eds.), Population status of California sea otters. Final report to the Minerals Management Service, U.S. Department of Interior 14-12-001-3003.

Riedman, M. L., and J. A. Estes. 1990. The sea otter (*Enhydra lutris*): Behavior, ecology, and natural history. U.S. Fish and Wildlife Service Biological Report 90(14). 126 p.

Rounds, P., S. Rice, M. M. Babcock, and C. C. Brodersen. 1993. Variability of *Exxon Valdez* hydrocarbon concentrations in mussel bed sediments. Pages 182–183, *in Exxon Valdez* oil spill symposium, 2–5 February 1993, Anchorage, Alaska. (Available, Oil Spill Public Information Center, 645 G Street, Anchorage, Alaska 99501.)

VanBlaricom, G. R. 1988. Effects of foraging by sea otters on mussel-dominated intertidal communities. Pages 48–91, *in* G. R. VanBlaricom and J. A. Estes (eds.), The community ecology of sea otters. Springer-Verlag, Berlin, West Germany.

Wade, T. L., E. L. Atlas, J. M. Brooks, M. C. Kennicutt II, R. G. Fox, J. Sericano, B. Garcia, and D. DeFreitas. 1988. NOAA Gulf of Mexico status and trends program: Trace organic contaminant distribution in sediments and oysters. Estuaries 11:171–179.

Wade, T. L., T. J. Jackson, T. J. McDonald, D. L. Wilkinson, and J. M. Brooks. 1993. Oysters as biomonitors of the APEX Barge oil spill, Galveston Bay, Texas. *In* Proceedings, 1993 international oil spill conference, 29 March-1 April 1993. Tampa, Florida.

Wild, P. W., and J. A. Ames. 1974. A report on the sea otter, (*Enhydra lutris* L.), in California. California Department of Fish Game. Marine Resources Technical Report 20. 93 p.

⋈ ⋈ ⋈

Chapter 12

Observations of Oiling of Harbor Seals in Prince William Sound

Lloyd F. Lowry, Kathryn J. Frost, and Kenneth W. Pitcher

INTRODUCTION

Harbor seals, *Phoca vitulina richardsi*, are among the most commonly seen marine mammals in Prince William Sound (PWS). They occur primarily in the coastal zone where they feed, haul out to rest, bear and care for their young, and molt (Pitcher and Calkins 1979). Haulout areas include floating glacial ice, intertidal reefs, rocky shores, mud bars, and gravel and sand beaches. Unlike northern fur seals (*Callorhinus ursinus*) and Steller sea lions (*Eumetopias jubatus*), harbor seals do not form distinct rookeries during the pupping and breeding season. Pups are born at the same general locations that are used as haulout sites at other times of year.

Prior to the *Exxon Valdez* oil spill (EVOS) there was only a limited understanding of how pinnipeds would interact with spilled oil. Their amphibious habits make them susceptible to contact with oil both in the water and on shore. Previous studies of seals in areas affected by oil spills (reviewed by St. Aubin 1990) have generally produced equivocal or inconclusive results. Deficiencies of these observations have been due largely to inadequate effort devoted to studies, logistical difficulties, and inherent problems with studying pinniped behavior, especially when they are at sea. Nonetheless, available information indicated that seals would contact spilled oil and the oil would stick to their pelage. Many questions remained, however, such as whether seals could detect and would avoid spilled oil (St. Aubin 1990).

When the *Exxon Valdez* ran aground, some of the largest harbor seal haulout sites in PWS, and waters adjacent to those sites, were directly impacted by substantial amounts of spilled crude oil. In the early weeks of the spill, seals surfaced in and swam through floating oil while feeding and moving to and from haulout sites. On haulout sites in oiled areas, they crawled over and rested on oiled

Marine Mammals and the *Exxon Valdez*   **209**

rocks and algae throughout the spring and summer. Pups were born in May and June, when some of the haulout sites still had oil on them, resulting in pups becoming oiled. Many pups were cared for by oiled mothers.

As part of the Natural Resources Damage Assessment program (NRDA) a study was designed to investigate and quantify, as possible, the effects of oil and the subsequent clean-up operations on the distribution, abundance, and health of harbor seals. One of the objectives was to describe the patterns of oiling and the behavior of seals in the spill area.

METHODS

All of the field work for this project was conducted in PWS because of the intensity of oiling of seals and their habitats, relatively easy access to major haulout sites, and availability of logistics. Initial efforts concentrated on trend-count sites in eastern and central PWS that had previously been selected and used for population monitoring (Pitcher 1989; Frost et al., Chapter 6). As the oil spread to western PWS, the study area expanded to include additional seal haulout sites in parts of that region.

Observations were made on harbor seals in PWS from 29 March through 4 September 1989. These observations were made at irregular intervals, except during 15 May–13 July when we conducted 2- to 5-day sampling periods at intervals of approximately 10 days. During initial efforts we visited as many haulout sites in both oiled and unoiled areas as was possible. Beginning in June, we limited our efforts to oiled haulout sites at Seal Island, Bay of Isles, and Herring Bay. Additional observations were made in these three areas in April and May 1990.

Seals were counted and observed from small boats using 7- to 10-power binoculars, or with a 25-power spotting scope from nearby locations on shore. Whenever possible each seal was classified as to the degree of pelage oiling. Seals were classified as "heavily oiled" when all visible parts of the body were coated with oil and appeared a uniform dark chocolate brown or black (Fig. 12-1). Seals that were oiled on some part of their body, but did not appear a uniform dark color were classified as "oiled" (Fig. 12-2). Early in the sampling period, this category was subdivided into lightly and moderately oiled groups, but as the season progressed this distinction was often difficult to make. Thus, the lightly and moderately oiled categories were combined. Seals were classified as "unoiled" if no oil could be seen on the body. Qualitative observations of behavior were recorded, especially if the behavior appeared different from what we considered normal for harbor seals.

Haulout sites were visually inspected for presence of oil on the substrate or algae. Haulout sites were classified as "oiled" if oil was visible on the rocks or algae,

⋈ ⋈ ⋈

Figure 12-1. A group of seals that were classified as "heavily oiled" in Herring Bay, Prince William Sound, 26 May 1989. Note the uniform, dark appearance of the entire body.

Figure 12-2. A harbor seal and its pup that were classified as "oiled" at Applegate Rocks, Prince William Sound, 8 June 1989. Note the dark coloration of the pelage behind the neck of the adult and on the flank of the pup.

✕◇ ✕◇ ✕◇

"intermediate" if no oil was detectable on the actual site but adjacent areas were oiled, or "unoiled" if there was no oil detectable on the haulout site or nearby.

RESULTS

Oil spread southwestward from the tanker grounded on Bligh Reef, and contacted harbor seal haulout sites in central and western PWS. Some haulout sites were completely covered with oil in layers and puddles, while others had only oily bands or a spattering of oil on the rocks. Oil on the water near haulout sites and offshore ranged from thin sheens to thick, heavy layers with mousse.

Characteristics of oil contamination in the seal habitats changed throughout the summer. After May, little floating oil remained except for thin sheens near heavily oiled shorelines. Most haulout sites became cleaner, either due to shoreline treatment or natural washing. Some major haulout sites were given high priority for cleanup and were treated by mid-May (Zimmerman et al., Chapter 2). However, in some heavily impacted areas a sticky oil residue remained on rocks and algae at least into September 1989.

When our observations began, seals were seen hauled out at some traditional locations even when those sites were heavily oiled. They were also seen swimming near haulout sites in water with an oil sheen and relatively heavy oil coverage. During the first 2 weeks after the spill, the pattern of oiling on their bodies often suggested contact with oil while in the water. Some seals had oil only on the top of the head, while others were oiled over the entire head and neck as if they were wearing a hood. Some were oiled on the head, neck, and anterior part of the body, as though they had risen vertically halfway out of the water through oil. After approximately mid-April these patterns were no longer recognizable: seals appeared to be unoiled, oiled on parts of the head, back, sides, or belly, or coated with oil over the entire body.

During April–July 1989, we made observations at 34 major haulout areas in PWS (Table 12-1). As many as 10 separate counts were made in some areas, so seal numbers and percentages that were oiled are therefore shown as ranges. In every oiled area we saw oiled seals, and the proportion of oiled seals was highest at heavily oiled haulout sites. Oiled seals were seen at only 3 of 12 unoiled sites, and those were at locations near areas that were oiled.

Systematic boat-based observations were conducted in eastern and central PWS in May 1989. Initially, work was conducted throughout eastern and central PWS. Data from individual haulout sites were combined into oiled, unoiled, and intermediate categories (Table 12-2). During May, only 1% of all seals observed in unoiled areas were oiled, while in intermediate areas 32% of the seals were oiled. In oiled areas, 81% of the observed seals were oiled, most of them heavily.

⋊⋉   ⋊⋉   ⋊⋉

Table 12-1. Oiling of harbor seals and harbor seal haulout sites in Prince William Sound, 1989. Data on oiling of seals are for animals older than pups.

Haulout	Degree of oiling on shoreline	Observation period	Number of seals	Percent oiled
Eshamy Bay	unoiled	June	3	0
Fairmount Island	unoiled	May	15	0
Gravina Island	unoiled	May	10-20	0
Gravina Rocks	unoiled	May	2-9	0
Little Green Island	unoiled	May	40	20
Lower Herring Bay	unoiled	May	3	0
Olsen Bay	unoiled	May	22-48	0
Olsen Island	unoiled	May	3	0
Payday	unoiled	May	3	0
Point Pellew	unoiled	May	4	0
Port Chalmers	unoiled	May	19	5
Stockdale Harbor	unoiled	May	1	100
Agnes Island	light	April-July	15-40	5-66
Channel Island	light	May	18-32	11-66
Chenega Island	light	June	12	8
Evans Island	light	June	43	35
Fleming Island	light	June	2	50
Green Island	moderate	April	10	60
Lone Island	moderate	July	4	25
Perry Island SE	moderate	July	22	23
Bay of Isles	moderate-heavy	May-July	5-42	87-100
Crafton Island	moderate-heavy	June-July	17-33	76-83
Junction Island	moderate-heavy	June-July	14-28	36-56
Rua Cove/Marsha Bay	moderate-heavy	May	5	75
Applegate Rocks	heavy	April-July	26-204	51-81
Disk Island	heavy	May-June	1-8	100
Foul Pass/Ingot Island	heavy	May	5-6	100
Herring Bay	heavy	April-July	10-58	98-100
Little Smith Island	heavy	April-July	12-23	83-100
Northwest Bay	heavy	April-July	1	100
Peak Island	heavy	July	7	14
Seal Island	heavy	May-July	15-74	33-77
Smith Island	heavy	April-July	10-25	25-56
Upper & Lower Pass	heavy	May-June	10-25	100

Table 12-2. Percent of seals older than pups that were oiled, as determined from boat-based observations in Prince William Sound, May 1989. Haulout sites included are shown in Table 12-1.

Area type	Dates	Number of seals classified	Percent in category			
			Heavily oiled	Oiled	Unoiled	
Unoiled	15-18 May	58	2	2	97	
	23-27 May	124	0	0	100	
	Combined	182	<1	<1	99	
Intermediate	15-18 May	24	8	0	92	
	23-27 May	72	18	22	60	
	Combined	96	15	17	68	
Oiled	15-18 May	177	85	11	4	
	23-27 May	408	45	29	26	
	Combined	585	57	24	19	

Table 12-3. Percent of seals and seal pups that were oiled at Seal Island, Bay of Isles, and Herring Bay in Prince William Sound, May-September 1989. For each sample, the total number of seals classified is given in parentheses. (--- = no data)

Date	Seal Island percent oiled		Bay of Isles percent oiled		Herring Bay percent oiled	
	non-pups	pups	non-pups	pups	non-pups	pups
16-18 May	89 (19)	--- (0)	86 (7)	50 (2)	98 (49)	--- (0)
24-26 May	62 (50)	50 (4)	92 (25)	91 (9)	100 (54)	100 (8)
8-9 June	70 (64)	80 (15)	91 (22)	90 (10)	100 (16)	100 (8)
16-19 June	54 (57)	58 (26)	91 (33)	100 (18)	100 (48)	100 (18)
24-28 June	49 (53)	43 (7)	97 (33)	100 (12)	100 (52)	100 (17)
11-13 July	56 (66)	100 (1)	87 (38)	88 (8)	100 (34)	100 (9)
4 September	---	---	15 (34)	---	16 (58)	100 (2)

Figure 12-3. A group of harbor seals on a haulout site in Herring Bay, Prince William Sound, 4 September 1989. Note that most of the animals have molted and appear unoiled.

Figure 12-4. A heavily oiled harbor seal pup at Applegate Rocks, Prince William Sound, 16 June 1989.

Subsequent boat-based observations focused on three oiled areas, Seal Island, Bay of Isles, and Herring Bay. These areas were particularly suitable because they contained adequate numbers of seals that could be approached close enough to examine and classify. The incidence of oiling of seals differed among areas (Table 12-3). From 49% to 89% of the seals older than pups were classified as oiled at Seal Island, with fewer seals oiled in late June and July than in May. From May through July the percentage of oiled seals ranged from 86% to 97% in Bay of Isles, while in Herring Bay virtually all seals seen during every observation period were oiled. On 4 September 1989, most of the seals in Bay of Isles and Herring Bay showed no signs of external oiling (Fig. 12-3). There were no September observations at Seal Island.

In the heavily oiled areas many pups became oiled (Fig. 12-4). Newborn pups were sometimes seen with oil only around their nose. Some pups only 1–2 days old (as evidenced by a bright pink umbilicus) were already heavily oiled over their entire body. In the three study areas, the proportion of the pups seen that were oiled ranged from 43% to 100% (Table 12-3). Every pup seen in Herring Bay was oiled. Pups do not molt during their first year of life and were therefore still oiled during the September observation period. During September, it was difficult to separate pups from yearlings, and some pups may have been included in the nonpup category in Table 12-3.

Small boat observations were again conducted in Herring Bay, Bay of Isles, and Seal Island during 10–14 April and 29–31 May 1990. None of the seals seen appeared to be oiled. Haulout sites were examined, and no substantial amounts of oil were detected on the surface of the rocks or on the algae.

Harbor seals are generally difficult to approach and commonly go into the water if aircraft fly over at low altitude (less than 200 m). In PWS, healthy seals will never stay hauled out if people on foot or in boats approach within 100–200 m. Following the EVOS there were many observations of unusual behavior reported by biologists accustomed to observing harbor seals (Table 12-4). Oiled seals were variously reported as sick, lethargic, or unusually tame. On several occasions, investigators were able to approach on foot to within a few meters of oiled seals without causing the animals to flee. During the weeks immediately following the spill, it was often possible to fly over or circle hauled out seals in a helicopter at less than 80-m altitude and not cause them to go into the water. In areas such as Herring Bay, seals continued to haul out despite very extensive boat and aircraft traffic. During field work in September 1989 and April–May 1990, harbor seals at the same sites were noticeably more wary and more difficult to approach than they were in May–July 1989.

We observed many harbor seal mother–pup pairs during May–July 1989. Even when both the mother and pup were heavily oiled their behavior appeared normal. Females were attentive to their pups, and pups seemed normally bonded to their

Table 12-4. Observations of unusual behavior by oiled harbor seals in Prince William Sound, April-June 1989.

Date	Location	Observer[a]	Number of seals	Observations
4-12-89	Agnes Island	KP	8	Some heavily oiled; did not go into water when approached at very close range by helicopter.
4-13-89	Smith Island	KP	14	Stayed on rocks through 2 low passes (60 m) by helicopter; landed 50 m away and walked to within 12 m without spooking seals.
4-15-89	Smith Island	LL	13	No reaction by seals when helicopter circled 4 times at 80 m; seals oiled.
4-17-89	Smith Island	LL	13	Seals heavily oiled; seals did not spook when helicopter landed; approached within 20 m on foot.
4-17-89	Green Island	LL	10	At least 6 oiled; very reluctant to go into the water; stayed on rocks until circled closely within 30 m at 25 m altitude.
4-19-89	Smith Island	LL	11	Reluctant to go into water; some heavily oiled.
4-19-89	Applegate Rocks	LL	59	Most heavily oiled; 2/3 of seals stayed hauled out when helicopter circled 5 times at 60 m.
4-21-89	Herring Bay	LL, KF	24	All heavily oiled; none went into water until circled with helicopter down to 60 m, 8 stayed up until circled down to 25 m.
4-21-89	Smith Island	KP	>10	Seals spooked by helicopter but rehauled immediately while helicopter was present; extremely tame, seals oiled.
4-27-89	Northwest Bay	RS	10	Did not move when helicopter flew to within 200 m at 30 m altitude.
5-10-89	S. Applegate Rocks	KP	30	Remained hauled out in presence of large cleanup crew and heavy helicopter traffic.
5-11-89	S. Applegate Rocks	LL	10	Seals remained hauled out in presence of circling helicopter and Twin Otter.
5-15-89	Herring Bay	LL, KF	1	Heavily oiled seal; squinty eyes; did not move when approached by boat.
5-24-89	Seal Island	LL, KF	2	Oiled pup of unoiled female; very lethargic.
5-26-89	Herring Bay	KF	>10	Heavily oiled seals; allowed approach on foot to within 3-5 m; another group stayed on rocks until boat within 20 m.

(Table continues)

Table 12-4. Continued

Date	Location	Observer[a]	Number of seals	Observations
6-08-89	Applegate Rocks	KF	1	Heavily oiled adult; hauled out very high on beach; allowed approach to within 2 m.
6-10-89	Herring Bay	KF, LL	13	Appeared very ill; mucous nasal discharge; tattered nostril edges. Two of the pups in this group not very responsive; walked to within 2 m of one lightly oiled pup.
6-24-89	Herring Bay	LL, KF	6	Stayed on rocks when large H3 helicopter flew over at 60 m.
6-26-89	Evans Island	LL, KF	1	Did not move when boat approached very close; very tame; left eye very runny.

[a]KP = K. Pitcher; LL = L. Lowry; KF = K. Frost; RS = R. Shideler

mothers. On several occasions we saw pups nursing on heavily oiled females. The hair around the mammary glands of nursing females was noticeably cleaner, appearing as two light circles on a dark abdomen (Fig. 12-5).

DISCUSSION

Oiling of Seals

Our observations showed that seals at many locations in PWS became oiled as a result of the EVOS. Oiling of seals was most severe in central PWS (Smith Island, Little Smith Island, Seal Island, and Applegate Rocks), the region from Eleanor Island through the north part of Knight Island (Northwest Bay, upper and lower passages, Bay of Isles, and Herring Bay), and the west side of Knight Island Passage (Crafton Island and Junction Island). Of 585 seals observed in oiled areas in May 1989, 81% were classified as oiled.

Some seals also became oiled in the region west of PWS, but the degree of contamination is less well documented. The National Park Service reported oiled seals at Pony Cove and Morning Cove on the east side of the Kenai Peninsula (Hoover-Miller 1989). Oil was found on seals collected as part of the NRDA harbor seal study at Perl Island (tip of the Kenai Peninsula) and in the Barren Islands (Frost and Lowry 1993).

Although we were not able to watch individual seals becoming oiled, it is apparent that there were several mechanisms involved. Seals that were in the water in areas with floating oil surfaced through the oil. The result was usually a heavy coating on part or all of the anterior portion of the body, with a distinct line between oiled and unoiled parts. Other seals were oiled in patches, often on the belly, sides or back, or they were oiled over the entire body. Much of that oiling probably resulted from contact with oil on haulout sites. By the time pupping began in mid-May, most of the floating oil had been picked up. Nonetheless, many pups became oiled. Some of this could have resulted from contact with their oiled mothers. For example, oil that we saw on the noses of young pups probably resulted from nursing. However, when pups were entirely coated with thick heavy tar, we presumed this had come from oil on the haulout rocks and algae. Mothers with pups often hauled out in the upper intertidal zone where *Fucus gardneri* was the dominant algae. *Fucus* remained coated with sticky oil long after other algae and rocks appeared clean (Lowry and Frost, personal observations). Heavily oiled pups were seen even at haulout sites that had supposedly been cleaned by mid-May, including Seal Island and Applegate Rocks (e.g., Fig. 12-4). Ekker et al. (1992) described the chronic oiling that resulted when gray seal (*Halichoerus grypus*) pups lay on tar patches that became melted by their body heat.

Figure 12-5. A heavily oiled harbor seal and her heavily oiled pup in Herring Bay, Prince William Sound, 25 May 1989. Note the two light circles around the mammary glands of the adult seal.

We saw no evidence that seals attempted to avoid oil either on their haulout sites or in the water.

There were some differences in the incidence of oiling of seals in our three principal study areas. Incidence of oiling of seals and pups at Seal Island was generally 50–80%, which was lower than the 90–100% observed in Herring Bay and Bay of Isles. Some of this difference may be due to shoreline treatment. Seal Island was identified as one of the high priority areas for cleanup, and some of the gross contamination was removed from seal haulout sites there prior to 15 May. Herring Bay and Bay of Isles were not treated until later in the season. While this may suggest a beneficial result of the treatment given to Seal Island, the oiling of pups born there clearly shows that the "cleanup" did not completely remove oil from the environment.

Observations at Seal Island suggest a slight decrease in the proportion of oiled seals from May through July 1989. Possible explanations for this include: (1) immigration of clean seals into the area; (2) emigration of oiled seals away from the area; (3) mortality of oiled seals; or (4) natural cleaning of oiled seals. Based on radiotagging studies in Alaska and elsewhere, harbor seals are thought to show considerable site fidelity (Pitcher and McAllister 1981; Yochem et al. 1987). If this is also true in PWS then it is unlikely that immigration or emigration of seals was responsible for the decrease in the percent of oiled seals. We saw almost no oiled seals at unoiled sites during May 1989 and have no reason to think that unoiled seals would have moved to oiled sites. We conducted a simple experiment by soaking a piece of heavily oiled seal skin in clean seawater. After 7 days of soaking the hair had become much cleaner, but it was still visibly discolored. However, at the distances from which most of our observations of live seals were made it might have been classified as unoiled. Since much of the heaviest oil on the Seal Island haulout sites was removed in May, it is possible that some seals may have become cleaner with time.

In Herring Bay all seal haulout sites were oiled and they were treated by clean-up crews at various times up until 15 September 1992. Through mid-July, 98–100% of all seals seen were oiled, suggesting that any natural cleaning was offset by continued exposure to oil on rocks and algae at haulout sites. Circumstances in Bay of Isles, where some but not all haulout sites were heavily oiled, were intermediate. Treatment by clean-up crews in Bay of Isles was not complete until August.

When observations were made in Bay of Isles and Herring Bay on 4 September, over 80% of the seals other than pups appeared unoiled. This was probably due to molting which occurs annually in July–August (Pitcher and Calkins 1979). That some seals apparently remained unoiled after molting suggests that their haulout areas were relatively clean at that time. Thus, the persistence of heavy oiling on pups that did not molt may indicate that natural cleaning was very slow. None of the seals that we examined in 1990 showed any signs of external oiling.

Effects of Oiling on Seals

One possible effect of fouling with oil is interference with locomotion. Davis and Anderson (1976) reported two gray seal pups that were so heavily oiled that they drowned because their flippers were stuck to their bodies and they could not swim. Coating and death were also observed in seals exposed to oil during the *Torrey Canyon*, *Arrow*, and *Kurdistan* spills (Engelhardt 1987). In PWS following the EVOS, we did not observe any seals in which external oiling appeared to physically interfere with locomotion. It is possible that during the period shortly after the spill, when there was thick oil near the haulout sites, some seals were fouled badly enough to inhibit locomotion, but they were not seen or reported.

Oiling of the hair reduces its insulative value, but in normal seals this is not likely to be a major problem since they rely primarily on blubber for insulation (St. Aubin 1990). However, the brain lesions that occurred in oiled harbor seals affected the part of the brain responsible for sensing the environment (Spraker et al., Chapter 17). These lesions may have interfered with the ability to register and control temperature and could have caused thermoregulatory problems for oiled harbor seals following the spill.

Figure 12-6. A heavily oiled seal at Applegate Rocks, Prince William Sound, 8 June 1989. Note the eroded tissued around the right nostril.

Contact with oil can irritate or damage sensitive tissues, especially mucous membranes (St. Aubin 1990). On occasion we noticed that heavily oiled seals appeared to have difficulty keeping their eyes open. During the first week after the spill, we also experienced significant irritation of the eyes while working in heavily oiled areas. Conjunctivitis was found in several of the seals that were found dead and collected after the spill (Spraker et al., Chapter 17). Geraci and Smith (1976) documented similar symptoms in the eyes of ringed seals (*Phoca hispida*) that were experimentally exposed to Norman Wells crude oil. In the seals that we collected, dry, scaly skin occurred significantly more often in animals that were oiled. Similar symptoms were seen in the skin of experimentally oiled polar bears (*Ursus maritimus*) (Oritsland et al. 1981). We observed an oiled seal on Applegate Rocks that had severely eroded tissue around the margins of the nostrils (Fig. 12-6), but that animal was not collected and could not be examined in detail.

Concern has been expressed that pinniped pups might be reluctant to nurse on oiled mothers (St. Aubin 1990). We observed oiled harbor seal pups nursing on oiled mothers, and pups of oiled mothers that we collected appeared to be in normal physical condition (Frost and Lowry 1993). This is consistent with observations by Davis and Anderson (1976) that showed that interactions between oiled gray seal mothers and their pups were normal. The cleaner areas around the nipples of nursing females suggest that pups ingested some oil from their mother's fur while nursing.

Oil from the EVOS clearly caused behavioral changes in harbor seals in the months immediately following the spill. The changes that were observed in the field were most likely symptoms of the pathology that was documented in the brains of collected seals (Spraker et al., Chapter 17). We do not know the relative contributions of surface contact, inhalation, and ingestion to the contamination of seals and the observed nerve damage. Whatever the mechanism of exposure, changes in behavior induced by oil toxicity probably contributed to the deaths of seals (Frost and Lowry 1993).

CONCLUSIONS

Oil that spilled from the *Exxon Valdez* impacted a considerable amount of harbor seal habitat in central and western PWS. Our observations showed that harbor seals continued to use oiled habitats after the EVOS. Their pelage became oiled when they used oiled haulout sites or contacted oil in the water. It appeared that seals became cleaner with time if they were not repeatedly exposed to oil. Seals older than pups molted their pelage in August, and no evidence of external oiling was seen in 1990.

Seals gave birth to and cared for their pups on oiled haulout sites during May–July. Many pups became oiled, sometimes within a few days of their birth. Shoreline treatment conducted prior to the start of pupping did not prevent the contamination of pups.

During April–June 1989 the behavior of oiled seals in oiled areas was sometimes very different from that of harbor seals in "normal" circumstances. The behavioral changes were most likely due to pathological damage to nerves of the brain caused by hydrocarbon toxicity.

It is not clear how oiled pelage and skin may have contributed to the overall damage to seals that resulted from the spill. The possible effects of external oiling require further study.

Because of the large area to be covered, limited personnel, and other response and research priorities, we were not able to watch undisturbed seals as they encountered oil. Also, we were not able to observe what happened when oil initially moved into and onto areas where seals were swimming or hauling out. Since we could not repeatedly identify and observe specific animals, it was not possible to follow how individuals became oiled and what happened to them. If circumstances allow in the future, research on seals in oil spill areas should include detailed behavior studies on individuals that are marked visually or telemetrically.

∞   ∞   ∞

ACKNOWLEDGMENTS

We thank the crews of the ADF&G research vessel *Resolution* and the NOAA vessel *1273* for their support during field studies, and those people who assisted in making observations of seals, especially D. McAllister, R. Shideler, and C. George. The manuscript was improved by comments from T. Loughlin and two anonymous reviewers. This study was conducted in cooperation with the National Marine Fisheries Service, National Marine Mammal Laboratory, as part of the Natural Resource Damage Assessment program, funded by the *Exxon Valdez* Oil Spill Trustee Council.

REFERENCES

Davis, J. E., and S. S. Anderson. 1976. Effects of oil pollution on breeding gray seals. Marine Pollution Bulletin 7:115–118.

Ekker, M., S.H. Lorentsen, and N. Rov. 1992. Chronic fouling of grey seal pups at the Froan breeding ground, Norway. Marine Pollution Bulletin 24:92–93.

Engelhardt, F. R. 1987. Assessment of the vulnerability of marine mammals to oil pollution. Pages 101–115, *in* J. Kiuper and W. J. Van Den Brink (eds.), Fate and effects of oil in marine ecosystems. Martinus Nijhoff Publishing, Boston, Massachusetts.

Frost, K. J., and L. F. Lowry. 1993. Assessment of injury to harbor seals in Prince William Sound, Alaska, and adjacent areas following the *Exxon Valdez* oil spill. Final Report, Marine Mammal Study Number 5, State-Federal Natural Resource Damage Assessment. 95 p.

Geraci, J. R., and T. J. Smith. 1976. Direct and indirect effects of oil on ringed seals (*Phoca hispida*) of the Beaufort Sea. Journal of the Fisheries Research Board of Canada 33:1976–1984.

Hoover-Miller, A. 1989. Impact assessment of the T/V *Exxon Valdez* oil spill on harbor seals in the Kenai Fjords National Park, 1989. Unpublished Report, Kenai Fjords National Park, Seward, Alaska. 21 p.

Oritsland, N. A., F. R. Engelhardt, F. A. Juck, R. J. Hurst, and P. D. Watts. 1981. Effect of crude oil on polar bears. Environmental Studies No. 24. Department of Indian and Northern Affairs Canada, Northern Affairs Program. 268 p.

Pitcher, K. W. 1989. Harbor seal trend count surveys in southern Alaska, 1988. Final Report Contract MM4465852-1 to U.S. Marine Mammal Commission, Washington, D.C. 15 p.

Pitcher, K. W., and D. G. Calkins. 1979. Biology of the harbor seal, *Phoca vitulina richardsi*, in the Gulf of Alaska. U.S. Department of Commerce, Environmental Assessment of the Alaskan Continental Shelf Final Reports of Principal Investigators 19(1983):231–310.

Pitcher, K. W., and D. C. McAllister. 1981. Movements and haulout behavior of radio-tagged harbor seals, *Phoca vitulina*. Canadian Field-Naturalist 95:292–297.

St. Aubin, D. J. 1990. Physiologic and toxic effects of oil on pinnipeds. Pages 103–127, *in* J. R. Geraci and D. J. St. Aubin (eds.), Sea mammals and oil: Confronting the risks. Academic Press, New York.

Yochem, P. K., B. S. Stewart, R. L. DeLong, and D. P. DeMaster. 1987. Diel haul-out patterns and site fidelity of harbor seals (*Phoca vitulina richardsi*) on San Miguel Island, California, in autumn. Marine Mammal Science 3:323–332.

Chapter 13

Health Evaluation, Rehabilitation, and Release of Oiled Harbor Seal Pups

Terrie M. Williams, George A. Antonelis, and Jennifer Balke

INTRODUCTION

Many pinniped species have been exposed to petroleum hydrocarbons in either accidental spills or experimental studies (St. Aubin 1990). With the exception of recent studies from the *Exxon Valdez* oil spill (EVOS) (Frost and Lowry 1993; Spraker et al., Chapter 17), the pathological effects of oil contamination on this group of marine mammals has remained inconclusive. This may be due to: (1) variability in the quality of reports from oil spills (St. Aubin 1990); (2) species-specific differences in response to oil contamination; (3) differences in the type of petroleum product spilled; (4) variability in the degree of oil weathering before contamination (Neff 1990); and (5) inconsistencies in necropsy protocols or tissue collection techniques.

In general, contact with fresh crude oil can lead to deleterious physiological, pathological, and behavioral responses depending on the degree and route of exposure. A disruption in behavioral patterns including unusual swimming postures and prolonged haulout periods has been observed for pinnipeds exposed to crude oil (T. M. Williams, personal observation; Lowry et al., Chapter 12). Laboratory studies on phocid seals have also demonstrated that petroleum hydrocarbons may be transported by the blood and distributed to many tissues including blubber, muscle, and liver (Geraci and Smith 1976; St. Aubin 1990). Pathological changes were noted for adult harbor seals (*Phoca vitulina*) captured within the EVOS area (Spraker et al., Chapter 17). Yet, the relationship between systemic exposure to petroleum hydrocarbons and mortality remains unclear for pinnipeds.

Neonatal and immature pinnipeds appear more vulnerable than adults to the effects of oil contamination (St. Aubin 1990). For example, undeveloped locomotor muscles will limit a pup's ability to swim through or out of oiled water. The thin blubber layer of most neonates may also provide a comparatively poor thermal barrier if the fur is fouled with oil. This latter problem may be exacerbated in the young of species that rely exclusively on their pelage for insulation (i.e., fur seal pups, pagophylic seal pups in lanugo pelage).

Interactions between mothers and pups, especially during suckling periods, may be disturbed by an oil spill (St. Aubin 1990). In addition, maternal pathways (i.e., transplacental transfer or lactation) represent potential routes of exposure to petroleum hydrocarbons for developing pups. Prenatal exposure is possible from females oiled during pregnancy. Accidental or incidental oil ingestion by the pups may occur during suckling (Klaassen and Rozman 1991; Lowry et al., Chapter 12). Each of these pathways could compromise the health of developing and neonatal pinnipeds (St. Aubin 1990; Williams and Davis in press; Lowry et al., Chapter 12).

To examine the responses of young pinnipeds to oil contamination, we conducted a longitudinal study of harbor seal pups placed in rehabilitation centers after the EVOS. Initial and prerelease health of the animals was compared by monitoring for clinical signs of disease, loss in body weight, and changes in blood chemistry, hematological parameters, and behavior. Based on our findings, we describe the criteria necessary for evaluating rehabilitated pups before their reintroduction into the wild and assess the risks and benefits of rehabilitation for pinnipeds exposed to crude oil.

CAPTURE, CLEANING, AND REHABILITATION OF OILED SEAL PUPS

Capture of Oiled Seal Pups

Following the EVOS, 18 harbor seal pups (9 males, 9 females) were brought to rehabilitation facilities in Alaska. The estimated age of the pups ranged from 1 to 7 days as determined from: (1) the timing of the annual pupping season for seals in Prince William Sound; (2) pup weight; and (3) the presence of the umbilicus. Based on age and weight, all of the pups appeared to be full term. With the exception of one pup captured on 14 April, the harbor seals were collected over a 5-week period from 2 May to 4 June 1989. Pups were captured in Herring Bay ($n=8$) and Bay of Isles ($n=1$) on Knight Island, on the shoreline of Green ($n=1$) and Chenega ($n=1$) Islands, in Rocky ($n=3$) and Windy ($n=2$) Bays on the Kenai Peninsula, and in Larsen Bay ($n=1$) on Kodiak Island. One pup arrived at the rehabilitation center with no capture record.

Assessing Oil Exposure

All seal pups were heavily coated with oil when captured. In most cases the pups were found lying on or swimming near heavily oiled beaches. Due to their age, we estimated that the pups had been exposed to oil for less than 7 days. Except for the animal captured in April, the pups first contacted the oil 5–10 weeks after the spill. Because many of the aromatic components of the oil had dissipated within 2 weeks of the spill (Neff 1990), it is likely that external contamination was limited primarily to straight chain hydrocarbons. The toxicity of these compounds is considered low in comparison to aromatic petroleum hydrocarbons.

To assess the degree of internal contamination, blood samples from nine seal pups were analyzed for total paraffinic hydrocarbon (TPH) content (Williams 1990; Williams and Davis in press). Samples from all nine seal pups were taken on the day of admission to rehabilitation facilities. Additional blood samples were taken from six pups at approximately 2-week intervals after admission. Capture records and transport time to the rehabilitation facility indicated that each animal had fasted approximately 3–6 hours before the initial samples were taken.

Despite similar levels of external oiling, paraffinic hydrocarbon contents of the blood were variable for the pups (Table 13-1). Initial values ranged from 22 to 260 ppm (mean=91 ppm ± 81 SD). Three pups showed paraffinic hydrocarbon concentrations that were above the calculated threshold for mortality in oiled sea otters (Williams and Davis 1990; Williams and Davis in press). Yet, all three pups survived and were released to the wild within 10 weeks of capture.

Rather than declining with time, the TPH concentrations did not change in a consistent manner during the serial blood samples. Both increases and decreases

Table 13-1. Changes in total paraffinic hydrocarbon concentration in whole blood samples of harbor seal pups. All pups were heavily oiled externally at the time of capture. (--- = no data)

	Sample draw date		
Seal	Admission day (30 May - 1 June)	17 June 1989	1 July 1989
G	70 ppm	114 ppm	86 ppm
H	22 ppm	73 ppm	36 ppm
I	54 ppm	50 ppm	62 ppm
J	22 ppm	57 ppm	64 ppm
K	60 ppm	65 ppm	69 ppm
L	260 ppm	95 ppm	148 ppm
S2	34 ppm	---	---
S3	180 ppm	---	---
S4	120 ppm	---	---

in TPH levels were observed between initial and final blood samples. The persistent and variable levels of TPH in the blood suggest that petroleum hydrocarbons were sequestered and subsequently released from the seals' tissues. Fat and blubber in particular may serve as a storage depot for lipophilic organic compounds in many mammals (Klaassen and Rozman 1991). Consequently, the long-term effects of oil contamination may be of greater concern for pinnipeds than for other marine species such as sea otters which have comparatively small fat reserves.

Cleaning Oiled Seal Pups

A dilute solution of dishwashing detergent (Dawn^{TM}) had been used successfully to remove crude oil from the contaminated pelts of sea otters following the EVOS (Williams and Davis 1990). This solution was ineffective on the oiled harbor seal pups. However, full strength Dawn^{TM} detergent followed by freshwater rinses removed the oil. Cleaning protocols for seals were similar to those established for sea otters (Davis et al., 1988; Williams et al., 1988). Briefly, each seal was lathered with detergent and rinsed with fresh water. The process was repeated until no trace of oil was observed on the seal or in the rinse water. Each seal pup took approximately 40 minutes to wash.

Holding Facilities

The pups were initially placed in dry pens (1 m X 1 m X 0.5 m high) with periodic access to saltwater pools during the day. Access to the pools was increased as the swimming and thermoregulatory capabilities of a pup improved. Within 1 week, the pups were transferred in pairs to individual cages supplied with saltwater pools (1 m X 1 m X 0.7 m deep). Each pool was plumbed separately to avoid disease transmission between animals. Cages, haulout spaces, and pools were washed and disinfected daily with detergent, a dilute chlorine bleach solution, and Nolvasan^{TM}. To prevent the introduction or transmission of domestic diseases, the rehabilitation facilities followed basic quarantine procedures. Domestic animals and the general public were excluded from the facilities. Footbaths and protective clothing were used by staff members that routinely handled the seals.

CLINICAL EVALUATION AND CARE OF OILED PUPS

All seal pups were alert on arrival at rehabilitation centers. No sedation was necessary for handling the animals during clinical evaluations and cleaning. Initial examination revealed the following clinical problems: corneal ulceration, conjunctivitis, and diarrhea. Treatments included prophylactic administration of antibiotics, vitamins (Vitamin B), eye flushes, and ophthalmic ointment. Core body temperature was recorded with a rectal thermometer for 11 seal pups upon

✄   ✄   ✄

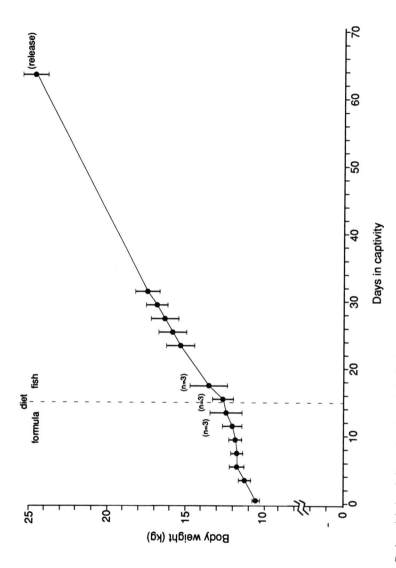

Figure 13-1. Body weight in relation to days in captivity for four harbor seal pups. The pups were captured approximately 2 months after the oil spill. Each point denotes the mean ± SD for N=4 pups, except where indicated by numbers in parentheses. All of the animals were placed on a fish diet after 14 days of receiving formula.

admission to the rehabilitation centers. Body temperatures ranged from 35.7° to 38.4°C (mean=37.0°C ± 0.7 SD) and were considered within the normal range of values for harbor seals (Dierauf 1990). Subsequent temperature measurements during the first days in rehabilitation centers were within this range and demonstrated that the pups were thermally stable. Only two of the seal pups exhibited clinical problems during rehabilitation or quarantine (see below).

Initial body weight of the 18 pups averaged 11.1 kg ± 2.2 SD (range=8.3–15.4 kg). All of the pups gained weight during rehabilitation. However, the rate of increase was affected by the method of feeding (Fig. 13-1). The pups were fed a formula diet (Multimilk^{TM} or Pedialyte^{TM}, plus minced fish) via stomach tube during the first 7–14 days of captivity. Increasing concentrations of chopped herring and cod liver oil were subsequently added to the formula. Whole herring or smelt was introduced into the diet after approximately 15 days of captivity. By the third and fourth weeks, the seal pups were eating primarily a fish diet.

The weight of the seal pups reflected each change in diet. In general, their weight remained stable or declined slightly on the formula diet. Body weight quickly increased once the pups started eating whole fish. Figure 13-1 illustrates the typical growth curve for four seal pups matched for weight, estimated age, and dietary regime. These pups had been captured 2 months after the spill. During the 15 days they were fed a formula diet their body weight increased only 18%. This contrasts with a 38% increase in weight for the subsequent 15 days after the pups were placed on a fish diet. Although the other pups differed in initial weight, they showed similar patterns in weight gain in response to changes in their diet.

Blood samples were obtained for 10 of the seal pups at the time of admission to rehabilitation centers (Table 13-2). Values for many of the parameters were within normal ranges for harbor seal pups (Bossart and Dierauf 1990). However, aspartate aminotransferase (AST), an indicator of hepatocellular function, was considered abnormally high for five of the seal pups examined (Table 13-2). Alanine aminotransferase (ALT) was elevated in two of these pups. Another exception was lactate dehydrogenase (LDH) which was consistently high for all of the pups. The mean value for the pups (1500 IU/l ± 458 SD) was 20% higher than the maximum normal value reported for harbor seals. However, LDH levels returned to normal values in subsequent blood samples of the seven pups retested. (Table 13-3; Fig. 13-2).

An elevation in LDH in mammals may be indicative of damage to a variety of organs including cardiac muscle, skeletal muscle, kidney, liver, and lungs (Kerr 1989). In oiled sea otters, 82% of the animals that subsequently died showed elevations in LDH. This elevation was attributed in part to skeletal and cardiac muscle damage associated with capture stress, shock, and hypothermia (Williams and Davis in press). Because the LDH concentration quickly returned to normal

✂ ✂ ✂

in the seal pups (Fig. 13-2), the original values were probably indicative of stresses associated with oiling, capture, and handling.

Anemia often developed in heavily oiled sea otters and persisted for 3–4 months during captivity (Williams et al. 1990). In contrast, anemia was not observed for the majority of harbor seal pups. In matched samples obtained for five pups, the mean packed cell volume (PCV) on admission, $58.2\% \pm 4.1$ SD, was not statistically different (at P> 0.05) from values determined 6 weeks later (Tables 13-2 and 13-3) nor did it differ from neonatal pinniped values (Bossart and Dierauf 1990).

Two seal pups eventually showed a marked decrease in PCV; these were the only seal pups to die during the rehabilitation program. Both pups had been captured relatively early (3 May 1989), housed for 24 hours in a rehabilitation center in Valdez, and then flown to a veterinary clinic in Anchorage. After 11 days in captivity the pups showed a decrease in PCV from approximately 61–48%. The animals were transferred to a prerelease facility where they died 2 weeks later. The results of a postmortem examination revealed severe gastric enteritis in one of the pups. Many factors may have contributed to the mortality of these pups including: (1) the stress associated with multiple transfers between facilities; (2) inconsistent rehabilitation or feeding protocols between different facilities; and (3) oil contamination early in the spill.

Survivorship of the remaining seal pups was high. In total, 15 of the 18 pups (83%) were transferred to prerelease facilities within 4–6 weeks after capture. The high survivorship of heavily oiled harbor seal pups is similar to that of fur seal pups contaminated during the *Sanko Harvest* spill in Australia (Gales 1991).

CRITERIA FOR REINTRODUCING REHABILITATED SEAL PUPS TO THE WILD

Two primary criteria had to be met before the rehabilitated seals could be released into their natural environments. First, the seals had to exhibit the behavioral maturity necessary for survival. Second, the rehabilitated harbor seals had to be clinically healthy and free of known disease(s) that could threaten the wild population. Clinical health parameters included a consistent increase in body weight, hematological and blood chemistry values within the normal range for harbor seal neonates, and normal respiratory patterns and heart rate.

Behavioral requirements included the ability to: (1) swim and orient normally in the water; (2) approach and consume fish independently; and (3) avoid contact with humans. Because the pups were removed from their mothers before weaning, imprinting behavior was a concern. Once a seal depends on humans for food or behaviorally imprints on humans, the likelihood of survival in the wild is greatly reduced. Therefore, the confirmation of independent behavior was especially important for these animals. The behavior of each seal was evaluated on a daily

∽ ∽ ∽

Table 13-2. Values for clinical blood chemistry and hematology for 10 heavily oiled harbor seal pups. All samples were taken within 1 day of admission to rehabilitation centers. Glu = glucose, BUN = blood urea nitrogen, AST = aspartate aminotransferase, ALT = alanine aminotransferase, CPK = creatine phosphokinase, LDH = lactate dehydrogenase, K = potassium, Phos = phosphorous, WBC = white blood cell count, and PCV = packed cell volume. (— = no data.

#	Glu mg/dl	BUN mg/dl	AST IU/l	ALT IU/l	CPK IU/l	LDH IU/l	K mEq/l	Phos mg/dl	Iron mg/dl	WBC x10³/mm³	PCV %
S6	210	42	91	81	100	1265	4.7	7.2	—	—	—
S7	150	45	65	46	1373	1030	5.4	7.9	360	—	—
D	138	27	57	58	—	1452	4.3	8.0	—	8.7	61
E	133	54	212	132	—	2144	4.5	8.0	—	—	61
F	238	73	131	137	158	1590	4.7	5.4	71	—	—
H	—	—	61	—	—	1110	—	—	—	8.0	61
I	162	30	147	—	—	—	4.7	6.6	250	8.1	56
J	—	—	171	—	—	2220	—	—	—	5.2	54
K	—	—	82	—	—	1190	—	—	—	6.8	64
L	88	37	256	—	—	—	4.6	8.6	95	10.1	60
n	7	7	10	5	3	8	7	7	4	6	7
x̄	160	44	127	91	554	1500	4.7	7.4	194	7.8	60
SD	50	16	69	42	719	458	0.3	1.1	136	1.6	3

234

Table 13-3. Values for prerelease clinical blood chemistry and hematology for heavily oiled harbor seal pups. All samples were taken within 1 week of release into the wild. Abbreviations are as defined in Table 13-2. (— = no data.)

#	Glu mg/dl	BUN mg/dl	AST IU/1	ALT IU/1	CPK IU/1	LDH IU/1	K mEq/1	Phos mg/dl	Iron mg/dl	WBC x10³/mm³	PCV %
S1	154	47	147	181	—	785	4.4	6.6	121	5.2	54
S2	154	53	155	82	300	—	4.4	8.6	156	6.8	52
S3	135	57	170	79	1163	1015	4.5	6.8	229	—	—
S6	140	69	530	174	690	1175	4.6	8.7	296	8.6	62
S7	128	46	96	58	1635	820	4.3	6.3	114	9.0	52
C	144	25	93	61	1121	735	4.0	7.8	172	7.7	67
G	145	40	227	91	104	—	4.5	8.3	455	6.7	60
H	137	33	—	78	1943	—	4.5	7.5	265	10.4	65
I	134	29	99	90	102	765	4.1	6.7	129	10.2	61
J	136	32	136	111	568	775	4.1	6.1	123	—	—
K	135	32	182	93	747	760	3.9	5.3	129	8.0	58
L	135	45	—	92	—	—	4.7	8.0	200	9.0	65
pw	152	29	117	70	284	900	4.1	6.7	106	13.2	62
n	13	13	11	13	11	9	13	13	13	11	11
x̄	140	41	177	97	787	859	4.3	7.2	192	8.6	60
SD	8	13	124	38	616	147	0.3	1.0	100	2.2	5

235

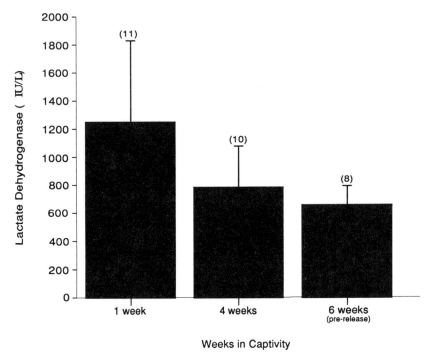

Figure 13-2. Plasma lactate dehydrogenase concentrations for harbor seal pups. Values for blood samples taken within 1 week of admission, at 4 weeks, and at 6 weeks in captivity are compared. Columns and vertical bars represent means +1 SD. Numbers in parentheses denote the number of pups.

basis with special attention to overall behavioral maturity during the final 2 weeks of captivity. Prerelease evaluations were made by experienced behaviorists and veterinarians.

Reducing the risk of disease transmission between the rehabilitated animals and wild population required the establishment of stringent quarantine conditions and testing protocols. Sanitary procedures for feeding and handling the seal pups were required in the rehabilitation facilities (see above discussion). In addition, direct or indirect exposure to domestic animals (e.g., feces or urine) was prohibited. Despite precautions, it was difficult to ensure that the seals were not exposed to pathogens while in the busy rehabilitation centers. Therefore, the pups were transferred to open-water quarantine pens once they were behaviorally ready for release. Transfers began on 1 July 1989, approximately 1 month after capture. The quarantine pens housed three seals each and were located in Halibut Cove, near Homer, Alaska. Each seal was kept in quarantine for a minimum of 14 days.

During the quarantine period, all seals were routinely observed for clinical signs of illness. Immediately before and 9–14 days into quarantine, blood samples

(15 cc) were collected for parvo- and canine-distemper-virus testing, and for standard hematology and blood chemistry analyses. Four to five cc of serum were also placed in long-term storage for baseline information.

Tests for exposure to parvo and distemper viruses were conducted by A. Osterhaus and I. Visser (National Institute of Public Health and Environmental Hygiene, Bilthoven, the Netherlands). The results were negative for all seals examined. Final values for hematology and blood chemistry (Table 13-3) were within the range for healthy newborn harbor seals (Bossart and Dierauf 1990), except for elevated AST in pups S6 and G. In addition, ALT levels were considered high for pups S1, S6, and J. Blood paraffinic hydrocarbon levels were below levels considered lethal for other marine mammals (Table 13-1).

Fifteen rehabilitated harbor seals were released into the wild near Halibut Cove on 5 August 1989. The remaining pup had escaped from the prerelease facility at Halibut Cove shortly after its arrival in early June. The estimated age of the seals was approximately 3 months. Weight at the time of release ranged from 24 to 29 kg. All of the harbor seals were flipper tagged prior to release. At the time of this writing none of the seals have been recovered or resighted since their release from Halibut Cove.

RISKS AND BENEFITS OF REHABILITATING HARBOR SEAL PUPS

The Marine Mammal Protection Act states that any wild marine mammal brought into captivity for rehabilitation must eventually be released into its natural habitat whenever feasible. During the EVOS, the feasibility of releasing rehabilitated harbor seal pups into the wild required special consideration of the potential risks and benefits to the individuals, their conspecifics, and the ecosystem in which they live. As custodians of harbor seals in all United States waters, the National Marine Fisheries Service was responsible for determining the guidelines to be followed prior to the release of rehabilitated harbor seals during the spill.

The short-term benefit of rehabilitation programs following an oil spill is the removal of individual animals from contaminated areas and a subsequent reduction in exposure to harmful petroleum hydrocarbons. Although important when crude oil is fresh, our results indicate that the benefits of removal are reduced as the more toxic, volatile hydrocarbon fractions dissipate with weathering (Williams and Davis in press). Most of the seal pups in this study were oiled more than 30 days after the spill; the relative toxicity of the oil seemed to have declined with few ensuing medical problems. Further study will be required to determine if medical problems eventually arise in oil-contaminated pinnipeds that were rehabilitated and released.

◇◇ ◇◇ ◇◇

A primary concern when marine mammals are returned to the wild following captivity is the risk of introducing undetected pathogens into the wild population. Often the transmission of these pathogens is accidental. However, once introduced, the pathogen is difficult, if not impossible, to eliminate (Sinderman 1993). The impact of a highly contagious pathogenic virus on marine mammal populations has recently been demonstrated by the mass mortality of Baikal seals (*Phoca sibirica*) in Siberia (Grachev et al. 1989) and by the death of more than 17,000 harbor seals in the North and Baltic Seas (Osterhaus et al. 1990). In view of these incidents, a responsible rehabilitation program involving pinnipeds demands extraordinary efforts to reduce the risk of disease transmission.

During the EVOS, the rehabilitation team had to balance the risk of disease transmission in crowded wildlife centers to risks posed by interactions between wild seals and rehabilitated pups placed in open-water pens. By choosing a remote location for the pens that was not frequented by wild seals, rehabilitators tried to reduce the chance of such interactions.

In future spills, the relative risk of transferring diseases to the wild population and the prognosis for long-term survival of the rehabilitated animal must be considered before seals are captured, treated, and released back into the wild. One solution to the problem may be to significantly reduce the number of marine mammals brought into captivity for rehabilitation. Alternatively, the rehabilitator selects to place the benefits to a few individuals over the potential risk to the population. This choice may be the most appropriate action for severely endangered populations or isolated stocks of marine mammals that are in danger of extinction (Kleiman 1989). For example, the Hawaiian monk seal (*Monachus schauinslandi*), Mediterranean monk seal (*M. monachus*), and declining sub-populations of Steller sea lions (*Eumetopias jubatus*) in Alaska warrant special consideration. A compromise recently used during a spill involving fur seals (*Arctocephalus forsteri*) (Gales 1991) involved establishing cleaning and short-term holding facilities near the affected rookeries. This option reduces the potential for disease transmission by avoiding transport to facilities, but must account for prolonged disruption of the rookery.

Assessing the risk of disease transmission is complicated because of the lack of tests specific for detecting pathogens in pinnipeds. Furthermore, little is known about the immune responses of marine mammals or the incidence of exposure to domestic disease and animals. Some populations of pinnipeds (i.e., seals and sea lions that haul out on public beaches or wharves) routinely encounter domestic animals.

To reduce the risks, a set of standards and guidelines must be established to protect the wild population from the accidental introduction of disease. Important components of a rehabilitation and release program for pinnipeds include: (1) rigorous marine disease research programs; (2) regional and national diagnostic centers to screen for known pathogens and parasites in marine mammals scheduled for release into the wild; (3) a national policy for rehabilitation criteria; and (4) an ongoing baseline data and monitoring program for diseases occurring in wild populations. Quarantine procedures in the rehabilitation facility, health screening, and testing for specific pathogens will also help to reduce the risks. Last, by releasing rehabilitated marine animals into the original population, the chance of introducing a subpopulation to a novel pathogen will be reduced (Sinderman 1993).

CONCLUSIONS

As reported by Geraci and Smith (1976) for harp seal (*Phoca groenlandica*) pups and Gales (1991) for Australian fur seal pups, we found few toxic or thermal consequences associated with external oiling of the harbor seal pups in this study. Internal exposure to oil was of greater concern for the seal pups due to the persistent nature of petroleum hydrocarbons in the blood. However, the animals did not display behavioral or clinical signs of organ dysfunction. Despite evidence of internal and external oil contamination, 15 of the 18 seals (83%) survived to release. It is not known if the sequestering or intermittent circulation of TPH will have a long-term detrimental effect on these animals. Clearly, further studies concerning the factors that control the turnover of petroleum hydrocarbons in marine mammal tissues are warranted.

Because the harbor seal pups showed few signs of clinical illness, the value of capturing and rehabilitating pinnipeds following an oil spill is questionable especially after the oil has weathered. Species that rely primarily on blubber for insulation will be comparatively resistant to the thermal effects of oil contamination. Conversely, if significant levels of petroleum hydrocarbons are transferred through milk to nursing pups, a rehabilitation program may avoid undue exposure of the pups to oil. In future spills, the relative risk to the wild population must be considered a primary factor when deciding if a rehabilitation program should be initiated.

ACKNOWLEDGMENTS

The authors thank A. Osterhaus and I. Visser for conducting the parvo and distemper virus testing for the seal pups. M. Beck and the Halibut Cove facility were invaluable for the rehabilitation of the seals. We also thank Exxon Company, USA for providing personnel and facilities through their Marine Mammal Rehabilitation Program. I. Barker, K. Frost, M. Haebler, L. Lowry, T. McIntyre, E. Sinclair, and T. Spraker shared their insights regarding marine mammal rehabilitation, release, and risks of disease transmission to wild populations.

✂ ✂ ✂

REFERENCES

Bossart, G. D., and L. A. Dierauf. 1990. Marine mammal clinical laboratory medicine. Pages 1–52, *in* L. A. Dierauf (ed.), CRC Handbook of marine mammal medicine: Health, disease, and rehabilitation. CRC Press, Boca Raton, Florida.

Davis, R. W., T. M. Williams, J. A. Thomas, R. A. Kastelein, and L. H. Cornell. 1988. The effects of oil contamination and cleaning on sea otters (*Enhydra lutris*). II. Metabolism, thermoregulation, and behavior. Canadian Journal of Zoology 66:2782–2790.

Dierauf, L. A. 1990. CRC Handbook of marine mammal medicine: Health, disease, and rehabilitation. CRC Press, Boca Raton, Florida. 735 p.

Frost, K. J., and L. F. Lowry. 1993. Assessment of damages to harbor seals caused by the *Exxon Valdez* oil spill. Pages 300–302, *in Exxon Valdez* oil spill symposium, 2–5 February 1993, Anchorage, Alaska. (Available, Oil Spill Public Information Center, 645 G Street, Anchorage, Alaska 99501.)

Gales, N. J. 1991. New Zealand fur seals and oil: An overview of assessment, treatment, toxic effects, and survivorship. Report to the West Australian Department of Conservation and Land Management. Mount Pleasant, Western Australia. 35 p.

Geraci, J. R., and T. G. Smith. 1976. Direct and indirect effects of oil on ringed seals (*Phoca hispida*) of the Beaufort Sea. Journal of the Fisheries Research Board, Canada 33:1976–1984.

Grachev, M. A., V. P. Kumarev, L. V. Mamaev, V. L. Zorin, L. V. Baranova, N. N. Denikina, S. I. Belikov, E. A. Petrov, V. S. Kolesnik, R. S. Kolesnik, V. M. Dorofeev, A. M. Beim, V. N. Kudelin, F. G. Nagiera, and V. N. Sidorov. 1989. Distemper virus in Baikal seals. Nature 338:209.

Kerr, M. G. 1989. Veterinary laboratory medicine. Blackwell Scientific Publications, Cambridge, Massachusetts. 270 p.

Klaassen, C. D. and K. Rozman. 1991. Absorption, distribution, and excretion of toxicants. Pages 50–87, *in* M. O. Amdur, J. Doull, and C. D. Klaassen (eds.), Toxicology: The basic science of poisons. Pergamon Press, New York, New York.

Kleiman, D. G. 1989. Reintroduction of captive mammals for conservation. Bioscience 39:152–161.

Neff, J. M. 1990. Composition and fate of petroleum and spill-treating agents in the marine environment. Pages 1–33, *in* J. R. Geraci and D. J. St. Aubin (eds.), Sea mammals and oil: Confronting the risks. Academic Press, San Diego, California.

Osterhaus, A. D. M. E., J. Groen, H. E. M. Spijkers, H. W. I. Broeders, F. G. C. M. UytdeHaag, P. deVries, J. S. Teppema, I. K. G. Visser, M. W. G. Van Bildt, and E. J. Vedder. 1990. Mass mortality in seals caused by a newly discovered mobillivirus. Veterinary Microbiology 23:343–350.

Sinderman, C. J. 1993. Disease risks associated with importation of nonindigenous marine animals. Marine Fisheries Review 54:1–10.

St. Aubin, D. J. 1990. Physiologic and toxic effects on pinnipeds. Pages 103–127, *in* J. R. Geraci and D. J. St. Aubin (eds.), Sea mammals and oil: Confronting the risks. Academic Press, Inc., San Diego, California.

Williams, T. M. 1990. Evaluating the long term effects of crude oil exposure in sea otters. Wildlife Journal 13:42–48.

Williams, T. M., and R. W. Davis. 1990. Sea otter rehabilitation program: 1989 *Exxon Valdez* Oil Spill. Report to Exxon Company, USA. International Wildlife Research. 201 p.

Williams, T. M., and R. W. Davis. In Press. Emergency Care and Rehabilitation of Oiled Sea Otters. University of Alaska Press, Fairbanks, Alaska.

Williams, T. M., R. A. Kastelein, R. W. Davis, and J. A. Thomas. 1988. The effects of oil contamination and cleaning on sea otters (*Enhydra lutris*). I. Thermoregulatory implications based on pelt studies. Canadian Journal of Zoology 66:2776–2781.

Williams, T. M., J. McBain, R. K. Wilson, and R. W. Davis. 1990. Clinical evaluation and cleaning of sea otters affected by the T/V *Exxon Valdez* oil spill. Pages 236–257, *in* K. Bayha and J. Kormendy (eds.), Sea otter symposium: Proceedings from a symposium to evaluate the response effort on behalf of sea otters after the T/V *Exxon Valdez* oil spill into Prince William Sound. U.S. Fish and Wildlife Service, Biological Report 90(12).

Chapter 14

Effects of Masking Noise on Detection Thresholds of Killer Whales

David E. Bain and Marilyn E. Dahlheim

INTRODUCTION

Killer whales (*Orcinus orca*) are toothed whales with a cosmopolitan distribution that includes Prince William Sound (PWS). They feed on a variety of prey ranging from small schooling fishes to the largest marine mammals, in addition to invertebrates, birds, and reptiles (Anonymous 1982). They are highly social animals that generally travel in schools of 3 to over 100 individuals. Killer whales communicate using a repertoire of pulsed calls and whistles (Dahlheim and Awbrey 1982; Ford 1984; Bain 1986; Morton et al. 1986) and have the ability to echolocate (Diercks et al. 1971).

Killer whale communication signals carry information regarding the geographic origin (Dahlheim and Awbrey 1982), individual identity (Hoelzel and Osborne 1986), pod membership (Ford 1991), activity level (Bain 1988), and location of the caller. Most calls consist of both low- and high-frequency components. The low-frequency component typically has a fundamental between 250 and 1500 Hz and has measurable harmonics ranging to about 10 kHz or higher. The high-frequency component has a fundamental ranging from about 5 to 12 kHz, with harmonics ranging to over 100 kHz. The high-frequency component is very directional, with most of its energy projected in a beam directly ahead of the animal. The low-frequency component is relatively omnidirectional, but more energy is projected ahead and to the side rather than directly behind the whale (Schevill and Watkins 1966). Because killer whales are more sensitive to the high-frequency component (Hall and Johnson 1972), and the high-frequency component contains more energy, it offers a greater potential for long-distance communication in the forward direction. However, killer whales often travel parallel to one another, and the low-frequency component is likely to be more important for whales traveling in that spatial arrangement.

During echolocation, killer whales emit a signal and listen for an echo that contains information about the surfaces that reflect the signal. Such information includes the location of the surface, its size, and reflectivity (which will correlate with what it is made of). Presumably, whales can use this information to locate and identify schools of fish or other prey and to navigate around obstacles such as rocks and islands.

Vessels engaged in cleanup and other related activities following the *Exxon Valdez* oil spill (EVOS) increased noise levels in PWS. The added noise may have impacted cetaceans such as killer whales because whales rely on sound to communicate, and species like killer whales that echolocate rely on sound to navigate and find food.

Masking is the term used to describe any impairment in an animal's ability to a detect a signal. The effects of masking sounds are complicated, and may depend on the structure, timing, and source location. Generally, the effects are strongest when the masking sound is near the signal source in space and time and is similar in structure to the signal to be detected (Scharf 1975).

The sound pressure of a signal declines as it propagates. There are two main factors to be considered. The first factor is that as the distance of transmission increases, the sound pressure decreases due to spreading losses. These losses are generally 10 to 20 times the log of the propagation distance. The second consideration is that water does not propagate sound perfectly because some of the sound energy is absorbed and scattered. This excess attenuation increases linearly with distance and increases irregularly with increasing frequency. The echoes an animal receives will be subject to these losses in both directions, as well as losses due to imperfect reflection (Au 1993).

As with killer whale calls, the noise from vessels does not propagate evenly in all directions. The noise from vessels is generally strongest directly behind the vessel and is minimal to the front.

In a related study, Bain et al. (unpublished data) used standard methods involving pure tones, white noise, and fixed signal and noise source locations to determine the range of frequencies that killer whales hear and found that their hearing could be masked by noise, similar to the response observed in other mammals. The whales were sensitive from about 500 Hz (the lowest frequency tested) to over 100 kHz. Critical ratios ranged from 20 dB at 10 kHz to 40 dB at 80 kHz. The increase in critical ratio with frequency is typical of mammals, so killer whale hearing appears to be impaired in a manner similar to that of terrestrial mammals.

In this study, we attempted to gather data that reflected real-world situations. First, we tested the effects of noise from different orientations. Then, we tested the response of killer whales using simulated vessel noise rather than white noise. Finally, we tested the ability of killer whales to detect natural communication and echolocation signals. Specifically the objectives of this study were to

≫   ≫   ≫

(1) determine the masking effects of vessel noise on the ability of killer whales to hear pure tones; (2) determine whether the direction to the masking noise source influences its effect; and (3) determine the masking effects of vessel noise on the ability to hear killer whale calls and echolocation clicks.

METHODS

The subjects were two captive female killer whales. "Yaka" was collected in December 1969 from the A5 pod in British Columbia and "Vigga" was collected from Icelandic waters in November 1980. Both were 2–3 years old at the time of collection. Yaka was 22–24 years old and Vigga was 12–14 years old at the time of the study. The whales were kept at Marine World/Africa USA, Vallejo, California, during all phases of this study.

The whales were trained to station against a bar submerged 1 m below the water's surface (Fig. 14-1). They were to remain there until released by the test stimulus, a trainer's whistle, or a recall signal. That is, a go/no-go paradigm was employed. A modified up–down staircase method was used. As was standard practice in their other trained behaviors, the whales were positively reinforced with variable quantities of whole fish for correct responses, and reinforcement was withheld following incorrect responses. Under this training regime, incorrect responses were rare (Bain et al., unpublished data).

Trials proceeded as illustrated in Figure 14-2. Once the whale was properly stationed, a random delay of from 1 to 10 seconds preceded the test stimulus. If the whale "responded" during this waiting period (a false alarm), she was not positively reinforced and the signal level was not changed. If the whale responded during the 2-second test stimulus or immediately following the 2-second response period (a correct detection), she was positively reinforced and the stimulus level was reduced for the next trial. Following the response period, there was an additional waiting period until a total of 15 seconds had elapsed since the beginning of the trial. "Responses" during this second waiting period (both a "miss" and a "false alarm") were not positively reinforced and the stimulus level was increased for the next trial. At 15 seconds, a recall signal was given. If the whale waited for the recall signal (a miss), she was positively reinforced and the stimulus level was increased for the next trial.

On occasion we performed catch trials. No stimulus was presented during some of these trials and the oscillator was set to very high or very low frequencies such that the transducer could not produce them during test trials. Personnel involved in the study were varied to ensure that inadvertent cuing did not develop. In addition, the whales' responses were judged by an observer who was blind to signal timing. The general method described above was varied slightly in different ways for each experiment as described below.

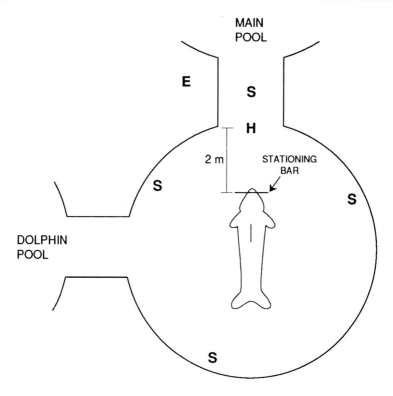

Figure 14-1. Layout of the experimental facility. H indicates position of hydrophone, S indicates positions of speakers producing masking noise, and E is experimenter's position.

Due to reverberation in the pool, received signal levels fluctuated. Thresholds are based on peak signal levels attained during a 2-second period and are probably accurate within 6 dB.

Experiment A

Vessel Noise and Pure Tones

Pure tones were generated with a General Radio Oscillator Model 1312 and gated by a photoelectric switch. Tones were transduced using a Celesco LC 32 hydrophone[1].

---

[1] Mention of trade names does not imply endorsement of specific products.

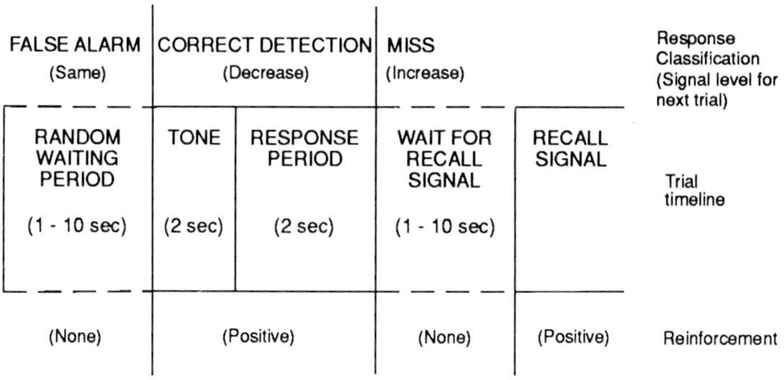

FALSE ALARM (Same)	CORRECT DETECTION (Decrease)	MISS (Increase)		Response Classification (Signal level for next trial)
RANDOM WAITING PERIOD (1 - 10 sec)	TONE (2 sec)	RESPONSE PERIOD (2 sec)	WAIT FOR RECALL SIGNAL (1 - 10 sec)	RECALL SIGNAL — Trial timeline
(None)	(Positive)	(None)	(Positive)	Reinforcement

Figure 14-2. Killer whale trial timeline. The middle row shows the different phases of the trial. Responses were scored as indicated on the upper line, and the subjects were reinforced as indicated on the lower line.

Vessel noise recorded in PWS during the cleanup of the EVOS was supplied by the National Marine Mammal Laboratory. A segment of this recording was digitized and stored on disk. It was played back continuously throughout the session using the Engineering Design Signal System operating on a Compaq Deskpro 486/25 and amplified using a Hafler 2400 professional amplifier and projected using a Lubell Labs 98 Speaker coupled by an AC 201 transformer (Fig. 14-3).

Masked thresholds were determined at three frequencies (8, 16, and 20 kHz) corresponding to the high end of the low-frequency component, and the range of frequencies typically constituting the strongest part of killer whale clicks and calls.

Experiment B

Directional Effects on Masking

Tones were generated with the General Radio Oscillator Model 1312 and sounds were projected using a B & K 8105 hydrophone. A low-frequency band of noise containing energy primarily from 500 Hz to 5 kHz was produced as described in Experiment A. The speaker was placed at one of five locations, directly in line with the test tone, approximately 30° above the test tone, approximately 30° below the test tone, approximately 90° to the side of the test tone, and approximately 150° behind the whale. These directions were chosen to cover a variety of conditions that might be experienced in the wild (e.g., vessels in front of, beside, or behind the whale).

To test the effects of orientation, detection thresholds were determined at three different frequencies (4, 8, and 20 kHz) at each orientation. These frequencies were chosen to include one frequency well within the band of noise, the high end

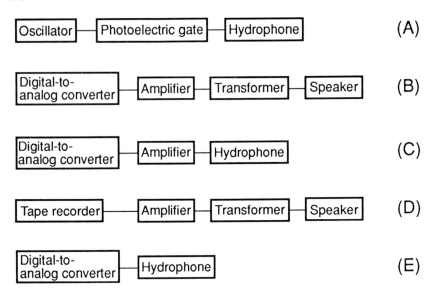

Figure 14-3. Schematic diagrams of the electronic equipment used for sound playback experiments. A=generation of pure tones, B=low-frequency masking noise, C=high-frequency masking noise, D=vessel noise to mask simulated whale sounds, and E=simulated whale sounds.

of the omnidirectional portion of the call, and the strongest portion of the directional component of calls.

Experiment C

Vessel Noise and Whale Sounds

The test stimuli used in this experiment consisted of either a simulated call [based on Call N32 (Ford, 1984) from killer whales that range from British Columbia to southeastern Alaska] or a click train. These sounds were recorded in British Columbia using a B & K 8104 hydrophone, B & K 2635 charge amplifier, and HP 3968A tape recorder and digitized using an Engineering Design Signal System running on a Compaq Deskpro 486/25. Low-frequency noise (ambient noise on the tape) was filtered from these sounds before replay. A faint 27-kHz tone was added to the call to facilitate transfer from pure tone tests with which the subjects were familiar (Fig. 14-4). In addition, all noise between clicks in the click

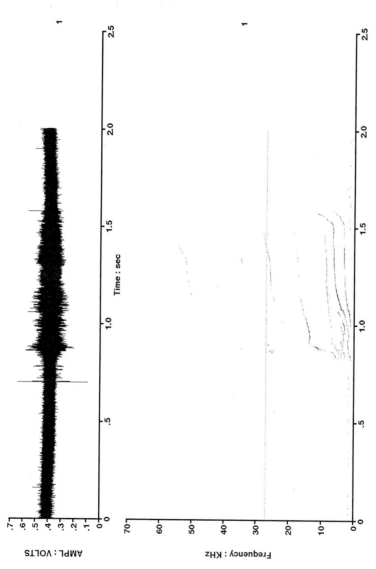

Figure 14-4. Killer whale call used as a test stimulus. The wave form is shown above and the spectrogram is shown below. This is the signal produced by the computer (the received level was too low to produce a suitable image). Note the pure tone at 27 kHz which was added to facilitate transfer from pure tone trials to pulsed stimuli.

train was digitally removed (Fig. 14-5). Output from the computer was directly transduced by a B & K 8105 hydrophone.

The masking noise used in Experiment A was synthesized using the Engineering Design Signal System running on a Compaq Deskpro 486/25 and recorded on a Sony WD-5M cassette recorder. The noise was played back from the cassette recorder and amplified using a Hafler 2400 professional amplifier and projected using a Lubell Labs 98 speaker coupled by an AC 201 transformer.

Sound Propagation

To determine potential impacts of vessel noise on killer whales, the following assumptions were made: Propagation losses averaged 15 log (range) + 3 dB per kilometer excess attenuation (see Malme et al. 1982; Au 1993); source levels of whale calls were typically 180 dB re 1 microPascal at 1 m for high-frequency components and 150 dB re 1 microPascal at 1 m for low-frequency components. Thresholds in nature in the absence of vessel noise were similar to those obtained under quiet conditions in the test pool, and the vessel noise used for testing was average in terms of its effects.

RESULTS

Vessel Noise and Pure Tones

Comparison of sensitivity to pure tones in the presence of vessel noise to that in a quiet background was as expected (Table 14-1). Masking was observed at frequencies ≤20 kHz (the range of peak energy in killer whale phonations). Higher levels of noise produced a stronger masking effect. We were unable to produce a loud enough tone to determine a threshold at 8 kHz with the highest noise level used.

Directional Effects and Masking

The relative effectiveness of masking noise varied with direction (Table 14-2). At all frequencies tested, the masking noise from in front of and slightly below the whale had the strongest effect. In general, masking noise from the side or behind had the least effect.

⋈   ⋈   ⋈

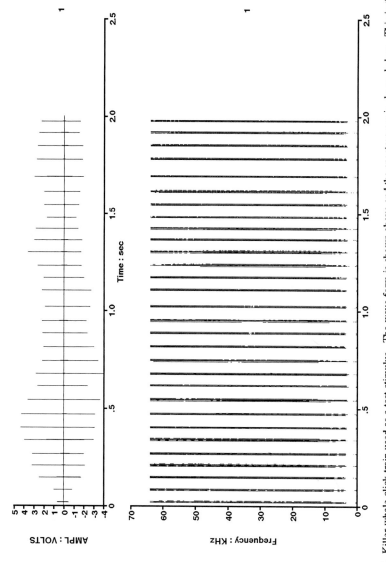

Figure 14-5. Killer whale click train used as a test stimulus. The wave form is shown above and the spectrogram is shown below. This is the stimulus as produced by the computer (the received level was too low to produce a suitable image).

Table 14-1. Detectability of pure tones in the presence of vessel noise. Thresholds in the presence of the tank ambient noise alone and two levels of vessel noise are shown. (--- = no data)

	No Noise	0 dB	20 dB
Yaka (Pacific Female)			
8 kHz	0	20	---
16 kHz	0	6	20
20 kHz	0	32	46
Vigga (Atlantic Female)			
8 kHz	0	14	---
16 kHz	0	14	23
20 kHz	0	17	37

Table 14-2. Relative effectiveness of masking noise from different directions. Position refers to the position of the noise source (see Figure14-1).

	Vigga			Yaka		
Frequency (kHz)	20	8	4	20	8	4
Position						
Low	0	0	0	0	0	0
High	-10	0	-1	-2	-2	-7
Side	-40	-16	-7	-8	-4	-7
Back	-24	-10	-12	-16	-10	-5

Table 14-3. Effects of vessel noise on detectability of whale sounds. Sensitivities in the presence of noise relative to its absence (in dB) are shown for an Atlantic female (Vigga) and a Pacific female (Yaka).

	Vigga	Yaka
Clicks	-10	0
Call	0	0

The variation in effectiveness of masking noise may be due to one or both of two factors. First, the whales may have a directional component to their hearing, as do bottlenose dolphins (*Tursiops truncatus*) (Johnson 1968; Au and Moore 1984), so that the perceived noise level changes with orienation. Second, they may be able to selectively attend to sounds from a particular direction and detect signals with a lower signal-to-noise ratio than when the noise and signal are coming from the same direction.

Vessel Noise and Whale Sounds

Table 14-3 shows the relative sensitivity of the whales to a click train and a call in the presence and absence of masking noise. There is little difference in the detectability of these test signals in the presence of masking noise.

Sound Propagation Model

Under the assumptions used here, at a distance of 10 km, sound would be attenuated by 90 dB [15 log (10,000) + 3(10)]. At this range, the high-frequency component should still be audible directly in front of the whale, but the low-frequency component would be barely audible in other directions.

If noise were to elevate thresholds by 20 dB, this would reduce detection ranges to 5 km.

DISCUSSION

This study shows that, under ideal conditions of low noise and killer whale sounds with significant high-frequency energy, boat noise can have little or no effect. However, it would be expected that higher levels of noise would have an effect, and that sounds without high-frequency energy would be more easily affected. Since the high-frequency component of killer whale sounds is relatively directional (Schevill and Watkins 1966), this suggests that there would be minimal effects straight ahead and maximal effects to the side or behind.

Figure 14-6 illustrates a situation that may have occurred in PWS during clean-up activities. Killer Whale Group A is swimming parallel to Killer Whale Group B. Killer Whale Group C is ahead of Group A, and Killer Whale Group D is ahead of Group B. Two vessels are close to Group A. Vessel 1 is not far away on the side, while Vessel 2 is directly in front of Group A. These vessels are distant from the other groups of killer whales. Based on this study, the following general predictions could be made about the effects of these vessels on killer whale communication.

Killer Whale Group A would be expected to have its ability to hear Group C impaired (Fig. 14-6). This prediction is based on Bain et al. (unpublished data) in

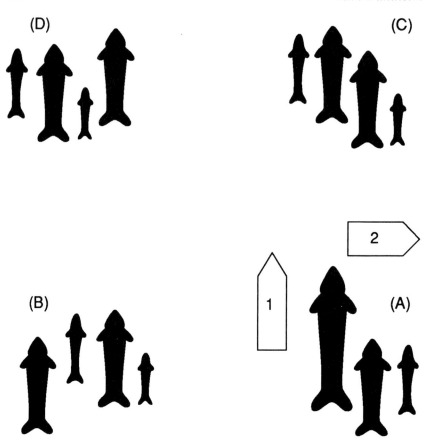

Figure 14-6. Conceptualized depiction of the impact of vessel noise on killer whales. A to D show positions of killer whale groups; 1 and 2 represent vessel positions. See text for discussion.

which high levels of low-frequency noise reduced sensitivity to all frequencies of sound tested. Likewise, its ability to hear Group B would be impaired. Vessel 2 would be the primary cause of the impairment since masking noise from in front of the whales has a stronger effect than noise from behind or the side (Experiment B).

Killer Whale Group B would also have its ability to hear Group A impaired. Although the low levels of noise from distant vessels only impairs the ability to hear low frequencies (Experiment A), only low-frequency components of killer whale sounds travel to the side (Schevill and Watkins 1966). Finally, Group D should be able to hear Group B unimpaired since the high-frequency components

⤜⤛ ⤜⤛ ⤜⤛

of Group B's calls will reach Group D, and these components will not be masked by the low levels of low-frequency noise (Experiment C).

Although each of the qualitative predictions are supported by data from this study, only the ability of Group D to hear Group B was actually tested in a quantitative manner. The other parts of the configuration could be tested easily using the paradigm developed in this study. However, due to the number of call types used by killer whales, the variety of vessel noises, and spatial arrangements to be considered, testing these combinations was beyond the scope of this study.

The impact of reduced detection distances is likely to be felt most strongly in increased difficulty in finding food. Individual killer whales will have a reduced area over which they can search for food. The shortened distance over which they can hear other whales reduces the potential for cooperative foraging (see Hoelzel 1993). The impact of this effect will depend on overall food availability.

There are a number of behavioral changes the whales might make that would mitigate the survival consequences of reduced food availability. These include turning to face other whales when phonating, increasing the time spent foraging, foraging in areas where there is a high probability of approaching prey closely so that prey can be detected even in the presence of noise, and adjusting activity cycles so that foraging occurs when noise is minimal. Although these are potential behavioral changes, it is unknown whether the whales have the behavioral flexibility to actually make any of them.

CONCLUSIONS

Vessel noise impairs the ability of killer whales to detect low-frequency (up to at least 20 kHz) signals. Since only low-frequency components of killer whale calls are omnidirectional, vessel noise would be expected to impair killer whale communication. However, high-frequency components which propagate well in the forward direction are not effectively masked by low levels of vessel noise. Thus, killer whales should be able to communicate unhindered with the small proportion of whales that they are likely to be oriented toward under natural conditions.

Due to substantial variation in the effects of masking noise of a particular level depending on orientation, estimating levels of noise exposure alone will not be adequate for assessing any potential damage.

ACKNOWLEDGMENTS

We would like to thank P. Cates, B. Kriete, and J. Olhiser for extensive help with data collection and analysis. D. Cameron, J. Lawrence, D. Marrin-Cooney, J. Mullen, P. Povey, and A. Traxler of Marine World all helped make this work

⋈ ⋈ ⋈

possible by training the whales and maintaining reliable responses during this work. We would like to thank S. Allen and M. Demetrios for their support. S. Klappenback of Sea World, and P. Moore and W. Au of the Naval Oceans Systems Center gave helpful advice. We thank several members of the National Marine Mammal Laboratory staff and two anonymous reviewers who reviewed an earlier draft of this manuscript. Research at Marine World was conducted under NMML contracts #43ABNF002499 and #43ABNF002500.

REFERENCES

Anonymous. 1982. Report of the workshop on identity, structure and vital rates of killer whale populations, Cambridge, England, June 23–25, 1981. Report of the International Whaling Commission 32:617–629.

Au, W. W. L. 1993. The sonar of dolphins. Springer-Verlag. New York, New York.

Au, W. W. L., and P. W. B. Moore. 1984. Receiving beam patterns and directivity indices of the Atlantic bottlenose dolphin. Journal of the Acoustical Society of America 75:255–282.

Bain, D. E. 1986. Acoustic behavior of *Orcinus*. Pages 335–371, *in* B. Kirkevold and J. Lockard (eds.), Behavioral biology of killer whales, Alan R. Liss, New York, New York.

Bain, D. E. 1988. An evaluation of evolutionary process. Ph.D. dissertation. University of California, Santa Cruz. 257 p.

Dahlheim, M. E., and F. Awbrey. 1982. A classification and comparison of vocalizations of captive killer whales (*Orcinus orca*). Journal of the Acoustical Society of America 72:661–670.

Diercks, K. J., R. T. Trochta, C. F. Greenlaw, and W. E. Evans. 1971. Recording and analysis of dolphin echolocation signals. Journal of the Acoustical Society of America 49:1729–1732.

Ford, J. K. B. 1984. Call traditions and dialects of killer whales (*Orcinus orca*) in British Columbia. Ph.D. dissertation. University of British Columbia, Canada.

Ford, J. K. B. 1991. Vocal traditions among resident killer whales (*Orcinus orca*) in coastal waters of British Columbia. Canadian Journal of Zoology 69:1454–1483.

Hall, J. D., and C. S. Johnson. 1972. Auditory thresholds of a killer whale *Orcinus orca* Linnaeus. Journal of the Acoustical Society of America 51:515–517.

Hoelzel, A. R. 1993. Foraging behaviour and group dynamics in Puget Sound killer whales. Animal Behavior 45:581–591.

Hoelzel, A. R., and R. W. Osborne. 1986. Killer whale call characteristics: Implications for cooperative foraging strategies. Pages 373–403, *in* B. Kirkevold and J. Lockard (eds.), Behavioral biology of killer whales, Alan R. Liss, New York, New York.

Johnson, C. S. 1968. Masked tonal thresholds in the bottlenosed porpoise. Journal of the Acoustical Society of America 44:965–967.

Malme, C. I., P. R. Miles, and P. T. McElroy. 1982. The acoustic environment of humpback whales in Glacier Bay and Frederick Sound/Stephens Passage, Alaska. Bolt, Beranek and Newman Inc. Report No. 4848. 187 p.

Morton, A., J. Gale, and R. Prince. 1986. Sound and behavioral correlations in captive *Orcinus orca*. Pages 303–333, *in* B. Kirkevold and J. Lockard (eds.), Behavioral biology of killer whales, Alan R. Liss, New York, New York.

Scharf, B. 1975. Audition. Pages 113–149, *in* B. Scharf and G. S. Reynolds (eds.), Experimental sensory psychology. Scott, Foresman and Co., Glenview, Illinois.

Schevill, W. E., and W. A. Watkins. 1966. Sound structure and directionality in *Orcinus* (killer whale). Zoologica. 51:71–76.

Chapter 15

Cetaceans in Oil

James T. Harvey and Marilyn E. Dahlheim

INTRODUCTION

Oil from natural and anthropogenic sources occurs mostly along coastlines. Only 7.7% of the petroleum hydrocarbons entering the marine environment come from natural sources. The anthropogenic sources of petroleum hydrocarbons include the atmosphere (9.2%), municipal and industrial wastes or runoff (36.3%), and production and transportation (46.8%) (Neff 1990). Large oil spills probably present the greatest short-term threat to coastal organisms, whereas wastes and runoff create chronic problems. Coastal cetaceans may contact petroleum during migration, feeding, or breeding. Usually cetaceans contact oil at the water's surface where they may inhale volatile hydrocarbons, the oil may adhere to their skin, eyes, or baleen, or prey may become contaminated. Because naturally occurring oil probably was not abundant during the evolution of cetaceans, there was little selective pressure to develop capabilities for detecting and evading oil. Consequently, there has been concern regarding the behavior of cetaceans near oil. Only a few researchers, however, have studied the possible evasive actions of cetaceans and the direct and indirect effects of oil.

There have been few quantitative studies of cetacean behavior in response to oil. Gray whales (*Eschrichtius robustus*) traveling off California breathed less frequently in the presence of naturally occurring oil than when they were outside oil slicks (Evans 1982). Fin whale (*Balaenoptera physalus*), humpback whale (*Megaptera novaeangliae*), bottlenose dolphin (*Tursiops truncatus*), and other cetaceans have been observed entering areas with oil and swimming and behaving normally (Geraci 1990).

On 24 March 1989, the T/V *Exxon Valdez* ran aground in Prince William Sound (PWS) spilling 11 million gallons of Prudhoe Bay crude oil. Numerous federal, state, local, and private personnel participated in the cleanup and conducted studies

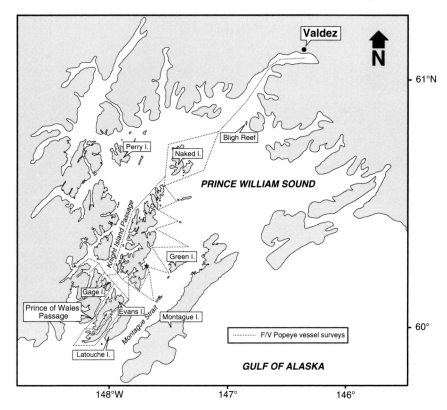

Figure 15-1. Study area in southwestern Prince William Sound where boat surveys were conducted during 1-9 April 1989. Dotted line represents the vessel tract. The * indicates the area near Snug Harbor where an oiled Dall's porpoise was observed.

of the effects of this oil spill on the environment (Zimmerman et al., Chapter 2). The purpose of this chapter is to report the limited observations of cetaceans in PWS made immediately after the *Exxon Valdez* oil spill (EVOS) and to compare these observations with past reports.

METHODS

From 1 to 9 April 1989, daily systematic surveys (Fig. 15-1) were conducted in PWS aboard the 8–m F/V *Popeye*. One observer scanned the starboard side (0°–90°) and one the port side (270°–360°) of the vessel whenever under way. Surveys were conducted primarily in areas with oil visible on the surface or in areas

Table 15-1. Date, times of effort, and sea state (B0 = Beaufort 0, B1 = Beaufort 1, etc.) during surveys of marine mammals conducted in Prince William Sound and the numbers of cetaceans counted, and numbers observed in oil. (--- = no data)

Date 1989	Time	Sea state	Number of cetaceans	Number in oil
1 April	1030-1630	B1	6	5
2 April	1040-1935	B0	21	21
3 April	0910-1610	B0	9	9
4 April	1608-1910	B0	2	2
6 April	0840-1820	B1-B2	4	---
7 April	0825-1645	B0	21	18
8 April	1245-1900	B0-B1	19	19
9 April	0805-1200	B1-B3	6	6

where oil had been observed. Upon sighting a marine mammal (except sea otters) the following data were recorded: species, location, time, number of individuals, approximate distance and angle of sighting from the vessel, sea state, behavior of animals (resting, traveling, or feeding), response of animals to oil on the surface (detection, avoidance, or attraction), presence or absence of oil on water, and qualitative assessment of amount of oil.

Behaviors of marine mammals were qualitatively scored into categories as fast or slow traveling, resting, or unusual behaviors (e.g., prolonged periods at the surface by cetaceans). We recorded any appearance of oil on the animals. Generally we did not stop when marine mammals were sighted, although we occasionally slowed to make more detailed observations of their behavior and appearance.

Observations of cetaceans in oil also were solicited from researchers working in PWS. These anecdotal observations provided additional information which increased the sample size. This anecdotal information included the species observed, location, behavior, and the presence of oil visible on the water and animals.

Unfortunately, we did not quantify the amount of survey effort in oil and outside oiled areas. Therefore, there was no ability to compare sighting frequencies of cetaceans between the two areas (i.e., we could not assess whether cetaceans avoided or were attracted to oil).

Difference in the frequencies of fast or slow travel of Dall's porpoise (*Phocoenoides dalli*) in areas of light or moderate to heavy oil was tested using a contingency table. Oil was considered light when there was a sheen on the water's surface, and was considered moderate to heavy when there were visible concentrations (such as mousse).

RESULTS

Nine surveys (72.8 hours) were completed, mostly in the southwest portion of PWS (Fig. 15-1). All survey effort was confined to periods with a sea state of Beaufort 3 or less (Table 15-1). One hundred cetaceans were sighted; 80 Dall's porpoise, 18 killer whales (*Orcinus orca*) in one pod, and 2 harbor porpoise (*Phocoena phocoena*) were located in one group. An average 2.7 (SE=0.25) Dall's porpoises were seen in the 30 groups.

Groups of Dall's porpoise were observed 21 times in areas with light sheen, seven occasions in areas with moderate to heavy surface oil, once in no oil, and once when the amount of oil was not recorded. When oil was present, Dall's porpoise were swimming rapidly on 13 occasions and swimming slowly 15 times. Swimming speed was independent of amount of oil present (χ^2=0.431, P>0.05). That is, Dall's porpoise were not observed to swim faster in areas with moderate to heavy amounts of oil when compared to areas with no oil or a light sheen.

We could determine whether cetaceans had oil on their skin only when we were close to the animal (usually within 50 m). Thus, only five Dall's porpoise groups (12 individuals) were confirmed not to have oil on their skin (3 groups in light sheen, 2 in moderate-heavy oil); for 25 groups (64 individuals) we could not confirm the presence or absence of oil on their skin; and only one solitary Dall's porpoise was observed with oil on its skin as it swam through an area covered with a light sheen. This oiled porpoise was seen slowly traveling south on 3 April, 6.4 km east of Snug Harbor (Fig. 15-1). The dorsal surface of the body, from the blowhole posterior to the dorsal fin, was covered with oil. Also, the upper third of the dorsal fin was oiled. This Dall's porpoise was about 1.2 m long and was initially sighted because it was blowing every 10 to 12 seconds and remaining at the surface for extended periods of time (1–2 minutes). We approached within 20 m of the animal before it dived, and it appeared stressed.

Eighteen killer whales were observed for about 30 minutes as they traveled from Gage Island to Prince of Wales Passage. We did not observe any oil on any individual (3 of which were calves) as they traveled at a moderate speed. At one point, they dove under the vessel, which was positioned ahead of a patch of light oil, and surfaced in the area with oil. In addition, the two harbor porpoise without visible oil on their dorsum were seen off Shelter Bay (north end of Evans Island) in an area with a light sheen and some droplets of oil on the water's surface.

Other researchers working in PWS also observed cetaceans in areas with and near oil on the water's surface. On 30 March, two killer whales were seen south of Green Island (L. Lowry and K. Frost, Alaska Department of Fish and Game, Fairbanks, Alaska, personal communication). These killer whales were headed toward an area with oil about 3.2 km away. Although there were reports of abnormal activity by killer whales in this area, no abnormal activity was confirmed

∽   ∽   ∽

by trained observers. Killer whales were observed on 31 March, 5 August, and 2 and 3 September in PWS by C. Matkin (Dahlheim and Matkin, Chapter 9). On the first date, AB pod was observed in lower Knight Island Passage remaining most of the day in an oil-free area before swimming through heavy slicks. On the latter three occasions, killer whales swam through light to moderate oil without apparent avoidance.

On 5 April, near Bishop Rock, Latouche Passage (southwestern PWS), five Dall's porpoise were observed swimming through water with a light sheen. All the individuals had a brown chevron on the top of their heads, apparently from surfacing in oil (C. Matkin, Homer, Alaska, personal communication). On 6 April, Matkin saw three Dall's porpoise bowriding 1.6 km northeast of Sleepy Bay (north tip of Latouche Island). Each individual had the same brown markings seen the day before. This brown chevron, however, may be natural coloration.

On 3 April 1989, gray whales were observed at the southwest entrance to PWS during aerial reconnaissance flights (J. Lentfer, Homer, Alaska, personal communication). Three gray whales were observed swimming northwest through a moderate amount of oil. An additional six gray whales were observed near Cape Puget (southwest entrance to PWS). During the 10 minutes of observation, the gray whales continually swam at the surface and appeared lethargic. Fumes from the oil could be detected in the plane at 100- to 200-m elevation (J. Lentfer, personal communication).

DISCUSSION

Most observations of cetaceans in oil are anecdotal or qualitative in nature. Of six reports of cetaceans swimming near or in oil (Geraci 1990), no animals avoided oil or behaved abnormally. Although gray whales did not avoid natural slicks of oil off southern California, gray whales spent less time at the surface by blowing less frequently (Evans 1982). With the limited data, lack of controls, and natural variability, it is difficult to test the effects of oil on cetaceans. Rarely do investigators have the opportunity to design experiments to test whether cetaceans detect and avoid oil.

In one controlled test, captive bottlenose dolphin, presumably using vision, could detect the difference between an oiled and uncontaminated water's surface (Geraci et al. 1983). Thin layers of oil, corresponding to oil 1-mm thick were not detected as well or at all compared to thicker layers of oil. Bottlenose dolphin also used tactile senses to detect and avoid a 1-cm-thick slick of mineral oil. Once the layer of oil was contacted the dolphins avoided the area. These data may indicate that odontocetes, and possibly baleen whales, cannot see a light sheen of oil on the water's surface, but may detect moderate to heavy amounts of oil using their visual or tactile senses.

〰〰〰〰〰〰

Although our data are limited, there was no apparent change in swimming speed of Dall's porpoise in light versus moderate to heavy amounts of oil. Either Dall's porpoise were unable to detect oil, detected oil but it did not affect their behavior, or our small sample size did not allow us to determine a difference. Only one cetacean was seen with oil on its skin. This observation suggests that cetaceans may avoid areas with oil, oil does not adhere well to their skin, or the presence of oil on cetaceans is difficult to see.

If cetaceans cannot detect petroleum products at some distance, a response may not occur until they directly contact oil. Cetaceans may reverse their path, thereby returning to oil-free water. This action may not be employed often because individual movements would be erratic. Cetaceans could reduce their contact with oil by increasing the duration of dives and decreasing the duration of time breathing at the surface. With increased duration of dives, cetaceans effectively decrease the number of breaths per time unit. Energetically, cetaceans require a given amount of oxygen per time depending on their activity level. Therefore, they may decrease the number of times they surface for a short period. This strategy will reduce the amount of contact with oil, but eventually the animal must breathe at the rate necessary for its metabolic requirements. If the oil spill is extensive, cetaceans cannot employ this tactic for long periods of time. Cetaceans generally expose a limited amount of their dorsum while breathing. It seems, therefore, that cetaceans may effectively reduce their exposure to oil for limited periods of time but cannot substantially reduce their exposure if a large amount of water is covered with oil.

The relatively few observations of cetaceans in or near oil may reflect: (1) the seasonally low density of cetaceans in PWS; (2) avoidance of the area because of the presence of oil; or (3) avoidance because of the increased human activity (e.g., aircraft, vessel, and on land). This is difficult to assess because there were no surveys conducted outside the impacted area (i.e., no controls). The paucity of observations of cetaceans in oil underscores the difficulty in obtaining quantitative information needed to assess the effects of oil on these animals.

Petroleum substances can damage mammalian skin, but experiments with captive cetaceans indicated little effect with exposures of 75 minutes (Geraci and St. Aubin 1985). Petroleum compounds did not significantly reduce epidermal cell proliferation or change lipid composition. The skin apparently was a barrier to allergenic hydrocarbons. These results are difficult to interpret, however, because sample sizes were small and results obtained from experiments with captive animals may not reflect natural processes.

To assess the effects of an oil spill on cetaceans, the following data are needed: abundance, distribution, behavior, year-classes, mortality rate, pollutant loads, and condition (e.g., nutritional, reproductive, and neurological) of a species before the

⋈   ⋈   ⋈

spill. These same variables should be measured during, within 1 year, and 2–10 years after the spill. This requires a great deal of effort in the area impacted before and after a spill. Some short-term (0–1 month) effects of oil may be: (1) changes in cetacean distribution associated with avoidance of aromatic hydrocarbons and surface oil, changes in prey distribution, and human disturbance; (2) increased mortality rates from ingestion or inhalation of oil; (3) increased petroleum compounds in tissues; and (4) impaired health (e.g., immunosuppression). A few long-term effects include: (1) change in distribution and abundance because of reduced prey resources or increased mortality rates; (2) change in age structure because certain year-classes were impacted more by oil; (3) decreased reproductive rate; and (4) increased rate of disease or neurological problems from exposure to oil.

Because data regarding abundance, distribution, health, and reproductive and mortality rates are rarely available before a spill, assessing the effects of oil on cetaceans requires concentrated effort at the time of a spill. One recommendation would be placing radiotags on cetaceans. Response of individuals in the presence of oil could be recorded. Certain hypotheses could be easily tested. For example, do cetaceans spend less time at the surface in the presence of oil, do they change direction to avoid oil, and does their behavior change through time? Because there are relatively few cetaceans in any particular area, we suggest collecting more data on a few identifiable individuals rather than surveying large amounts of water searching for the few animals in an area. It seems the ultimate answers regarding the effects of oil on cetaceans require information obtained from individuals monitored for prolonged periods of time.

CONCLUSIONS

Daily vessel surveys of PWS were conducted from 1 to 9 April 1989 to determine relative abundance and behavior of cetaceans in response to the EVOS. Two observers scanned the forward quarter of a vessel and recorded species, number, location, time, behavior, and presence or absence of oil for each marine mammal sighting. Observations were also obtained from other researchers working in the area. During nine surveys (72.8 hours), 80 Dall's porpoise, 18 killer whales, and 2 harbor porpoise were observed. Swimming speed of the Dall's porpoise was independent of the amount of oil present. We observed oil only on one individual, a Dall's porpoise. This individual had oil on the dorsal half of its body and appeared stressed because of its labored breathing pattern. In no case did cetaceans alter their behaviors when in areas with oil. These observations are consistent with other reports of cetaceans behaving normally in the presence of oil.

➤ ➤ ➤

ACKNOWLEDGMENTS

We sincerely thank K. Frost, J. Lentfer, L. Lowry, C. Matkin, and C. Monnett for providing their observations of cetaceans in oil. G. Maykowskyj skippered the F/V *Popeye* used for the surveys. S. Mizroch critiqued an earlier version of this chapter. This work was supported by the National Marine Mammal Laboratory and the *Exxon Valdez* Oil Spill Trustee Council.

REFERENCES

Evans, W. E. 1982. A study to determine if gray whales detect oil. Pages 47–61, *in* J. R. Geraci and D. J. St. Aubin, (eds.), Study of the effects of oil on cetaceans. Final Report, Contract No. AA551-CT9-29, U.S. Department of Interior, Bureau of Land Management, Washington, D.C.

Geraci, J. R. 1990. Physiologic and toxic effects on cetaceans. Pages 167–197, *in* J.R. Geraci and D. J. St. Aubin (eds.), Sea mammals and oil: Confronting the risks. Academic Press, San Diego, California.

Geraci, J. R., and D. J. St. Aubin. 1985. Expanded studies of the effects of oil on cetaceans. Contract No. 14-12-0001-29169, U. S. Department of Interior, Minerals Management Service, Washington, D.C. 144 p.

Geraci, J. R., D. J. St. Aubin, and R. J. Reisman. 1983. Bottlenose dolphin, *Tursiops truncatus*, can detect oil. Canadian Journal of Fisheries and Aquatic Science 40:1515–1522.

Neff, J. M. 1990. Composition and fate of petroleum and spill-treating agents in the marine environment. Pages 1–33, *in* J. R. Geraci and D. J. St. Aubin (eds.), Sea mammals and oil: Confronting the risks. Academic Press, San Diego, California.

Chapter 16

Pathology of Sea Otters

Thomas P. Lipscomb, Richard K. Harris, Alan H. Rebar,
Brenda E. Ballachey, and Romona J. Haebler

INTRODUCTION

In the months following the *Exxon Valdez* oil spill (EVOS), 994 sea otters (*Enhydra lutris*) from oil-spill-affected areas died (Doroff et al. 1993). Carcasses collected from these areas and otters that died in rehabilitation centers are included in this number. The actual number that died was probably much greater.

Within days of the spill, the Exxon Company (USA) funded an effort to rehabilitate oil-contaminated sea otters (Davis 1990). Initially, clinical veterinarians working on the rehabilitation effort performed partial necropsies on some of the sea otters that died. Soon, veterinary pathologists from the University of Alaska and the U.S. Environmental Protection Agency provided assistance. Later, rehabilitation centers were constructed and other veterinarians with special training in pathology were hired by Exxon to provide diagnostic support.

In late April 1989, veterinary pathologists from the U.S. Fish and Wildlife Service (USFWS) assumed responsibility for pathologic evaluation of oil-spill-affected sea otters. The USFWS requested assistance from veterinary pathologists of the Armed Forces Institute of Pathology (AFIP) in June 1989. Eventually, as part of the Natural Resources Damage Assessment program, AFIP veterinary pathologists were asked to carry out histopathological studies of the tissue specimens collected by all parties and to perform necropsies on carcasses that had been collected and frozen. A veterinary clinical pathologist was requested to assess hematology and clinical chemistry findings in otters that had been held in the rehabilitation centers.

In spite of the best efforts of many dedicated people working under extremely difficult conditions, there are significant limitations in the pathological studies. The absence of a detailed necropsy protocol and of full documentation of necropsy findings during the first several weeks after the spill resulted in important data being lost. Often, samples of all major organs were not collected. In some cases, no necropsy report was available. Specimens for toxicologic analysis for petroleum hydrocarbons were not consistently collected. The absence of a detailed toxicology protocol suggests that the samples may not have been collected properly. The lack of a consistent numbering system for identification of specimens caused major problems: some samples were useless because they could not be identified. Many blood samples could not be transported to laboratories quickly enough to prevent significant deterioration of the samples because of inclement weather and the remote locations of the rehabilitation centers. Thus, in a number of cases, data could not be used because significant deterioration of the specimens was considered likely to have occurred. Since more than one laboratory was used to analyze blood samples, problems with comparability of results were encountered. The laboratory tests were performed to aid the clinical veterinarians in diagnosis and treatment of individual animals, not as part of a consistent protocol; thus, there is variation in the amount and type of laboratory data available for each otter. These problems illustrate the need for development of contingency plans that include detailed protocols before disasters occur.

SEA OTTERS THAT DIED IN REHABILITATION CENTERS

Following the oil spill, sea otters that appeared oil contaminated, were considered to be in danger of becoming oil contaminated, or that behaved abnormally were captured and transported to rehabilitation centers (Bayha and Hill 1990). On arrival, oil exposure was assessed by visual examination. Degree of oil contamination was graded according to the following criteria: oil covering greater than 60% of the body (heavily contaminated); oil covering 30–60% of the body (moderately contaminated); oil covering less than 30% of the body or a light sheen on the fur (lightly contaminated); or, no visible oil (uncontaminated).

Clinical, Hematologic, and Blood Chemistry Studies

Clinical records and laboratory data for 21 oil-contaminated sea otters that died within 10 days of arrival at the rehabilitation centers were examined. Seven were heavily contaminated, five moderately contaminated, and nine lightly contaminated. Selection of these otters was based on completeness of clinical records, availability of laboratory data from acceptable samples, and availability of results of histopathologic examination.

Clinically, shock was the most common terminal syndrome and was characterized by hypothermia, lethargy, and often hemorrhagic diarrhea. This syndrome was rarely observed on arrival at the centers but generally developed within 48 hours in heavily and moderately oil-contaminated otters. Lightly contaminated otters generally developed shock during the second week after arrival. A high

✂ ✂ ✂

proportion of otters in all three grades of oil contamination had seizures at or near the time of death. Anorexia was also relatively common.

Blood values were compared to reference ranges established for normal sea otters from southeastern Alaska. Abnormalities were interpreted according to conventions used for the interpretation of laboratory data in dogs and other carnivores/omnivores (Duncan and Prasse 1986).

The most common hematologic abnormality in sea otters showing all grades of oil contamination was leukopenia (decreased white blood cell count) characterized by decreased mature neutrophils with increased numbers of immature neutrophils (degenerative left shift) and decreased lymphocytes. Degenerative left shifts indicate severe inflammation. Terminal endotoxemia is a possible cause of the leukopenia in these otters. The lymphopenia reflects either systemic stress or the influence of glucocorticoids administered by clinical veterinarians. Severe stress could also result in the release of large amounts of glucocorticoids from the adrenal cortices which could cause sequestration and possible destruction of circulating lymphocytes. Anemia was also relatively common but could not be further characterized from the available data.

The main clinical chemistry abnormalities were azotemia, hyperkalemia, hypo-proteinemia/hypoalbuminemia, increased serum transaminases indicative of hepatocellular leakage, and hypoglycemia.

Azotemia (the accumulation of nitrogenous waste products in the blood) was the most common clinical chemistry abnormality, and its prevalence was similar in otters showing all three grades of oil contamination. Azotemia indicates inadequate kidney function, which may result from a primary kidney problem (renal azotemia), insufficient blood and oxygen supply to the kidney (prerenal azotemia), or obstruction of the lower urinary tract (postrenal azotemia). Unfortunately, urine-specific gravities were not available to aid in differentiation of prerenal from renal azotemia. Clinical findings did not support postrenal azotemia. The clinical histories suggest that shock and hemorrhagic diarrhea may have caused a poor renal blood supply resulting in prerenal azotemia. In the few animals that developed true renal azotemia (serum urea nitrogen values greater than 200 mg/dl), the long-term reduced renal blood supply probably led to primary terminal renal injury.

Hyperkalemia (abnormally high blood potassium) and hypoproteinemia/hypoalbuminemia (abnormally low blood protein and albumin) were found less commonly than azotemia but were probably also related to diarrhea and shock. Possible causes of hyperkalemia in these otters include release of potassium from dying cells and acidosis. Acidosis is a condition that causes decreased blood pH. Severe diarrhea is a frequent cause of acidosis. Protein loss from diarrhea is the most likely cause of the hypoproteinemia/hypoalbuminemia.

Increased hepatocellular leakage-associated serum transaminases were only slightly less common than azotemia. Hepatocellular leakage might have been

⋙   ⋙   ⋙

caused by primary hepatotoxicity, but may also have been caused by anorexia. In fasting associated with anorexia, tissue stores of fat are mobilized and transported to the liver resulting in increased cell membrane permeability of hepatocytes with leakage of transaminases into the blood. Anorexia is also the most likely cause of the hypoglycemia; hypoglycemia probably caused seizures.

Anatomic Pathology Studies

Histopathologic examinations were performed on 51 oil-contaminated and on 6 uncontaminated sea otters that died in rehabilitation centers. Pups were excluded because of the small number available. Samples from six apparently healthy adult sea otters that were killed by gunshot in an area not affected by an oil spill as part of unrelated research were used as normal controls. Of the 51 oil-contaminated otters that died in rehabilitation centers, 16 were heavily contaminated, 13 were moderately contaminated, and 22 were lightly contaminated. Complete sets of tissues were not available from all otters. A more detailed description of this study is reported elsewhere (Lipscomb et al. 1993).

Among oil-contaminated otters that died in the centers, interstitial pulmonary emphysema was the most common lesion. It was found in 11/15 (73%) heavily contaminated, 5/11 (45%) moderately contaminated, and 3/20 (15%) lightly contaminated otters. Histologically, the lesion appeared as expanded areas of clear space within the interlobular septa of the lung (Fig. 16-1).

Gastric erosions were found in 2/14 (14%) heavily contaminated, 7/9 (78%) moderately contaminated, and 4/17 (24%) lightly contaminated otters. Microscopically, the erosions consisted of discrete, 1- to 3-mm-diameter areas of coagulative necrosis that affected superficial to midlevel gastric mucosa (Fig. 16-2).

Hepatic lipidosis occurred in 8/16 (50%) heavily contaminated, 5/12 (42%) moderately contaminated, and 1/19 (5%) lightly contaminated otters. The lesion was characterized by variably sized, single to multiple, round, sharply delineated, unstained intracytoplasmic vacuoles in hepatocytes (Fig. 16-3). The distribution of the lesion was predominantly periportal, but in severe cases it was diffuse. Renal lipidosis was found only in otters that also had hepatic lipidosis. It was present in 10/42 (24%) oil-contaminated otters that died in rehabilitation centers. The microscopic appearance was characterized by single or multiple, variably sized, round, discrete, unstained intracytoplasmic vacuoles within proximal and distal renal tubular epithelium (Fig. 16-4). The intracytoplasmic vacuoles in both liver and kidney stained red with oil red O which indicated the presence of lipids.

Centrilobular hepatic necrosis was seen in 4/16 (25%) heavily contaminated, 3/12 (25%) moderately contaminated, and 4/19 (21%) lightly contaminated otters. In affected livers, centrilobular hepatocytes exhibited nuclear pyknosis, karyorrhexis, karyolysis, and increased eosinophilia of cytoplasm with preservation of basic cell shape (Fig. 16-5).

⋊⋉   ⋊⋉   ⋊⋉

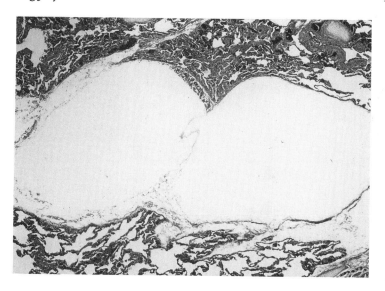

Figure 16-1. Lung; interstitial emphysema. Interlobular septum is expanded by gas bubbles. Adjacent alveoli are atelectatic.

Among the six uncontaminated otters that died in rehabilitation centers, one had gastric erosions, one had hepatic lipidosis and multifocal hepatic necrosis, and one had focally extensive hepatic necrosis. The remaining otters included one with peritonitis caused by perforation of the small intestine. Another had mild acute enteritis and mild subacute hepatitis of undetermined cause. The remaining otter in this group had no significant histologic lesions and no necropsy report was available.

The six apparently healthy otters collected from an area that had not been affected by an oil spill had no significant histologic lesions. Various incidental lesions were found in otters in each of the three oiled groups.

Possible Mechanisms of Lesion Development and Clinicopathologic Correlations

The most common serious complication of petroleum hydrocarbon ingestion in human beings and animals is aspiration pneumonia (Richardson and Pratt-Thomas 1951; Rowe et al. 1951; Eade 1974). Aspiration pneumonia occurs when large amounts of foreign material enter the lungs through the airways. Oil-contaminated sea otters attempt to remove the oil by grooming with the mouth (Siniff et al. 1982), which would seem to provide ample opportunity for aspiration. However, no evidence of aspiration pneumonia was found in oil-contaminated sea otters.

⋙   ⋙   ⋙

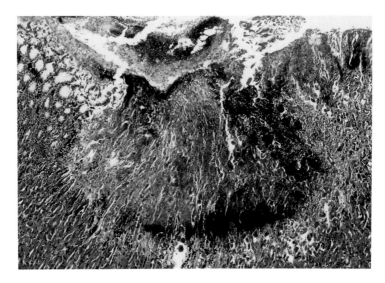

Figure 16-2. Gastric mucosa; erosion. Focal area of coagulative necrosis. (Reprinted by permission of Veterinary Pathology from Lipscomb, T. P., Harris, R. K., Moeller, R. B., Pletcher, J. M., Haebler, R. J., and Ballachey, B. E. 1993. Histopathologic lesions in sea otters exposed to crude oil. Veterinary Pathology 30.)

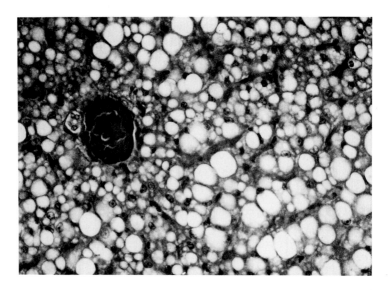

Figure 16-3. Liver; lipidosis. Diffuse vacuolation of hepatocytes caused by lipid accumulation. Portal area is at left.

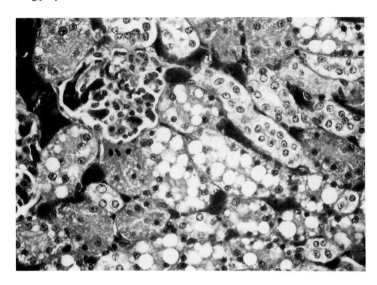

Figure 16-4. Kidney; lipidosis. Vacuolation of tubular epithelium caused by lipid accumulation.

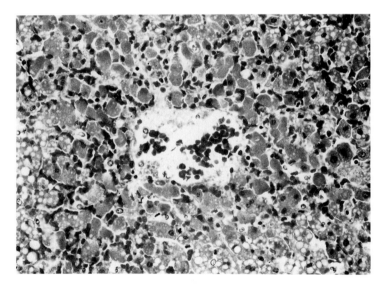

Figure 16-5. Liver; centrilobular necrosis and midzonal lipidosis. Central vein is at center. (Reprinted by permission of Veterinary Pathology from Lipscomb, T. P., Harris, R. K., Moeller, R. B., Pletcher, J. M., Haebler, R. J., and Ballachey, B. E. 1993. Histopathologic lesions in sea otters exposed to crude oil. Veterinary Pathology, 30.)

Interstitial pulmonary emphysema, which is the accumulation of bubbles of air within the supportive connective tissues of the lungs, was remarkably common in oil-contaminated sea otters. The mechanism by which exposure to crude oil causes sea otters to develop this lesion is unclear. Dyspnea was prevalent in oil-contaminated sea otters presented to rehabilitation centers; many of these otters also had subcutaneous emphysema (Williams et al. 1990), which forms by extension of pulmonary emphysema through the mediastinum into muscle fascia and subcutaneous tissue. Rupture of alveolar walls is the usual mechanism by which air enters the pulmonary interstitium. Alveoli may rupture when there is a combination of forced expiration or coughing and bronchiolar obstruction that produces greatly increased pressures within alveoli (Cotran et al. 1989). Sea otters have well-developed pulmonary interlobular septa and thus may be predisposed to the development of interstitial emphysema (Dungworth 1985). Interstitial emphysema has been reported in sea otters with pneumonia (Mattison and Hubbard 1969; Cornell et al. 1979). Pneumonia was not found in the oil-contaminated sea otters. Although not previously reported, inhalation of volatile components of crude oil such as benzene might have damaged alveolar septa and caused the lesion, but other lesions, such as interstitial pneumonia, likely to result from inhalation of an irritant vapor, were not found. Another possibility is that small amounts of oil were aspirated and caused coughing and dyspnea that ruptured alveolar walls and forced air into the interstitium, yet the volume of oil that entered the lungs was small enough to be removed without development of pneumonia. It is likely that, by undetermined mechanisms, exposure to crude oil caused the sea otters to become dyspneic; the combination of dyspnea and anatomical predisposition probably caused the development of interstitial pulmonary emphysema in a high proportion of oil-contaminated sea otters. Pressure changes that occur during diving may have exacerbated the emphysema.

Gastric erosions were common in oil-contaminated sea otters that died in the centers and were also found in one-sixth of the uncontaminated sea otters that died in the centers. The most likely mechanisms by which oil-contaminated sea otters might develop gastric erosions are either severe stress or a direct effect of the oil on the gastric mucosa. Gastrointestinal erosion/ulceration has been reported in sea otters that died in captivity and in the wild and has been attributed to stress (Stullken and Kirkpatrick 1955; Mattison and Hubbard 1969). Erosions caused by ingestion of corrosive liquids are extensive, but the erosions in these sea otters were small, discrete, and largely confined to the stomach. Thus, the erosions were probably caused by stress. It is unclear whether gastric erosions in oil-contaminated sea otters that died in captivity developed because of stress of capture and captivity or because of stress associated with oil exposure. It is likely that all sources of stress contributed to development of gastric erosions.

⋈   ⋈   ⋈

Hepatic lipidosis (the accumulation of lipids within hepatocytes) was present frequently in oil-contaminated sea otters that died in the centers and also was found in one-sixth of the uncontaminated sea otters that died in a center. Renal lipidosis (the accumulation of lipids within the tubular epithelial cells of the kidney) was somewhat less common and was found only in otters that also had hepatic lipidosis. Potential causes of hepatic and renal lipidosis include toxicity, mobilization of stored fat due to inadequate food consumption, and hypoxia. Hepatic lipidosis caused by hypoxia is primarily centrilobular (Kelly 1985), but the lipidosis in these sea otters was predominantly periportal, indicating that hypoxia is unlikely to be the cause. Experimentally oil-contaminated sea otters had marked increases in activity and metabolic rate with unchanged or decreased time devoted to feeding (Costa and Kooyman 1982; Siniff et al. 1982). Thus, mobilization of stored fat is likely to occur in oil-contaminated sea otters. Anorexia and elevated hepatocellular leakage-associated serum transaminases were common in oil-contaminated sea otters that died in the centers. Some otters with elevated transaminases had hepatic lipidosis. The hepatocellular leakage may have resulted from accumulation of lipids in hepatocytes because of fat mobilization. A direct or metabolite-associated toxic effect may also have caused the hepatorenal lipidosis. Hepatic lipidosis has been reported in rats (Bogo et al. 1982), mice (Gaworski et al. 1982), cattle (Winkler and Gibbons 1973), sheep (Adler et al. 1992), and a ringed seal (Smith and Geraci 1975) exposed to petroleum hydrocarbons, but mechanisms were not determined. Lipidosis of renal tubular epithelium of undetermined cause has been reported in hydrocarbon-exposed rats (Bogo et al. 1982). In the oiled sea otters, no renal lesions were found that would be expected to cause azotemia, and there was no evidence of urinary tract obstruction. Therefore, the azotemia is considered pre-renal. Centrilobular hepatic necrosis (death of hepatocytes that surround central veins of the liver) was relatively common in oil-contaminated sea otters that died in rehabilitation centers and was not found in uncontaminated sea otters that died in the centers. Potential causes of centrilobular hepatic necrosis include toxins and conditions that cause hepatic ischemia such as anemia, heart failure, and shock. Other lesions that would support heart failure were not found. Shock was a common syndrome in oil-contaminated otters that died in rehabilitation centers. Anemia was also common. Crude oil ingestion (Leighton 1986) and gastric erosion with hemorrhage are possible causes of anemia, but gastric erosions and centrilobular hepatic necrosis rarely occurred in the same otters, so anemia due to gastric hemorrhage was not a common cause of centrilobular necrosis. It is likely that shock, and in some cases anemia, contributed to centrilobular hepatic necrosis. Toxin-induced hepatic necrosis is also possible. Centrilobular hepatic necrosis of undetermined cause was found by researchers who gave crude oil orally to birds (Leighton 1986). Hepatic necrosis probably contributed to the increases in hepa-

≫ ≫ ≫

tocellular leakage transaminases observed in some otters in the rehabilitation centers.

The histopathologic studies failed to identify the cause of the inflammatory stimulus responsible for the degenerative left shift identified in the hematology results.

SEA OTTERS THAT DIED IN THE WILD

Histopathologic Studies

Tissues from five oil-contaminated sea otters that were found dead in oil-spill-affected areas were examined histologically. One had interstitial pulmonary emphysema and hepatorenal lipidosis. Two others had hepatorenal lipidosis. The remaining two otters had no significant histologic lesions and no necropsy reports were available. The presence of interstitial pulmonary emphysema and hepatorenal lipidosis in noncaptive oil-contaminated sea otters and the absence of these lesions in normal controls suggest that these lesions were caused by oil exposure rather than captivity.

Necropsy Studies

Following the EVOS, sea otter carcasses were collected from oil-spill-affected areas, placed in plastic bags, and frozen. In summer 1990, carcasses were thawed and complete necropsies were performed. Specimens were collected for toxicologic analysis in accordance with an established protocol. Results of toxicologic testing are reported elsewhere in this volume (Mulcahy and Ballachey, Chapter 18). Histologic examinations were not performed because of the artifacts produced by freezing and thawing of tissues.

A total of 214 carcasses examined were judged adequately preserved; too few pups were available to be included. Of the 214 carcasses, 152 were externally oil contaminated and 62 had no detectable external oil. Among oil-contaminated otters, 100/152 (66%) had interstitial pulmonary emphysema and 83/152 (55%) had gastric erosion and hemorrhage; 64/152 (42%) had both of these lesions. In the uncontaminated group, 13/62 (21%) had interstitial pulmonary emphysema and 4/62 (6.5%) had gastric erosion and hemorrhage; all four that had gastric erosion and hemorrhage also had interstitial pulmonary emphysema. Among uncontaminated sea otters, two had lesions that contributed to their deaths; one had vegetative valvular endocarditis and the other had a gunshot wound in the thorax. Lungs of both of these otters had interstitial emphysema. A variety of incidental lesions were found in many of the otters.

Emphysematous lungs were characterized by generally diffuse expansion of interlobular septa by clear, round to oblong, gas-filled bubbles that ranged up

⋙   ⋙   ⋙

to 5 cm in diameter (Fig. 16-6). Pulmonary parenchyma adjacent to affected septa was compressed. Occasionally, the emphysema extended into mediastinum, muscle fascia, and subcutaneous tissues of the neck and back.

Stomachs with erosions contained small to abundant amounts of dark red to black blood (Fig. 16-7). Numbers of erosions varied from a few to about 50. The pylorus was affected most frequently, but in many cases all regions of the gastric mucosa were affected. The amount of hemorrhage in the lumen generally correlated with the number of erosions. Rarely, one or two erosions were present in the duodenal mucosa. The erosions were punctate, round to oval mucosal defects with bright to dark red bases. The diameter ranged from 2 mm to 1 cm and was usually from 2 to 4 mm. Other lesions found microscopically in oil-contaminated sea otters, such as hepatic and renal lipidosis and centrilobular hepatic necrosis, could not be conclusively identified by gross examination alone.

This study confirms the association of interstitial pulmonary emphysema with exposure to crude oil in sea otters. The incidence of emphysema was threefold higher in oil-contaminated versus uncontaminated otters. There are several possible explanations for the presence of emphysema in some of the uncontaminated otters. These otters may have been lightly oil contaminated but were able to remove the oil prior to death. They may have breathed volatile components of crude oil while not coming in contact with liquid crude oil, or their interstitial pulmonary emphysema may have had an unrecognized cause other than crude oil exposure.

The high incidence of gastric erosion and hemorrhage in oil-contaminated sea otters is particularly interesting. This lesion was commonly identified histologically in oil-contaminated sea otters that died in rehabilitation centers, but it was unclear whether stress from oil exposure or from capture and captivity was the cause. This study indicates that gastric erosion and hemorrhage are associated with exposure to crude oil in the absence of capture and captivity.

Data from the necropsies were studied in an attempt to determine cause of death. Sea otters that were oil contaminated, had one or both lesions associated with crude oil exposure (interstitial pulmonary emphysema or gastric erosion and hemorrhage), and that did not have lesions indicative of another possible cause of death were considered to have strong evidence of death caused by oil exposure. Of the 214 sea otters examined, 123 (57%) fit these criteria. Those otters that were oil contaminated and had neither lesions associated with crude oil exposure nor lesions indicative of another possible cause of death were considered to have evidence of death caused by oil exposure. Sea otters that conformed to these criteria comprised 29/214 (14%). Carcasses that did not have detectable external oil, did not have lesions associated with oil exposure, and did not have lesions indicative of another possible cause of death were considered to have an undetermined cause of death. This group consisted of 49/214 (23%). Carcasses that had neither detectable external oil nor lesions indicative of another possible cause of death, but that did

⚬⚬⚬   ⚬⚬⚬   ⚬⚬⚬

Figure 16-6. Lung; interstitial emphysema. Interlobular septa are expanded by trapped air (arrows). H=heart and L=liver. (Photograph courtesy of Dr. Terrie M. Williams.)

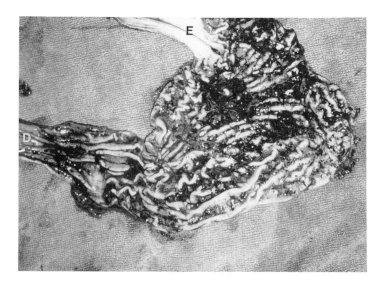

Figure 16-7. Opened stomach; gastric hemorrhage. Mucosal surface is covered by dark, partially digested blood. E=esophagus and D=duodenum.

⋈   ⋈   ⋈

have emphysema and/or gastric erosions included 11/214 (5%) and were also considered to have an undetermined cause of death. The uncontaminated sea otter with the gunshot wound in the thorax and emphysema and the uncontaminated sea otter with vegetative valvular endocarditis and emphysema comprised 2/214 (1%) and were considered to have died primarily because of conditions unrelated to oil exposure.

CONCLUSIONS

In spite of significant limitations, these studies represent the largest and most detailed investigation of the pathological effects of an oil spill on a marine mammal species. The findings support some long-held assumptions and bring to light much new information.

Because they lack a thick layer of subcutaneous fat, sea otters rely on their pelage for protection from the cold water. Thus, it was suspected that sea otters would be highly vulnerable to oil spills because contact with crude oil would dramatically decrease the insulating properties of their fur, resulting in hypothermia and death (Williams et al. 1988). Indeed, the oil spill had a devastating effect on the sea otters of Prince William Sound, and hypothermia was a major problem in sea otters presented to rehabilitation centers (Williams et al. 1990). Death caused by hypothermia can occur without distinctive gross or histologic lesions.

Clinical, hematologic/blood chemistry, and postmortem findings, combined with previous research, suggest the following scenario. Oil-contaminated sea otters rapidly become hypothermic. They devote themselves to a life-or-death struggle to remove the oil by grooming. Feeding is drastically curtailed, and energy stores are rapidly depleted. Grooming is marginally effective at best and results in ingestion of crude oil. By unknown mechanisms, exposure to the oil causes interstitial pulmonary emphysema which compromises respiration. Their desperate situation causes a powerful stress reaction. Gastric erosions form as the physiologic effects of stress reach a critical level. Hemorrhage into the gut begins. The combined effects of these factors overwhelm the otters; shock ensues, followed by death. Some sea otters succumb to hypothermia rapidly, and no lesions form. Others live long enough to develop some or all of the morphological markers that characterize this syndrome: interstitial pulmonary emphysema, gastric erosion and hemorrhage, hepatic and renal lipidosis, and centrilobular hepatic necrosis. Otters that are captured and taken to rehabilitation centers are subjected to additional stressors but are given medical and supportive care.

⋙　⋙　⋙

REFERENCES

Adler, R., H. J. Boermans, J. E. Moulton, and D. A. Moore. 1992. Toxicosis in sheep following ingestion of natural gas condensate. Veterinary Pathology 29:11–20.

Bayha, K., and K. Hill. 1990. Overall capture strategy. Pages 41–60, *in* Sea Otter Symposium: Proceedings of a symposium to evaluate the response effort on behalf of sea otters after the T/V *Exxon Valdez* oil spill into Prince William Sound, Anchorage, Alaska, 17–19 April 1990, U.S. Fish and Wildlife Service Biological Report 90 (12).

Bogo, V., R. W. Young, T. A. Hill, C. L. Feser, J. Nold, G. A. Parker, and R. M. Cartledge. 1982. The toxicity of petroleum JP5. Pages 46–66, *in* Proceedings of the symposium, the toxicology of petroleum hydrocarbons, The American Petroleum Institute, Washington, D.C.

Cornell, L. H., K. G. Osborn, J. E. Antrim, and J. G. Simpson. 1979. Coccidioidomycosis in a California sea otter (*Enhydra lutris*). Journal of Wildlife Diseases 15: 373–378.

Costa, D. P., and G. L. Kooyman. 1982. Oxygen consumption, thermoregulation, and the effect of fur oiling and washing on the sea otter, *Enhydra lutris*. Canadian Journal of Zoology 60:2761–2767.

Cotran, R. S., V. Kumar, and S. L. Robbins. 1989. The respiratory system. Pages 771–772, *in* Robbins Pathologic Basis of Disease, 4th ed., W. B. Saunders, Philadelphia, Pennsylvania.

Davis, R. W. 1990. Facilities and organization. Pages 3–58, *in* T. M. Williams and R. W. Davis (eds.), Sea Otter Rehabilitation Program, 1989 *Exxon Valdez* oil spill. International Wildlife Research.

Doroff, A., A. R. DeGange, C. Lensink, B. E. Ballachey, J. L. Bodkin, and D. Bruden. 1993. Recovery of sea otter carcasses following the *Exxon Valdez* oil spill. Pages 285–288, *in Exxon Valdez* oil spill symposium, 2-5 February 1993, Anchorage, Alaska. (Available, Oil Spill Public Information Center, 646 G Street, Anchorage, Alaska 99501.)

Duncan, J. R., and K. W. Prasse. 1986. Reference values. Pages 227–234, *in* Veterinary Laboratory Medicine, Clinical Pathology, 2nd ed, Iowa State University Press, Ames, Iowa.

Dungworth, D. L. 1985. The respiratory system. Pages 443–447, *in* K. V. F. Jubb, P. C. Kennedy, and N. Palmer (eds.), Pathology of Domestic Animals, Academic Press, Orlando, Florida.

Eade, N. R. 1974. Hydrocarbon pneumonitis. Pediatrics 54:351–356.

Gaworski, C. L., J. D. MacEwen, E. H. Vernot, R. H. Bruner, and M. J. Cowan. 1982. Comparison of the subchronic inhalation toxicity of petroleum and oil shale JP-5 jet fuels. Pages 67–75, *in* Proceedings of the symposium, the toxicology of petroleum hydrocarbons, The American Petroleum Institute, Washington, D.C.

Kelly, W. R. 1985. The liver and biliary system. Pages 253–255, *in* K. V. F. Jubb, P. C. Kennedy, and N. Palmer (eds.), Pathology of domestic animals, Academic Press, Orlando, Florida.

Leighton, F. A. 1986. Clinical, gross, and histologic findings in herring gulls and Atlantic puffins that ingested Prudhoe Bay crude oil. Veterinary Pathology 23:255–263.

Lipscomb, T. P., R. K. Harris, R. B. Moeller, J. M. Pletcher, R. J. Haebler, and B. E. Ballachey. 1993. Histopathologic lesions in sea otters exposed to crude oil. Veterinary Pathology 30:1–11.

Mattison, J. A., and R. C. Hubbard. 1969. Autopsy findings on thirteen sea otters (*Enhydra lutris nereis*) with correlations with captive animal feeding and behavior. Pages 99–101, *in* Proceedings of the 6th annual conference on sonar and diving mammals, Stanford Research Institute, Menlo Park, California.

Richardson, J. A., and H. R. Pratt-Thomas. 1951. Toxic effects of varying doses of kerosine administered by different routes. American Journal of Medical Science 221:531.

Rowe, L. D., J. W. Dollahite, and B. J. Camp. 1951. Toxicity of two crude oils and of kerosine to cattle. Journal of the American Veterinary Medical Association 162:61–66.

Siniff, D. B., T. D. Williams, A. M. Johnson, and D. L. Garshelis. 1982. Experiments on the response of sea otters *Enhydra lutris* to oil contamination. Biology and Conservation 23:261–271.

Smith, T. G., and J. R. Geraci. 1975. The effect of contact and ingestion of crude oil on ringed seals of the Beaufort Sea. Pages 1–67, *in* Beaufort Sea project, Technical Report No. 5, Institute of Ocean Science, Sidney, British Columbia, Canada.

Stullken, D. E., and C. M. Kirkpatrick. 1955. Physiological investigation of captivity mortality in sea otters (*Enhydra lutris*). Pages 476–494, *in* Transactions of the Twentieth North American Wildlife Conference, Wildlife Management Institute, Washington, D.C.

Williams, T. M., R. A. Kastelein, R. W. Davis, and J. A. Thomas. 1988. The effects of oil contamination and cleaning on sea otters (*Enhydra lutris*). I. Thermoregulatory implications based on pelt studies. Canadian Journal of Zoology 66:2776–2781.

Williams, T. M., R. Wilson, P. Tuomi, and L. Hunter. 1990. Critical care and toxicologic evaluation of sea otters exposed to crude oil. Pages 82–100, *in* T. M. Williams and R. W. Davis (eds.), Sea otter rehabilitation program, 1989 *Exxon Valdez* oil spill, International Wildlife Research, Galveston, Texas. 201 p.

Winkler, J. K., and W. J. Gibbons. 1973. Petroleum poisoning in cattle. Modern Veterinary Practice (Nov):45–46.

Chapter 17

Gross Necropsy and Histopathological Lesions Found in Harbor Seals

Terry R. Spraker, Lloyd F. Lowry, and Kathryn J. Frost

INTRODUCTION

When the T/V *Exxon Valdez* grounded in Prince William Sound (PWS) and spilled approximately 11 million gallons of crude oil, substantial oiling of harbor seals (*Phoca vitulina richardsi*) and their habitat resulted. In the early weeks of the spill, harbor seals were exposed to oil in the water and on land. They swam through oil slicks and breathed aromatic hydrocarbons at the air/water interface. At harbor seal haulout sites in oiled areas, the seals crawled through and rested on oiled rocks and algae throughout the spring and summer. Pups were born on haulout sites in May and June when some sites still had oil on them resulting in extensive external oiling of pups (Lowry et al., Chapter 12). Toxicological analyses of tissues confirmed that many seals had been exposed to and had assimilated toxic hydrocarbon contaminants (Frost et al., Chapter 19).

Following the *Exxon Valdez* oil spill (EVOS), oiled seals were observed to be lethargic and were described as being sick or unusually tame (Lowry et al., Chapter 12). Excessive lacrimation, squinting, and disorientation were also observed in oiled seals.

As part of the Natural Resources Damage Assessment program, our study was designed to investigate and quantify the effects of oil and disturbance associated with cleanup on the distribution, abundance, and health of harbor seals. One objective was to document pathological conditions that may have occurred in oiled seals. This chapter describes gross necropsies and histological lesions found in harbor seals following the EVOS.

281

METHODS

Wildlife rescue personnel, beach clean-up crews, and harbor seal researchers located and collected dead and debilitated harbor seals. Whenever possible, the date found, collection location, sex, and relative age were noted. The degree of external oiling was recorded as heavy (a thick coating of oil over the entire body), moderate (some oil on most of the body), light (a thin layer of oil on some part of the body), or unoiled (no oil visible). Carcasses were kept as cool as circumstances permitted. The time between death and sampling usually was not known, but probably ranged from 1 to several days.

To evaluate pathological changes associated with the spilled oil, 27 free-ranging harbor seals were collected by researchers and necropsied. Animals were killed by shooting them in the neck with a high-powered rifle. Seals were collected on six field trips: April 1989—PWS; June 1989—PWS; July 1989—Gulf of Alaska (GOA); October/November 1989—PWS and GOA; April 1990—PWS; and August 1990—George Inlet, Ketchikan (southeastern Alaska). Collected animals were weighed, measured, and photographed; time, date, location, and circumstances of collection were noted; and all gross abnormalities were recorded.

Tissue samples were collected from shot animals as soon after death as possible; the postmortem interval ranged from 10 to 30 minutes. Blood samples were taken from the extradural intervertebral vein and serum, plasma, and whole blood were frozen for later analysis. Multiple tissues from organ systems were taken for histopathology (nervous, cardiovascular, respiratory, lymphohematopoietic, urogenital, integument, digestive, endocrine, musculoskeletal and special senses (Appendix 17-A). Tissue samples from each seal were preserved in 10–15 liters of 10% neutral buffered formalin, with the fixative exchanged three times over a 4- to 5-day period. Tissues were transported to the Colorado State Diagnostic Laboratory, Fort Collins, Colorado, for processing. Tissues were imbedded in paraffin and sectioned at 5–6 μ. Most tissue sections were stained with hematoxylin and eosin. Brain, spinal cord, gasserian ganglion, and major nerves of vibrissae were stained with Bodian's nerve fiber stain and cresyl violet/luxol fast blue. Heart muscle was stained with Mallory's phosphotungstic acid, hematoxylin, and Masson's trichrome stain. Approximately 75–115 slides were examined from each collected seal, including multiple tissues from each organ system.

Due to degradation that occurred prior to sampling and fixation, relatively few tissues from animals that were found dead or that died in captivity were in a condition suitable for histopathology. Wherever possible, tissues were collected, processed, and examined using the procedures described above.

⊃○⊂   ⊃○⊂   ⊃○⊂

RESULTS

Seals Found Dead

Nineteen seals were obtained from PWS and the adjacent GOA and necropsied between early April and early July 1989 (Table 17-1). Fifteen were found dead, three died in captivity, and one was shot by a native hunter. Of the 19 seal carcasses, 9 were heavily oiled, 6 were lightly or moderately oiled, and 4 were unoiled. Thirteen were pups, including two oiled pups that were captured alive in early May and died after approximately 1 month in a rehabilitation facility, four dead premature pups found during April, and seven dead pups found during the normal pupping period (mid-May through early July). Two of the last group were unoiled, one was lightly oiled, and four were heavily oiled. Not all recovered carcasses were suitable for complete necropsies and histopathology. They were either scavenged, with major parts of the body and internal organs missing, or decomposed. Results of the examination of these carcasses are summarized in Table 17-2.

The seal from the native hunter (MH-HS-4) was unoiled and appeared unremarkable. Two other seals (KP-HS-1 and MH-HS-8) had probably died due to traumatic impact. Both had fractured ribs and extensive damage to various organs; they were lightly to moderately oiled. One heavily oiled adult female (MH-HS-6) was captured alive by a wildlife rescue crew and died on the way to a rehabilitation facility. This seal had severe pyometra and peritonitis that were probably associated with an *in utero* infection and abortion. It also had acute suppurative pneumonia and bilateral conjunctivitis. Another heavily oiled adult seal (MH-HS-7) had conjunctivitis and mild intramyelinic edema in the brain. A lightly oiled subadult seal (GA-HS-1) had internal hemorrhage.

Two pups (AF-HS-2 and MH-HS-10) may have been stillborn. One (AF-HS-2) showed no abnormalities, and the other (MH-HS-10) had free blood in the body cavity. Two pups that apparently died shortly after birth (MH-HS-2 and 3) appeared normal on examination. Two other dead pups (MH-HS-5 and 14) were severely emaciated perhaps due to malnutrition; one (LL-HS-1) had mild hepatitis and encephalitis; one (MH-HS-9) had gastrointestinal hemorrhage and possible nerve damage; and one (MH-HS-11) had been too badly scavenged for detailed examination. One heavily oiled pup (MH-HS-15) had severe purulent dermatitis and may have succumbed to a bacteremia; however, bacteria from this animal were not cultured.

Two of the oiled pups had been brought to rehabilitation centers, cleaned, and then held for approximately 4 weeks. One (MH-HS-12) had severe dermatitis, congested lungs, and hemorrhage in the stomach and small intestine. The cause of death may have been stress and septic shock. The other (MH-HS-13) was emaciated and had congestion and hemorrhage of the small intestine. The likely cause of death was hypotensive shock associated with emaciation and debilitation.

∢ ∢ ∢

Table 17-1. Harbor seals that were found dead in Prince William Sound (PWS) and the Gulf of Alaska (GOA), or that died in captivity, during *Exxon Valdez* oil spill response and damage assessment.

Specimen number	Date found	Location	Degree of oiling	Comments
no number	9 April 1989	Eleanor Island, PWS	heavy	premature pup in lanugo
AF-HS-2	16 May 1989	Herring Bay, PWS	unoiled	pup
GA-HS-1	25 June 1989	Dutch Group, PWS	light	subadult
KP-HS-1	20 May 1989	Raspberry Cape, GOA	light	subadult
LL-HS-1	15 May 1989	Herring Bay, PWS	light	pup
MH-HS-2	12 April 1989	Eleanor Island, PWS	moderate	premature pup
MH-HS-3	19 April 1989	Green Island, PWS	unoiled	premature pup
MH-HS-4	20 April 1989	Tatitlek Narrows, PWS	unoiled	subsistence kill, juvenile
MH-HS-5	21 April 1989	Applegate Rocks, PWS	heavy	premature pup, scavenged
MH-HS-6	1 May 1989	Herring Bay, PWS	heavy	captured alive and died, adult
MH-HS-7	28 April 1989	Windy Bay, GOA	heavy	predated or scavenged, adult
MH-HS-8	11 May 1989	Axel Lind Island, PWS	light	adult
MH-HS-9	25 May 1989	Drier Bay, GOA	unoiled	pup, scavenged
MH-HS-10	30 May 1989	Herring Bay, PWS	heavy	pup
MH-HS-11	30 May 1989	Herring Bay, PWS	heavy	pup, scavenged
MH-HS-12	2 May 1989	Herring Bay, PWS	moderate	in lanugo when caught, rehabilitated pup, died 31 May in captivity
MH-HS-13	3 May 1989	PWS	heavy	rehabilitated pup, died 31 May in captivity
MH-HS-14	22 June 1989	Chugach Bay, GOA	heavy	pup, badly autolyzed
MH-HS-15	9 July 1989	Herring Bay, PWS	heavy	pup

Table 17-2. Results of examinations of harbor seals that were found dead or that died in captivity during *Exxon Valdez* oil spill response and damage assessment.

Specimen number	Necropsy and histopathology results	Comments
no number	not examined	aborted
AF-HS-2	lungs not inflated, no lesions found	probably stillborn
GA-HS-1	hemorrhage in mesenteries, intestine, and trachea	autolysis
KP-HS-1	fractured ribs, ruptured organs, possible nerve damage	died due to blunt trauma
LL-HS-1	hepatitis and encephalitis	died shortly after birth
MH-HS-2	no significant lesions found	died shortly after birth
MH-HS-3	no significant lesions found	died shortly after birth
MH-HS-4	no significant lesions found	subsistence kill
MH-HS-5	inspissation of bile, hepatic atrophy	may have died due to malnutrition
MH-HS-6	severe pneumonia, chronic pyometra, peritonitis, conjunctivitis	had aborted or resorbed a fetus
MH-HS-7	conjunctivitis, nerve damage	samples moderately autolyzed
MH-HS-8	fractured ribs, ruptured organs, mild pneumonia, and hepatitis	died due to blunt trauma
MH-HS-9	hemorrhagic gastroenteritis, possible nerve damage	autolysis
MH-HS-10	peritonitis, hemorrhagic kidneys, blood in body cavity	probably stillborn
MH-HS-11	none	organs scavenged
MH-HS-12	severe dermatitis, hemorrhage in lungs and small intestine, mild nerve damage	died due to stress and septic shock
MH-HS-13	emaciated, hemorrhage in small intestine, possible nerve damage	died due to emaciation and shock
MH-HS-14	depletion of lymphoid and adipose tissue	samples autolyzed, possibly malnourished
MH-HS-15	severe dermatitis and septicemia	probably due to bacterial infection

In summary, hemorrhage of internal organs, sometimes with free blood in the body cavity, was found in four seals; severe dermatitis in two; conjunctivitis in two; and emaciation in three. In three seals, histopathologic examination suggested the presence of nerve damage in the brain, including intramyelinic edema and neuronal degeneration; these lesions were difficult to confirm due to the degree of autolysis. In four seals, no significant lesions were found.

Seals Collected

Of the 27 animals collected, 14 were females and 13 were males (Table 17-3). Seventeen were adults, seven were subadults, and three were pups of the year. Eleven were collected from oiled haulout areas in PWS during April–June 1989; six were collected from adjacent areas in the GOA in June–July 1989; two were collected from PWS and the GOA in October-November 1989; and six were collected from PWS in April 1990. Two "control" animals were collected over 1000 km south of the spill area, near Ketchikan, Alaska, in August 1990.

Gross Pathology of Collected Seals by Body System

INTEGUMENT: TS-HS-1 was heavily oiled. The oil on its pelage was heavy, sticky, and quickly stained the gloves of the examiner. The pelage of seals TS-HS-2, 3, 4, 5, 6, 7, 8, 9, 10, 11, 14, and 17 was also oiled but the texture was dry and less tacky. Oil did not rub off the pelage of these seals as easily as from TS-HS-1. Seals TS-HS-2, 3, 4, 5, 6, 7, 8, 9, 10, 11, and 17 had oil on their heads and trunks, whereas TS-HS-14 had oil only on its trunk. Pectoral blubber thickness varied from 1.3 to 4.3 cm in all animals except TS-HS-1 which had a blubber thickness of 6.5 cm. TS-HS-13 had several small focal areas of alopecia, probably due to the beginning of molt. Three animals (TS-HS-12, 15, and 19) had scars and lacerations caused by bite wounds.

CARDIOVASCULAR SYSTEM: The heart was normal in size and shape in all 27 animals. All animals had a small amount of adipose tissue around the coronary vessels both at the base of the heart and between the ventricles. Gross evidence of anemia was not found. Heartworms (*Dipetalonema spirocauda*) were found in seven animals (TS-HS-10, 12, 13, 18, 20, 25, and 27).

RESPIRATORY SYSTEM: A mild infection of nasal mites (*Halarachne miroungae*) occurred in 15 animals. Mites were not observed in one pup (TS-HS-6). The nasal turbinates were not examined in three animals. Small white foci were found both under and within the parenchyma of the lungs in seven animals (TS-HS-2, 12, 16, 17, 19, 20, and 21). One animal (TS-HS-17) had several small hyperplastic lymphoid nodules on the laryngeal folds.

Table 17-3. Harbor seals collected in Prince William Sound (PWS) and the Gulf of Alaska (GOA) during *Exxon Valdez* oil spill response and damage assessment.

Specimen number	Date	Location	Degree of oiling	Sex	Weight (kg)	Age
TS-HS-1[a]	29 April 1989	Herring Bay, PWS	very heavy	F	98	adult
TS-HS-2	16 June 1989	Bay of Isles, PWS	very heavy	M	190	adult
TS-HS-3[b]	16 June 1989	Seal Island, PWS	heavy	F	150	adult
TS-HS-4[b]	16 June 1989	Seal Island, PWS	heavy	F	25	pup
TS-HS-5	17 June 1989	Bay of Isles, PWS	very heavy	F	100	adult
TS-HS-6	17 June 1989	Applegate Rocks, PWS	light	F	25	pup
TS-HS-7	17 June 1989	Bay of Isles, PWS	very heavy	F	70	adult
TS-HS-8[c]	17 June 1989	Bay of Isles, PWS	very heavy	M	25	pup
TS-HS-9	18 June 1989	Herring Bay, PWS	very heavy	M	100	adult
TS-HS-10	18 June 1989	Herring Bay, PWS	very heavy	F	56	subadult
TS-HS-11	18 June 1989	Herring Bay, PWS	very heavy	F	55	subadult
TS-HS-12	25 June 1989	Afognak Island, GOA	unoiled	F	55	subadult
TS-HS-13	25 June 1989	Afognak Island, GOA	unoiled	F	59	subadult
TS-HS-14	29 June 1989	W Amatuli Island, GOA	moderate	M	95	adult
TS-HS-15	30 June 1989	Ushagat Island, GOA	unoiled	M	90	adult
TS-HS-16	30 June 1989	Ushagat Island, GOA	unoiled	F	80	adult
TS-HS-17	6 July 1989	Perl Island, GOA	light	F	65	adult
TS-HS-18	26 October 1989	Big Fort Island, GOA	unoiled	M	43	subadult
TS-HS-19	1 November 1989	Agnes Island, PWS	unoiled	M	87	adult
TS-HS-20	11 April 1990	Herring Bay, PWS	unoiled	M	55	subadult
TS-HS-21	12 April 1990	Herring Bay, PWS	unoiled	M	30	subadult
TS-HS-22	12 April 1990	Herring Bay, PWS	unoiled	M	110	adult
TS-HS-23[a]	12 April 1990	Eleanor Island, PWS	unoiled	F	79	adult
TS-HS-24	12 April 1990	Herring Bay, PWS	unoiled	M	75	adult

(Table continues)

Table 17-3. Continued

Specimen number	Date	Location	Degree of oiling	Sex	Weight (kg)	Age
TS-HS-25	13 April 1990	Bay of Isles, PWS	unoiled	M	55	adult
TS-HS-26	15 August 1990	Ketchikan	unoiled	F	60	adult
TS-HS-27	16 August 1990	Ketchikan	unoiled	M	65	adult

[a] pregnant
[b] pup of TS-HS-3
[c] pup of TS-HS-7

DIGESTIVE SYSTEM: One animal (TS-HS-24) had extremely worn teeth, several of which were missing. Three other animals (TS-HS-2, 9, and 22) had a mild to moderate degree of wear to the teeth. The remaining 23 had normal, relatively sharp teeth. Lesions were not found in the oral cavity or esophagus in any animal. Three pups (TS-HS-4, 6, and 8) did not have any gastrointestinal parasites. All animals 1 year and older had two species of nematodes on and within the mucosa of the stomach (*Anisakis* spp. and *Contracaecum osculatum*). These parasites were usually associated with ulcers within the mucosa that varied from 1.0 to 1.5 cm in diameter. Usually the ulcerated area was in the center of a firm raised nodule. Active ulcers contained parasites, whereas healed ulcers did not. Small firm nodules could be palpated within the submucosa of the stomachs of all animals with parasites. Ulcers were found in 18 of 24 animals over 1 year of age. Two of the pups (TS-HS-4 and 8) had milk in their stomachs, and three adults (TS-HS-11, 16, and 18) had remains of octopus or fish in their stomachs. Two seals (TS-HS-16 and 19) had yellow fibrous plaques within the serosa of the pylorus. These did not penetrate through the muscular wall of the stomach. Acanthocephala (*Corynosoma* spp.) were found in the small intestine of 21 of 24 animals over 1 year of age. Approximately 20–25% of the small intestine was opened and examined, therefore some of the intestinal parasites may have been missed. Small white foci were found under the capsule or within the parenchyma of the liver in 23 of 24 adult and juvenile seals and 1 of 3 pups. Pancreas and salivary tissue were normal in all animals.

LYMPHOHEMATOPOIETIC SYSTEM: Peripheral, mediastinal, and mesenteric lymph nodes were normal in all animals. The bone marrow of the ribs and vertebrae was red. The spleen was normal in size and shape in all animals and the thymus was of normal size considering the age in each animal. Tonsils were considered normal in all animals.

UROGENITAL SYSTEM: Kidneys, urinary bladder, and ureters were normal in all animals. One adult male (TS-HS-2) had lesions on the penis and prepuce. These lesions were characterized by multifocal raised, firm, proliferative plaques 5 mm in diameter. One male (TS-HS-25) had a fibrous cord of tissue attached to the glans and to the midregion of the dorsal aspect of the prepuce. The cord caused a dorsal deviation of the penis and would have interfered with breeding. Testes and prostate glands were normal in all male animals. The uterus, ovaries, and cervix were considered normal in all females except one. This animal (TS-HS-12) had a small 1.5- to 2.0-cm red-brown, amorphous mass in the left horn of the uterus. The remainder of the reproductive tract of this animal was normal. Two of the females were pregnant (TS-HS-1 and 23) and each contained a single normal fetus.

✕✕✕ ✕✕✕ ✕✕✕

MUSCULOSKELETAL SYSTEM: Muscle mass, color, and texture were normal in all animals. Trunk muscles were a dark black, whereas neck and head muscles were a lighter gray. Ribs and femurs from all animals had a normal snap when broken.

ENDOCRINE SYSTEM: The pituitary gland was normal in all 25 animals in which it could be examined. Thyroid and adrenal glands were normal in size and shape in all animals.

NERVOUS SYSTEM: The brain was examined in 25 of 27 animals collected. Subdural hemorrhages associated with gunshot wounds were present in four animals (TS-HS-8, 11, 15 and 18). Spinal cord, brachial plexus, and peripheral nerves were examined in all animals and were considered normal.

SPECIAL SENSES: The cornea of TS-HS-1 appeared to be slightly opaque. Mildly reddened and injected conjunctiva were observed in six animals (TS-HS-1, 3, 9, 10, 14, and 19). A small 3- to 4-mm white circular mass (ball of organized fibrin) was found in the anterior chamber of the right eye of TS-HS-12. A 1-cm abscess containing a light brown semisolid material was present under the dorsal conjunctiva of the right eye in TS-HS-19. The ears and ear canals appeared normal in most animals.

Histopathology of Collected Seals by Body System

INTEGUMENT: Ten animals (TS-HS-2, 3, 4, 5, 7, 8, 10, 11, 17, and 18) had mild acanthosis and orthokeratotic hyperkeratosis of the epidermis (Figs. 17-1 and 17-2). This was likely due to a surface irritant; all of these animals except TS-HS-18 were oiled. Three seals (TS-HS-1, 9, and 27) had mild ulceration of the skin. The cause of these small ulcers was not determined. A larval nematode (probably *Peloderma* sp.) was found in hair follicles of the skin in the ventral trunk of eight animals (TS-HS-1, 3, 5, 9, 19, 22, 24, and 26). These infected hair follicles usually were not surrounded by any type of inflammatory reaction. Mites (*Demodex* sp.) were found in four animals (TS-HS-3, 23, 24, and 25). *Demodex* sp. were most common within the necks of sebaceous glands within the follicles of vibrissae and adjacent hair follicles. Inflammation was not associated with mite infection.

SPECIAL SENSES: Sixteen animals had a mild to moderate lymphoplasmacytic conjunctivitis (TS-HS-1, 2, 5, 8, 9, 10, 11, 15, 16, 17, 18, 19, 20, 21, 22, and 23), which could have been a nonspecific response to irritation or mild infection. The cause of the small mass of fibrin in the anterior chamber of TS-HS-12 was probably due to a previous anterior uveitis, the cause of which was

❯◇ ❯◇ ❯◇

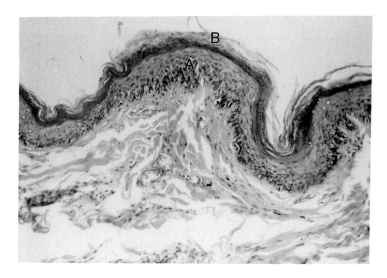

Figure 17-1. Photomicrograph of normal skin from the ventral trunk region of an unoiled harbor seal (TS-HS-26). Note the thickness of the cellular layer (A) and the cornified layer (B) of the epidermis. H&E 40X

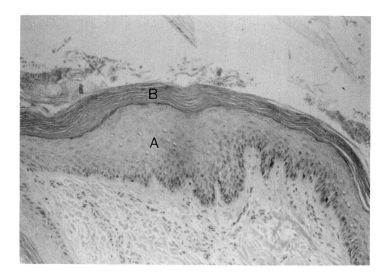

Figure 17-2. Photomicrograph of skin from the ventral trunk region of an oiled harbor seal (TS-HS-7). This section is characterized by mild acanthosis (A) and orthokeratotic hyperkeratosis (B) of the epidermis. Inflammation is not observed within the dermis. H&E 40X

not determined. The cause of the subconjunctival abscess in TS-HS-19 was also not determined. Lesions were not found within the posterior segment of the eye in any animal.

DIGESTIVE SYSTEM: The majority of the lesions found in the digestive system (gastritis, enteritis, colitis, multifocal granulomatous, or lymphoplasmacytic hepatitis) were associated with parasites and/or parasitic migration. The gross lesions described in the stomach and small multifocal white lesions within the liver corresponded to these parasitic lesions found during histopathological examination. These lesions are common in marine mammals. Two animals (TS-HS-24 and 27) had a coccidia similar in morphology to *Eimeria leuckarti* located within the lamina propria of the small intestine. The only lesions that may have been associated with crude oil toxicity included mild hepatocellular necrosis (TS-HS-1) and hepatocellular swelling with mild bile inspissation within canaliculi (TS-HS-1, 2, 3, and 8). These lesions were found in heavily oiled animals. This damage may have been associated with the exhaustion of the mixed function oxidase system during detoxification of toxins, such as the toxic components of crude oil. These hepatic lesions were mild and reversible and probably of little clinical significance.

RESPIRATORY SYSTEM: Mild to moderate lymphoplasmacytic rhinitis was found in 15 animals (TS-HS-2, 3, 5, 7, 9, 10, 13, 16, 17, 18, 20, 22, 24, 25, and 26). These lesions were probably associated with infection of nasal mites. The cause of the fibrous polyp in TS-HS-16 was probably due to a focal area of pharyngitis and was of little importance. Ten animals had a mild ulcerative tracheitis (TS-HS-1, 3, 6, 12, 14, 18, 23, 24, 25, and 26). This lesion and the mild lymphoplasmacytic laryngitis/pharyngitis (TS-HS-7, 17, and 20) were suggestive of an upper respiratory viral infection; however, intracellular inclusion bodies were not found within any of these lesions. These small lesions also may have been associated with nasal mites. Small foci of eosinophilic and pyogranulomatous pneumonia were found in 17 animals (TS-HS-1, 2, 3, 10, 11, 12, 13, 16, 17, 18, 19, 20, 21, 22, 24, 26, and 27). These lesions corresponded to the multifocal white lesions observed during the gross examination of the lungs. Lungworms were associated with these foci of pneumonia in all but two animals (TS-HS-24 and 26). The lungworms were compatible in size and shape with *Parafilaroides decorus*. Emphysema was not found in any of the animals.

UROGENITAL SYSTEM: Lesions were not found in any of the glomeruli, tubules, collecting ducts or juxtaglomerular apparatus of the kidneys in any of the animals. Two interesting lesions were found in males. One was multifocal, proliferative, squamous epithelial plaques on the penis and prepuce of TS-HS-2.

⋈ ⋈ ⋈

These hyperplastic squamous epithelial plaques contained cells with Cowdry type A intranuclear inclusion bodies typical of a herpes virus. Electron microscopy was inconclusive due to postmortem autolysis of the specimen. The second lesion (in TS-HS-25) was a congenital anomaly characterized by a thin (3 mm) fibrous band extending from the dorsal aspect of the anterior one-third of the penis to the prepuce, resulting in a dorsal deviation of the penis. One female (TS-HS-12) had a 1.5- to 2.0-cm amorphous mass within the uterus. The identity of this amorphous acellular mass was not determined but it may have been the remains of an embryo, an old intrauterine fibrin clot, or a structure similar to an amorphous acellular mass, called a hypomane, occasionally found within the uterus of horses. Considering the gestation period of harbor seals and the size of this mass, and if this mass was a dead embryo, the embryonic death would have occurred between January 1989 and early March 1989, before the EVOS.

MUSCULOSKELETAL SYSTEM: A few individual myocytes were necrotic in three animals (TS-HS-2, 20, and 21). The cause of the mild necrosis was not determined. Two animals (TS-HS-13 and 18) had mild eosinophilic myositis that was probably associated with parasitic migration and activity. These lesions are common in marine mammals. One animal had sarcocysts in skeletal muscle of the tongue (TS-HS-2).

CARDIOVASCULAR SYSTEM: An acute, mild, myocardial degeneration was found in one animal (TS-HS-10), the cause of which was not determined. Four animals (TS-HS-3, 14, 18, and 24) had mild lesions in the heart that were suggestive of parasitic activity. Ten animals had mild to moderate fibromuscular intimal proliferation of medium-sized arteries (TS-HS-4, 5, 8, 10, 13, 15, 17, 18, 20, and 25). Two of these (TS-HS-13 and 20) also had thrombosis of medium-sized arteries. These lesions were probably due to the activity of heartworms.

LYMPHOHEMATOPOIETIC SYSTEM: Tonsillitis was found in eight animals (TS-HS-3, 6, 8, 12, 16, 18, 22, and 23). This lesion is common in marine mammals and is probably associated with bacteria that are ingested. Ten animals had pyogranulomatous, eosinophilic adenitis of mesenteric lymph nodes (TS-HS-2, 3, 11, 12, 18, 19, 21, 22, 23, and 24). These lesions are due to parasitic migration and are common in marine mammals (Spraker, unpublished data).

ENDOCRINE SYSTEM: Three lesions were found in the endocrine organs. One animal had small multiple brachial cysts of the thyroid (TS-HS-3). These cysts are not uncommon in mammals. One animal had slightly hyperactive thyroid follicles (TS-HS-9) and one animal had a mild hyperplasia of the parathyroid glands (TS-HS-14). These three lesions were considered incidental.

⋙   ⋙   ⋙

NERVOUS SYSTEM: Two general categories of lesions were found in the nervous system, one that was probably associated with crude oil toxicity and the other not. The lesions considered not to be associated with oil toxicity included multifocal, mild, fibrous plaques within the meninges that occurred in eight animals (TS-HS-2, 5, 7, 9, 19, 23, 24, and 27). The cause of these lesions was not determined, but they have been observed in Steller sea lions (*Eumetopias jubatus*) and California sea lions (*Zalophus californianus*) (Spraker, unpublished data). The etiology and significance of this lesion are undetermined. Mineralized foci within the choroid plexus were found in three animals (TS-HS-7, 9, and 16). The cause of these lesions was not determined. Three animals had several small holes or microcavities within the brain stem (TS-HS-20, 22 and 24). This microcavitary degeneration was primarily located in the cuneate nucleus and trapezoid body. These foci were believed to represent neuronal degeneration and were extremely mild. This condition has been observed in adult Steller sea lions (Spraker, unpublished data). Eight animals had mild lymphocytic or lymphoplasmacytic cuffing of vessels (TS-HS-10, 13, 14, 15, 18, 19, 20, and 23), primarily in the obex and near or within the spinal tract of the trigeminal nerve. These lesions may have been associated with a latent viral infection such as a herpes virus. Sixteen animals had single to multiple small, round to oval, pale eosinophilic intracytoplasmic droplets in neurons, especially within the thalamus (TS-HS-2, 3, 7, 8, 9, 10, 11, 13, 16, 20, 21, 22, 23, 24, 25, and 27). These inclusions were found in intact and necrotic neurons. The etiology of these inclusions was undetermined, but did not appear to be paramyxovirus inclusions of phocine or canine distemper.

Lesions that were considered to be of greatest significance included intramyelinic edema of the large myelinated axons of the midbrain, neuronal swelling, neuronal necrosis, and axonal swelling and degeneration (Table 17-4; Appendix 17-B). Intramyelinic edema was found in seven animals (TS-HS-1, 2, 3, 7, 10, 11, and 17). It was severe in TS-HS-1 and was most prominent in the ventral caudal lateral and ventral caudal medial nuclei of the thalamus and within large myelinated fibers of the thalamus, corpus callosum, crus cerebri, and internal capsule (Fig. 17-3). Intramyelinic edema was mild in the other animals. The degree of neuronal swelling with loss of Nissl substance was also most severe in TS-HS-1. This lesion was primarily in the thalamus (Figs. 17-4 and 17-5). Mild neuronal swelling was found in TS-HS-2, 3, 5, 10, 11, 14, 16, 17, and 19. Neuronal necrosis was most evident in the ventral caudal lateral and ventral caudal medial nuclei of the thalamus (Fig. 17-6). These lesions were moderate in one animal (TS-HS-17) and mild in eight (TS-HS-1, 3, 5, 7, 10, 11, 14, and 16). Axonal swelling and degeneration may be associated with neuronal degeneration, secondary lesions following myelin damage, or primary lesions. Axonal swelling and

≫⊂ ≫⊂ ≫⊂

Table 17-4. Summary of lesions found in the brains of harbor seals collected in Prince William Sound, the Gulf of Alaska, and near Ketchikan, 1989-1990.

Specimen number	Degree of oiling	Intramyelinic edema	Neuronal swelling	Neuronal necrosis	Axonal swelling and degeneration
TS-HS-1	very heavy	+++	+++	+	+++
TS-HS-2	very heavy	+	+	-	-
TS-HS-3	heavy	+	+	+	-
TS-HS-5	very heavy	-	+	+	-
TS-HS-6	light	+	-	-	-
TS-HS-7	very heavy	+	-	+	-
TS-HS-8	very heavy	-	-	-	-
TS-HS-9	very heavy	-	-	-	-
TS-HS-10	very heavy	+	+	+	+
TS-HS-11	very heavy	+	+	+	+
TS-HS-13	unoiled	-	-	-	-
TS-HS-14	moderate	-	+	+	-
TS-HS-15	unoiled	-	-	-	-
TS-HS-16	unoiled	-	+	+	+
TS-HS-17	light	+	+	++	-
TS-HS-18	unoiled	-	-	-	-
TS-HS-19	unoiled	-	+	-	-
TS-HS-20	unoiled	-	-	-	-
TS-HS-21	unoiled	-	-	-	-
TS-HS-22	unoiled	-	-	-	-
TS-HS-23	unoiled	-	-	-	-
TS-HS-24	unoiled	-	-	-	-

(Table continues)

Table 17-4. Continued

Specimen number	Degree of oiling	Intramyelinic edema	Neuronal swelling	Neuronal necrosis	Axonal swelling and degeneration
TS-HS-25	unoiled	-	-	-	-
TS-HS-26	unoiled	-	-	-	-
TS-HS-27	unoiled	-	-	-	-

+ + + = severe; + + = moderate; + = mild; - = negative

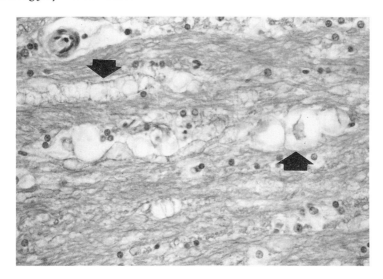

Figure 17-3. Photomicrograph of the ventral caudal lateral nucleus of the thalamus of an oiled harbor seal (TS-HS-1). Note the moderate degree of intramyelinic edema (swelling and separation of the myelin sheaths) and axonal degeneration (arrows). H&E 400X

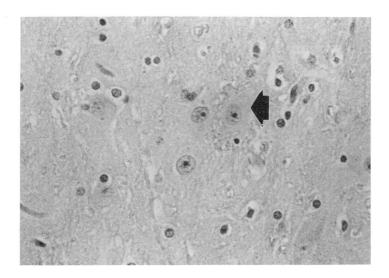

Figure 17-4. Photomicrograph of a well-fixed normal neuron (arrow) and neuropil within the ventral caudal lateral nucleus of the thalamus in an unoiled harbor seal (TS-HS-26). H&E 400X

degeneration were found within the thalamus, corpus callosum, internal capsule, and crus cerebri in three animals (Fig. 17-6). TS-HS-1 had severe axonal swelling and degeneration, whereas the other three animals had a mild degree of neuronal swelling and degeneration (TS-HS-10, 11 and 17).

DISCUSSION

Petroleum hydrocarbons may be taken into the bodies of seals through surface contact, ingestion, and inhalation (Engelhardt et al. 1977; Engelhardt 1987). Mammals are able to metabolize hydrocarbons through the production of mixed function oxidases that convert the hydrocarbons to metabolites that are excreted in urine and bile (Addison et al. 1986). The mechanisms used by seals for detoxification and excretion were described by Engelhardt et al. (1977). As they noted, at some hydrocarbon concentration, it is likely that the detoxifying and excretory mechanisms would cease to function, but the concentration at which that would occur is not known. Presence of hydrocarbon metabolites in the bile and of aromatic hydrocarbons in the blubber confirm that many of the seals examined in this study had been exposed to and assimilated crude oil (Frost et al., Chapter 19). This was especially true for the animals that were collected from oiled haulout areas within PWS during April–June 1989 and for some of the animals found dead.

Most field studies of the effects of oil on marine mammals have not included detailed pathological examinations, partly because it is difficult to obtain sufficiently fresh material from dead animals, and the results of examinations are usually equivocal. In spite of the substantial effort made to retrieve and sample animals found dead after the EVOS, we encountered similar problems in this study. For example, three seals showed signs of nerve damage but the diagnosis was not clear. Two of the seals found dead had died due to blunt trauma, but the source of the trauma was not determined. A reasonable hypothesis may be that nerve damage caused behavioral changes that made these animals more prone to accidents, such as being hit by a boat. Otherwise, while a variety of pathological conditions were found in the dead seals, the factors responsible for their deaths were not determined.

Thirteen of the carcasses found were pups, most of which either died shortly after birth or sometime later; their deaths were associated with emaciation and perhaps stress. Six of them were born well before the beginning of the normal pupping period in mid-May. Most were found in oiled areas, and six came from Herring Bay where all the seals were heavily oiled for several months after the spill (Lowry et al., Chapter 12). One adult female from Herring Bay that died in captivity had resorbed or aborted a fetus. Hoover-Miller (1989) reported a heavily oiled seal off the Kenai Peninsula that died while giving birth. These observations suggest that stress and/or toxic effects from the EVOS may have been associated with an

><> ><> ><>

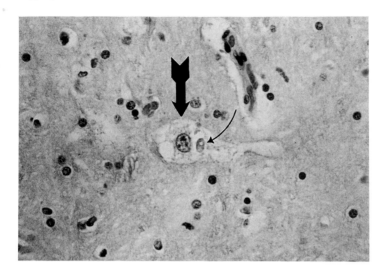

Figure 17-5. Photomicrograph of a large swollen degenerated neuron (bold arrow) containing an oval intracytoplasmic inclusion (thin arrow) surrounded by normal neuropil within the thalamus of an oiled harbor seal (TS-HS-1). H&E 400X

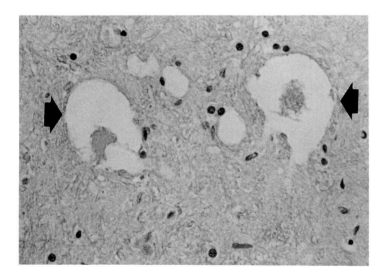

Figure 17-6. Photomicrograph of two large necrotic neurons (bold arrows) surrounded by a relatively normal neuropil within the ventral caudal lateral nucleus of the thalamus of an oiled harbor seal (TS-HS-17). H&E 400X

unusually high rate of abortions, premature births, and deaths of harbor seal pups and adults in heavily oiled areas. However, the specific nature of the causative factors were not identified.

The collected seals provided a better opportunity to document lesions that may have been associated with oil toxicity. However, there are major limitations due to the timing of sample collections. The first seal was available for examination and sampling 35 days after the spill. Most of the seals were not collected until June, almost 3 months after the *Exxon Valdez* grounded. Therefore, seals acutely affected by the more toxic volatile compounds in Prudhoe Bay crude oil may have died prior to our collections. The animals that we collected and examined were in relatively good condition compared to normal harbor seals (Pitcher and Calkins 1979; Frost and Lowry 1993). They were examples of animals that had survived the effects of the oil spill, at least in the short term. Pathological findings might have been much different if seals had been collected and properly sampled in the days immediately following the spill.

The lesions found in collected seals that were considered to be associated with oil toxicity are summarized in Table 17-5. Seventeen of the 27 animals had mild to moderate lymphoplastic conjunctivitis. The occurrence of moderate conjunctivitis was higher in seals that were oiled when collected than in unoiled seals (7 of 13 versus 3 of 14), but the difference was not statistically significant (χ^2=3.04, P>0.05). Ten animals had mild acanthosis and orthokeratotic hyperkeratosis of the epidermis. Of the 10 animals, all but one of the seals was oiled; the condition was milder in the unoiled seal. This difference in occurrence of epidermal lesions is statistically significant (9 of 13 versus 1 of 14, χ^2=11.14, P<0.01). Hepatocellular swelling and necrosis with mild bile inspissation within canaliculi occurred in the livers of four seals. All were heavily oiled animals collected in PWS in June 1989. Neurological lesions that may have been associated with oil toxicity were found in 9 of 12 oiled seals and 1 of 13 unoiled seals, a statistically significant difference (χ^2=11.78, P<0.01).

Other studies have shown that contact with oil can irritate or damage sensitive tissues, especially mucous membranes (St. Aubin 1990). Severe conjunctivitis and corneal abrasions occurred in the eyes of ringed seals (*Phoca hispida*) that were experimentally exposed to Norman Wells crude oil (Geraci and Smith 1976). The symptoms disappeared when the seals were put in clean water. Experimentally oiled polar bears (*Ursus maritimus*) showed signs of skin irritation (Oritsland et al. 1981).

Exposure to oil has been shown to affect liver function in seals, although most studies have been short term and the resulting damage minor and transient (Geraci and Smith 1976). The lesions in the livers of heavily oiled seals collected 3 months after the EVOS also suggested only minor and reversible damage. Standard blood chemistries run on sera from seals collected in 1989, 1991, and 1992 did not indicate

⋈   ⋈   ⋈

Table 17-5. Occurrence of pathology in tissues of harbor seals collected in Prince William Sound, the Gulf of Alaska, and near Ketchikan, 1989-1990.

Specimen number	Degree of oiling	Lymphoplastic conjunctivitis	Acanthosis and hyperkeratosis	Hepatocellular swelling and bile inspissation	Neuronal damage in brain
TS-HS-1	very heavy	++	-	++	+++
TS-HS-2	very heavy	++	+	+	+
TS-HS-3	heavy	-	+	+	+
TS-HS-4	heavy	-	+	-	NE
TS-HS-5	very heavy	++	+	-	+
TS-HS-6	light	-	-	-	-
TS-HS-7	very heavy	-	+	-	++
TS-HS-8	very heavy	++	+	+	-
TS-HS-9	very heavy	++	-	-	-
TS-HS-10	very heavy	+	+	-	+
TS-HS-11	very heavy	++	+	-	++
TS-HS-12	unoiled	+	-	-	NE
TS-HS-13	unoiled	-	-	-	-
TS-HS-14	moderate	-	-	-	++
TS-HS-15	unoiled	++	-	-	-
TS-HS-16	unoiled	+	-	-	++
TS-HS-17	light	++	+	-	++
TS-HS-18	unoiled	+	+	-	-
TS-HS-19	unoiled	+	-	-	+
TS-HS-20	unoiled	++	-	-	-
TS-HS-21	unoiled	+	-	-	-
TS-HS-22	unoiled	+	-	-	-

(Table continues)

Table 17-5. Continued

Specimen number	Degree of oiling	Lymphoplastic conjunctivitis	Acanthosis and hyperkeratosis	Hepatocellular swelling and bile inspissation	Neuronal damage in brain
TS-HS-23	unoiled	++	-	-	-
TS-HS-24	unoiled	-	-	-	-
TS-HS-25	unoiled	-	-	-	-
TS-HS-26	unoiled	-	-	-	-
TS-HS-27	unoiled	-	-	-	-

+ + + = severe; + + = moderate; + = mild; - = negative; NE = not examined

any significant abnormalities when compared to standard values for dogs, cats, and other seals (Frost and Lowry 1993). Tests for kidney and liver function showed no evidence of organ damage. Serum minerals and indicators of fat metabolism were also within normal limits.

The most significant histopathologic findings in collected seals were lesions in the midbrain. These occurred significantly more frequently in oiled seals and were not present in two control seals collected near Ketchikan. The lesions included intramyelinic edema, axonal degeneration and neuronal swelling, and necrosis. Intramyelinic edema, a sensitive and reversible indicator of neuronal damage, occurs when there is swelling within the lipid-rich myelin sheaths of the nerve axons. The swelling causes diffusion of the electrical impulses and reduces the ability of the axon to transmit neural signals. Neuronal swelling is also an extremely sensitive and acute, but reversible change caused by neurotoxins. Neuronal necrosis and dropout is a severe, nonreversible change. Axonal degeneration can be a primary lesion or can be secondary to intramyelinic edema or neuronal necrosis.

The thalamic nuclei (where the most severe lesions were located) relay impulses of the sensory system to the cerebrum. Damage to this area could result in failure of impulses to reach the cerebral or cerebellar cortex. The specific thalamic nuclei affected are primarily sensory to the head and body, with some influence on respiration. The thalamus may be the critical area for reception of some types of sensation, with the sensory cortex functioning to give the finer detail (Chusid 1982). Lesions that occurred in the ventral caudal lateral and ventral caudal medial nuclei of the thalamus would primarily alter peripheral proprioception. They could account for behavioral changes observed in oiled harbor seals such as decreased flight distance, disorientation, and increased tendency to haul out. These neuronal lesions, if severe, could have caused the affected seals to have extreme difficulty in performing normal tasks such as swimming, feeding, and diving. Ringed seals that were experimentally exposed to oil for 24 hours showed body quivering and uncontrolled body movements (Geraci and Smith 1976).

Effects of volatile petroleum hydrocarbons that have been documented in other mammals include abnormal nervous system function, anesthesia, respiratory failure, and death (Engelhardt 1987). The highly volatile C_5–C_8 hydrocarbons may cause central nervous system damage, axonal degeneration, and cerebral edema (Cornish 1980). The neuronal lesions that we found in harbor seals are somewhat comparable to those found in humans and rats exposed to low-molecular-weight aliphatic hydrocarbons. Axonal swelling and degeneration of large myelinated axons, with retraction of the myelin sheaths of peripheral nerves and long ascending and descending axons of spinal cord, cerebellar white matter, optic nerve, and mammillary bodies have been described in rats (Stolenburg-Didinger and Altenkirch 1988). Peripheral neuropathy has been described in humans following

⋙   ⋙   ⋙

n-hexane toxicity (Pleasure and Schotland 1984), but it was not observed to any significant degree in any of the seals that we examined.

It is likely that the primary impact of crude oil exposure on harbor seals was due to inhalation of short-chain aromatic hydrocarbons (Geraci and St. Aubin 1987). This hypothesis is supported because no lesions were found in the brain of a heavily oiled pup (TS-HS-8), but significant lesions were found in her mother (TS-HS-7; Table 17-4). Since most of the aromatic hydrocarbons would probably have dissipated by May when pups were born, pups should have had a much lower level of exposure to those compounds. If inhalation was a primary route of exposure, that could explain why lesions were found in some seals that were collected near areas affected by the spill but showed little or no evidence of external oiling.

Oil toxicity resulted in nerve damage, including intramyelinic edema, neuronal swelling, neuronal necrosis, and axonal swelling and degeneration within the midbrain. These lesions were probably most severe within the first 3–4 weeks after the oil spill. The first seal that was properly sampled was collected 5 weeks after the spill, and that animal had severe neuronal lesions that may have led to its death. Milder neuronal lesions were found in oiled animals collected during June to August 1989. By that time the animals most drastically affected by crude oil toxicity had already died and were no longer available for examination. Milder neuronal lesions found in animals collected 3–5 months after the spill suggests that some exposed animals recovered. Neuronal lesions were not found in seals collected in 1990. In humans, short-term exposure to low-molecular-weight aliphatics, such as *n*-hexane or methyl-*n*-butyl ketone, usually causes symptoms for several weeks, then recovery occurs (Pleasure and Schotland 1984).

CONCLUSIONS

This study documented pathological conditions observed in harbor seals collected after the EVOS. Nineteen seals were examined that were found dead in Prince William Sound and the Gulf of Alaska, or that had died in captivity. Fifteen of the 19 were oiled, but because of carcass conditions and the lengthy time between death and sampling, it was difficult to ascertain any causes of death. Thirteen of the 19 were pups, and stress or toxic effects of oil may have contributed to their deaths. Two older seals likely died due to blunt trauma. Twenty-seven additional seals, both oiled and unoiled, were collected, examined, and sampled within minutes after death. Gross examination and detailed histopathology of these 27 documented several types of lesions that were probably caused by crude oil. Skin irritation, conjunctivitis, and liver lesions occurred more frequently in oiled seals. This damage was relatively mild, and probably reversible, in most cases. Four types of lesions characteristic of hydrocarbon toxicity were found in the brain of oiled seals. These lesions occurred principally in the thalamus and probably explain disorientation and lethargy observed in seals immediately following the spill. The lesions were most acute in a seal collected 35 days after the spill and milder in seals collected later. It is likely that severely affected seals died prior to our sampling, and the animals that we collected were those that survived the short-term effects of the oil spill.

ACKNOWLEDGMENTS

We thank the crews of the ADF&G research vessel *Resolution* and the NOAA vessel *1273* for their assistance and support in field studies and all the people who assisted in collecting and handling seals, especially D. McAllister, D. Calkins, J. Lewis, M. Bates, C. Armisted, R. Haebler, K. Pitcher, A. Franzmann, and G. Antonelis. This study was conducted in cooperation with the National Marine Fisheries Service, National Marine Mammal Laboratory, as part of the Natural Resource Damage Assessment Study, funded by the *Exxon Valdez* Oil Spill Trustee Council. Helpful comments on the draft manuscript were provided by T. Loughlin and two anonymous reviewers.

REFERENCES

Addison, R. F., P. F. Brodie, A. Edwards, and M. C. Sadler. 1986. Mixed function oxidase activity in the harbour seal (*Phoca vitulina*) from Sable Is., N.S. Comparative Biochemical Physiology 85C(1):121–124.

Chusid, J. G. 1982. The thalamus. Pages 37–41, *in* J. G. Chusid (ed.), Correlative neuroanatomy and functional neurology, 19th edition. Lange Medical Publications, Los Altos, California.

Cornish, H. H. 1980. Solvents and vapors. Pages 468–496, *in* J. Doull, C. D. Klaassen, and M. O. Amdur (eds.), Casarett and Doull's Toxicology. MacMillan Publishing Company, New York, New York.

Engelhardt, F. R. 1987. Assessment of the vulnerability of marine mammals to oil pollution. Pages 101–115, *in* J. Kiuper and W. J. Van Den Brink (eds.), Fate and effects of oil in marine ecosystems. Martinus Nijhoff Publishing, Boston, Massachusetts.

Engelhardt, F. R., J. R. Geraci, and T. J. Smith. 1977. Uptake and clearance of petroleum hydrocarbons in the ringed seal, *Phoca hispida*. Journal of the Fisheries Research Board of Canada 34:1143–1147.

Frost, K. J., and L. F. Lowry. 1993. Assessment of injury to harbor seals in Prince William Sound, Alaska, and adjacent areas following the *Exxon Valdez* oil spill. Final Report, Marine Mammal Study Number 5, State-Federal Natural Resource Damage Assessment. 95 p.

Geraci, J. R., and T. J. Smith. 1976. Direct and indirect effects of oil on the ringed seals (*Phoca hispida*) of the Beaufort Sea. Journal of the Fisheries Research Board of Canada 33:1976–1984.

Geraci, J. R., and D. J. St. Aubin. 1987. Effects of offshore oil and gas development on marine mammals and turtles. Pages 587–617, *in* D. F. Boesch and N. N. Rabalais (eds.), Long-term environmental effects of offshore oil and gas development. Elsevier Applied Science, New York, New York.

Hoover-Miller, A. 1989. Impact assessment of the T/V *Exxon Valdez* oil spill on harbor seals in the Kenai Fjords National Park, 1989. Unpublished Report Kenai Fjords National Park, Seward, Alaska. 21 p.

⋙ ⋙ ⋙

Oritsland, N. A., F. R. Engelhardt, F. A. Juck, R. J. Hurst, and P. D. Watts. 1981. Effect of crude oil on polar bears. Environmental Studies No. 24. Department of Indian and Northern Affairs Canada, Northern Affairs Program, Ottawa, Canada. 268 p.

Pleasure, D. E., and D. L. Schotland. 1984. Acquired neuropathies. Pages 484–498, *in* L. P. Rowland (ed.) Merritt's textbook of neurology. 7th edition. Lea & Febiger, Philadelphia, Pennsylvania.

Pitcher, K. W., and D. G. Calkins. 1979. Biology of the harbor seal, *Phoca vitulina richardsi*, in the Gulf of Alaska. U.S. Department of Commerce/NOAA/OCSEAP, Environmental Assessment of the Alaskan Continental Shelf, Final Reports of Principal Investigators 19(1983):231–310.

St. Aubin, D. J. 1990. Physiologic and toxic effects of oil on pinnipeds. Pages 103–127, *in* J. R. Geraci and D. J. St. Aubin (eds.), Sea mammals and oil: Confronting the risks. Academic Press, San Diego, California.

Stolenburg-Didinger, G. and H. Altenkirch. 1988. Neurotoxic effects of hexacarbons (*n*-hexane, methyl-*n*-butylketones; 2,5-hexane-dione; 1,4-diketones). Pages 32–40, *in* T. C. Jones, U. Mohr, and R. D. Hunt (eds.), Nervous system: Monographs on pathology of laboratory animals. Springer-Verlag.

Appendix 17-A. List of tissues examined in each organ system of harbor seals collected following the *Exxon Valdez* oil spill.

NERVOUS SYSTEM: Frontal lobe, olfactory peduncles, corpus striatum, caudate nucleus, palladium, putamen, fornix, internal capsule, corpus callosum, optic chiasm, optic tracts, parietal cortex, infundibulum, stria habenularis thalami, thalamic nuclei, hypothalamic nuclei, pyriform lobe, amygdaloid body, corticopontine-nuclear-spinal projection fibers, mammillary bodies, subthalamic nuclei, zona incerta, crus cerebri, hippocampus, cingulate gyrus, oculomotor nerve, habenular nucleus, lateral geniculate body, thalamocortical projection fibers, choroid plexus, substantia nigra, medial geniculate, central grey substance, parasympathetic nucleus of oculomotor nerve, red nucleus, multiple areas of cerebral and cerebellar cortex, pons, rostral and caudal calyculus, reticular formation, medial longitudinal fasciculus, transverse and longitudinal fibers of the pons, pontine nuclei, trochlear nerve, cerebellar peduncles (rostral, middle, and caudal), lateral lemniscus and nucleus of lateral lemniscus, facial nerve and nucleus pyramids, trapezoid body, motor nucleus of the fifth nerve, nucleus of the spinal tract of the trigeminal nerve, fifth nerve, cochlear nucleus, vestibulocochlear nerve, vestibular nucleus, olivary nucleus, hypoglossal motor nucleus, parasympathetic nucleus of vagus, solitary tract and nucleus of solitary tract, nucleus ambiguous, nucleus gracilis, medial and lateral cuneate nucleus, spinal cord (cervical, thoracic, and lumbar) with spinal rootlets and dorsal root ganglia, gasserian ganglia, facial nerve, brachial plexus and peripheral nerves.

CARDIOVASCULAR SYSTEM: Myocardium - left papillary, atrium, valves, pulmonary artery, aorta, peripheral vessels.

DIGESTIVE SYSTEM: Soft palate, epiglottis, tongue, esophagus, stomach, duodenum, jejunum, ileum, caecum, rectum/colon, salivary tissue, pancreas, liver, gall bladder.

ENDOCRINE SYSTEM: Pituitary gland, parathyroid gland, thyroid gland, adrenal gland.

RESPIRATORY SYSTEM: Nose, nasal turbinates, trachea, lung.

LYMPHOHEMATOPOIETIC SYSTEM: Tonsil, spleen, lymph nodes, thymus.

UROGENITAL SYSTEM: Kidney, bladder, female (ovary, oviduct, uterus, cervix, and vagina) or male (testis, penis, and prostate gland).

INTEGUMENT: Skin (several parts of body, head, trunk and flippers) and vibrissae.

MUSCULOSKELETAL SYSTEM: Skeletal muscle of nose, head/neck, trunk, bone/bone marrow (ribs and vertebra).

SPECIAL SENSES: Eyes with conjunctiva, eyelids, external ear canal.

⋈ ⋈ ⋈

Appendix 17-B. Summary of histopathological lesions found in harbor seals collected in Prince William Sound and the Gulf of Alaska following the *Exxon Valdez* oil spill.

INTEGUMENT:

1. Acanthosis, mild, with orthokeratotic hyperkeratosis, skin: TS-HS-2, 3, 4, 5, 7, 8, 10, 11, 17, 18.
2. Dermatitis, mild, ulcerative, skin: TS-HS-1, 9 (with scars), 27.
3. Panniculitis, focal, mild, suppurative, subcutaneous tissues: TS-HS-1.
4. Inspissated hair follicles, moderate, skin of eyelids: TS-HS-11.
5. Dermatitis, microabscesses with folliculitis, skin: TS-HS-26.
6. Dermatitis, ulcerative with ballooning degeneration, skin of trunk and flippers: TS-HS-27.
7. Parasitism, larvae, nematodes, hair follicles, skin of trunk, suggestive of *Peloderma* sp.: TS-HS-1, 3, 5, 9, 19, 22, 24, 26.
8. Parasitism, demodex, hair follicles, skin: TS-HS-3, 23, 24, 25.
9. Acarinosis, mites, unidentified, skin of face and vibrissae: TS-HS-24 (maybe a *Halarachne miroungae* from the nasal cavity).

SPECIAL SENSES:

1. Conjunctivitis, mild to moderate, lymphoplasmacytic, conjunctiva: TS-HS-1, 2, 5, 8, 9, 10, 11, 12, 15, 16, 17, 18, 19, 20, 21, 22, 23.
2. Inspissated sebaceous glands with bacterial infection, ear canal: TS-HS-9.

DIGESTIVE SYSTEM:

1. Cheilitis, mild, ulcerative with microabscesses, lips: TS-HS-16.
2. Glossitis, mild, multifocal, lymphoplasmacytic, tongue: TS-HS-2, 6, 18, 20.
3. Stomatitis, mild, lymphoplasmacytic, ulcerative, oral cavity: TS-HS-15.
4. Parasitism, sarcocyst, few, tongue: TS-HS-2.
5. Esophagitis, mild, lymphoplasmacytic, esophagus: TS-HS-12, 26.
6. Gastritis, ulcerative, chronic active, pyogranulomatous multifocal, stomach (associated with *Anisakis* sp. and *Contracaecum osculatum*): TS-HS-1, 2, 3, 5, 7, 9, 10, 12, 11, 13, 14, 15, 17, 18, 19, 20, 21, 22, 24, 25, 26, 27.
7. Gastritis, chronic active, pyogranulomatous, multifocal, stomach (associated with *Anisakis* sp. and *Contracaecum osculatum*): TS-HS-16, 23.
8. Plaques, fibrous, chronic, serosa of pylorus: TS-HS-15.
9. Duodenitis, ulcerative, pyogranulomatous/eosinophilic, duodenum (associated with *Anisakis* sp. and *Contracaecum osculatum*): TS-HS-3, 7, 9, 18, 19, 20, 21, 22, 25, 26.
10. Duodenitis, pyogranulomatous/eosinophilic, duodenum (associated with *Anisakis* sp. and *Contracaecum osculatum*): TS-HS-12, 16, 23, 24, 27.
11. Enteritis, ulcerative, mild, multifocal, granulomatous/ eosinophilic, small intestine due to *Corynosoma* sp.: TS-HS-2, 10, 19, 20, 21, 22, 26, 27.
12. Parasitism, mild, small intestine, coccidia, compatible morphologically with *Eimeria leuckarti*: TS-HS-24, 27.
13. Colitis, pyogranulomatous, multifocal, mild, colon probably associated with parasites: TS-HS-11, 13, 18, 19, 20, 22, 24, 27.

(Appendix continues)

∛ ∛ ∛

Appendix 17-B. Continued

---

14. Colitis, ulcerative, pyogranulomatous, mild, multifocal, colon, probably associated with parasites: TS-HS-9, 10.
15. Hepatitis, multifocal, mild, pyogranulomatous/eosinophilic, liver, probably associated with parasitic larval migration: TS-HS-2, 3, 5, 8, 9, 10, 11, 12, 13, 15, 17, 18, 20, 21, 22, 23, 24, 25, 26, 27.
16. Necrosis, hepatocellular, individual cell, diffuse, mild with hepatocellular swelling, with inspissation of bile within canaliculi, moderate, liver: TS-HS-1.
17. Hepatocellular swelling, diffuse, mild with inspissation of bile within canaliculi, diffuse, liver: TS-HS-2, 3, 8.
18. Peritonitis, mild, focal, eosinophilic, peritoneal cavity, probably due to parasitic migration: TS-HS-17.
19. Sialoadenitis, mild, lymphoplasmacytic, multifocal, salivary tissue: TS-HS-1, 3, 20, 22.
20. Sialoadenitis, mild, chronic, lymphoplasmacytic with moderate fibrosis, salivary tissue: TS-HS-24, 26.

RESPIRATORY SYSTEM:

1. Rhinitis, mild to moderate, multifocal, nasal turbinates, due to nasal mites (*Halarachne miroungae*): TS-HS-2, 3, 5, 7, 9, 10, 13, 16, 17, 18, 20, 22, 24, 25, 26.
2. Fibrous, inflammatory, polyp, posterior aspect nasal cavity: TS-HS-16.
3. Pharyngitis/laryngitis, mild, lymphoplasmacytic, subacute, larynx/pharynx: TS-HS-7, 17, 20.
4. Tracheitis, mild, ulcerative, subacute, trachea (suggestive of a mild viral tracheitis): TS-HS-1, 3, 6, 12, 14, 18, 23, 24, 25, 26.
5. Pneumonia, multifocal, granulomatous and pyogranulomatous, lung (etiology lungworms): TS-HS-1, 2, 3, 10, 11, 12, 13, 16, 17, 18, 19, 20, 21, 22, 27.
6. Pneumonia, multifocal, suppurative, mild, lung: TS-HS-24, 26.
7. Bronchitis, mucoid with bronchial gland hyperplasia and excessive mucous in air passages, lung: TS-HS-20, 23, 24.
8. Bronchitis, focal, suppurative, lung due to unidentified amorphous foreign body: TS-HS-7.

NERVOUS SYSTEM:

1. Intramyelinic edema, acute, thalamus, corpus callosum, crus cerebri and internal capsule, brain, severe: TS-HS-1; mild: TS-HS-2, 3, 7, 10, 11, 17.
2. Neuronal swelling, acute with loss of Nissl substance, thalamus, severe: TS-HS-1; mild: TS-HS-2, 3, 5, 10, 11, 14, 16, 17; very mild 19.
3. Necrosis, subacute, neuronal, thalamus, brain, mild: TS-HS-1, 3, 5, 7, 10, 11, 14, 16; moderate: TS-HS-17.
4. Axonal swelling and degeneration, thalamus, corpus callosum, internal capsule, crus cerebri, brain, severe: TS-HS-1; mild: TS-HS-10, 11, 17.
5. Intracytoplasmic inclusions, multiple, eosinophilic, neurons, thalamus, brain: TS-HS-2, 3, 7, 8, 9, 10, 11, 13, 16, 18, 19, 20, 21, 22, 23, 24, 25, 27.
6. Encephalitis, perivascular cuffing, lymphocytic, mild to moderate, brain stem, gasserian ganglion and medulla oblongata: TS-HS-10, 13, 14, 15, 18, 19, 20, 23.

(Appendix continues)

⋈ ⋈ ⋈

7. Neuropathy, microcavitation, with mild axonal degeneration, trapezoid body and medulla oblongata: TS-HS-20, 22, 24, 25.
8. Plaques, fibrosus, mild, multifocal, chronic, meninges, brain: TS-HS-2, 5, 7, 9, 19, 23, 24, 27.
9. Mineralization, mild, multifocal, choroid plexus, brain: TS-HS-7, 9, 16.
10. Meningitis, mild, eosinophilic meninges, spinal cord probably due to *Dipetalonema spirocauda*: TS-HS-18.
11. Sclerosis, mild, multifocal, cerebellar peduncles: TS-HS-24.

UROGENITAL SYSTEM:

1. Plaques, proliferative, multifocal, epithelial with intranuclear (Cowdry type A) inclusion bodies, penis and prepuce compatible with a herpes infection: TS-HS-2.
2. Urethritis, mild, lymphoplasmacytic, diffuse, urethra: TS-HS-2, 15.
3. Mineralization, mild, multifocal, tubules, kidney: TS-HS-7.
4. Cervicitis/vaginitis, mild, lymphoplasmacytic, cervix and vagina, probably associated with parturition: TS-HS-7, 11.
5. Nephritis, mild, interstitial, subacute, lymphoplasmacytic with mild hydronephrosis, kidney: TS-HS-15.
6. Congenital anomaly, fibrous band, penis to prepuce: TS-HS-25.
7. Atrophy, moderate, diffuse, prostate gland: TS-HS-24.

MUSCULOSKELETAL SYSTEM:

1. Necrosis, myocytes, mild, multifocal, skeletal muscle of head and neck: TS-HS-2, 20, 21.
2. Myositis, mild, multifocal, eosinophilic, parasitic, diaphragm: TS-HS-13, tongue and trunk muscle: TS-HS-18.
3. Rhabdomyolyis, minimal, acute, skeletal muscle, diaphragm: TS-HS-4.

CARDIOVASCULAR SYSTEM:

1. Myocarditis, mild, multifocal, eosinophilic, lymphoplasmacytic, atrium and ventricle probably associated with parasitic migration: TS-HS-3, 18, 24.
2. Cardiomyopathy, myocardial degeneration, moderate, acute, multifocal, heart: TS-HS-10.
3. Endocarditis, valvular, mild, focal, lymphoplasmacytic, left atrioventricular valve, heart: TS-HS-14.
4. Fibromuscular intimal proliferation, mild to moderate, arteries (medium sized), vessels of mesentery, spleen, coronary, lung and periphery, probably associated with *Dipetalonema spirocauda*: TS-HS-4, 5, 8, 10, 13, 15, 17, 18, 20, 25, 27.
5. Fibromuscular intimal proliferation, moderate with thrombosis, lung: TS-HS-13, 20.

LYMPHOHEMATOPOIETIC SYSTEM:

1. Tonsillitis, mild, suppurative, deep crypts, tonsil: TS-HS-3, 6, 8, 12, 16, 18, 22, 23.
2. Cysts, brachial clefts, mild, thymus: TS-HS-3.
3. Abscesses, eosinophilic, mild, thymus: TS-HS-25.

(Appendix continues)

Appendix 17-B. Continued

4. Adenitis, pyogranulomatous/eosinophilic, subacute to chronic, active, multifocal, mild, mesenteric lymph nodes, associated with parasitic activity: TS-HS-2, 3, 11, 12, 18, 19, 21, 22, 23, 24.

ENDOCRINE SYSTEM:

1. Cysts, mild, multifocal, brachial clefts, thyroid: TS-HS-3.
2. Hyperactive follicles, mild, thyroids: TS-HS-9.
3. Hyperplasia, minimal, parathyroid glands: TS-HS-14.

Chapter 18

Hydrocarbon Residues in Sea Otter Tissues

Daniel M. Mulcahy and Brenda E. Ballachey

INTRODUCTION

On 24 March 1989, the T/V *Exxon Valdez* ran aground in Prince William Sound (PWS), eventually releasing 11 million gallons of Prudhoe Bay crude oil. The subsequent oil slick extended from PWS southwest along the Kenai Peninsula, past Kodiak Island to the Alaska Peninsula (Galt and Payton 1990). The spill encompassed extensive areas of sea otter (*Enhydra lutris*) habitat. Estimates of sea otter mortality due to the oil spill run between 3000 and 5000 animals, although only 878 carcasses were actually recovered (Bayha and Kormendy 1990). Some of the recovered sea otters may have died quickly from hypothermia, or inhalation and ingestion of oil, while others may have survived for varying lengths of time before succumbing. An unknown and presumably very small number of animals may have died from causes unrelated to the oil spill and drifted into the oil slick to become coated with oil.

The effects of petroleum exposure on sea otters and other marine mammals have been reviewed (Geraci and Smith 1977; Geraci and St. Aubin 1980, 1990; Engelhardt 1983; Engelhardt 1985; Waldichuk 1990). However, very little has been published on the concentrations of hydrocarbons found naturally in marine mammal tissues or in animals exposed to oil.

We report the results of hydrocarbon analyses of tissues taken from ten sea otters found dead in western PWS following the *Exxon Valdez* oil spill (EVOS). All of the carcasses were covered with oil when found, and evidence of the involvement of oiling in the deaths of these animals was provided by necropsy observations. However, the patterns of hydrocarbon prevalence and concentration varied among individual animals. We used those variations to divide the ten sea otters into three groups. In addition, we compare hydrocarbon residues in oiled sea otters to those in sea otters collected from an area in southeastern Alaska that has not experienced a crude oil spill.

METHODS

Animals

Shortly after the EVOS occurred, sea otter carcasses were recovered mostly from beaches and nearshore areas, placed in plastic bags with identifying tags, and held on the recovery boats sometimes for several days before transfer to collection centers and storage in freezer vans at -20°C. The ten sea otters (Table 18-1) were recovered dead in western PWS, and were frozen between 5 and 11 April 1989. These otters were selected for study because they were early victims of the spill and because they were heavily (60% of the pelage) or moderately (30-60% of the pelage) oiled (Williams et al. 1990).

Samples

The oiled sea otters were necropsied in the summer of 1990, and gross pathological lesions were noted (Lipscomb et al. 1993, Chapter 16). At necropsy, the carcasses were weighed and measured, the upper first premolar tooth was removed for age determination, the gender was determined, and samples of liver, muscle, kidney, brain, intestine, fat, and testes, when applicable, were taken. Samples were taken using instruments that were cleaned with hot, soapy water and rinsed with acetone and *n*-hexane. Testes were sampled from only two animals. Fat was obtained from only six animals due to the poor body condition of the rest. The abdomen of the animal was opened with care to avoid transferring oil into the body cavity. Jejunum was stripped of obvious amounts of lumenal material before the intestinal sample was obtained. Muscle tissue was taken mostly from the internal obturator muscle. The head was skinned and disarticulated from the body. The calvarium was opened using a reciprocating saw and brain tissue removed. Bile could not be obtained because it was generally not present in gall bladders of the frozen and thawed animals. The samples were placed in glass jars (Eagle Picher Environmental Services, Miami, Oklahoma) and frozen at -20°C in the dark. Samples were analyzed within 9 months of collection.

Hydrocarbon Analysis

Hydrocarbon analyses were done by the Geochemical and Environmental Research Group (GERG), Texas A&M University, College Station, Texas. The tissue extraction method used was initially developed by McLeod et al. (1985) and modified by Wade et al. (1988, 1993) and Jackson et al. (in press). Most of the tissue samples used for analyses ranged from 0.5 to 1.0 g wet weight. Aliphatic

Table 18-1. Histories of sea otters sampled for hydrocarbons. All animals were dead when discovered and all had moderate to heavy oiling of their pelage. Ages were determined by counting dental annuli.

Otter number	Sex	Recovery location	Age (y)	Weight (kg)	Length (cm)
VD015	M	Smith Island	9	21.8	137
VD018	M	Unknown	1	18.6	137
VD028	M	Eleanor Island	6	22.7	137
VD056	M	Unknown	5	22.1	119
VD059	F[a]	Knight Island	6	21.8	127
VD065	F	Knight Island	5	18.1	124
VD068	M	Knight Island	4	21.8	122
VD074	M	Knight Island	1	9.9	97
VD141	F	Knight Island	1	8.3	99
VD165	F[a]	Disk Island	9	22.8	ND

[a] Pregnant

hydrocarbons were separated by gas chromatography using a flame ionization detector. Aromatic hydrocarbons were separated and quantified using gas chromatography/mass spectrometry. The mean minimum detection limit (MDL) was 34.6 ng/g for the aliphatic hydrocarbons and 18.5 ng/g for the aromatic hydrocarbons.

Data Analysis

All values reported by the analytical laboratory were included in the analysis of data. The hydrocarbon data were converted to dry weights and transformed using a $\log_{10}(i+1)$ formula, where i was the hydrocarbon concentration. Means and standard deviations were determined on the transformed data. Data were back-transformed for the purpose of discussion in this paper.

Principal components analyses were used to compare the wet-weight hydrocarbon residue data from the 10 oiled otters to hydrocarbon data from 12 unoiled sea otters from southeastern Alaska (Ballachey and Mulcahy, unpublished data). Classification of the oiled and unoiled sea otters was based on the covariance matrix of the standardized, transformed data and a varimax rotation. Aliphatic and aromatic hydrocarbons were analyzed separately. Only the data for kidney, liver, and muscle were included in this analysis because these were the only tissues consistently sampled from both the oiled and unoiled sea otters. Because all three tissues were taken from each sea otter, and therefore were not independent samples, a separate principal components analysis was done for each tissue type.

⌖   ⌖   ⌖

RESULTS

Hydrocarbon Concentrations

A general pattern of aliphatic hydrocarbons in sea otter tissues could be described based on the data from the ten oiled sea otters in this study. The concentrations of the compounds in the low range of the n-alkane hydrocarbons, from C_{10} to about C_{16} or C_{17}, were variable from tissue to tissue, and all compounds were not necessarily found in each of the tissues. The C_{10} alkane was particularly variable. From C_{18} to C_{20}, the compounds generally were present in the tissues and their concentrations varied little between tissues from a single animal; concentrations progressively declined over this range of n-alkane hydrocarbons. From C_{20}, concentrations gradually increased, peaking at about C_{25} or C_{26}, followed by a gradual decline to about C_{29} or C_{30}. There was limited variation in concentrations between tissue types over this middle range of n-alkane hydrocarbons. From C_{30} to C_{34}, the incidence and concentration of compounds varied between tissue types. Pristane was generally present at higher concentrations than phytane. Occurrence of the unresolved complex mixture (UCM) fraction varied between tissues and in concentration.

With the aromatic hydrocarbons, a general pattern was not discernible. Most constant among the individual animals and tissue types was the presence of at least a low concentration of naphthalene and C_1-naphthalene.

Hydrocarbons in Oiled Sea Otters

The ten oiled sea otters could be separated into three groups based on the patterns of aliphatic and aromatic hydrocarbons detected in their tissues. Sea otters VD015, VD018, VD141, and VD165 had a relatively low mean concentration of total aliphatic (6000 ng/g) and total aromatic (700 ng/g) hydrocarbons in their tissues.

Figure 18-1. Aliphatic and aromatic hydrocarbons in tissues of sea otter VD141. Abbreviations: C_{10} through C_{34}: n-alkanes (the subscript is the number of carbon atoms); PRI: pristane; PHY: phytane; UCM: unresolved complex mixture; TOT: total aliphatic (not including the UCM) or total aromatic hydrocarbons; NAP: naphthalene; C_1N: C_1-naphthalene; C_2N: C_2-naphthalene; C_3N: C_3-naphthalene; C_4N: C_4-naphthalene; BIP: biphenyl; ANP: acenaphthylene; ANH: acenaphthene; FLU: fluorene; C_1F: C_1-fluorene; C_2F: C_2-fluorene; C_3F: C_3-fluorene; ANT: anthracene; PHE: phenanthrene; C_1P: C_1-phenanthrene; C_2P: C_2-phenanthrene; C_3P: C_3-phenanthrene; C_4P: C_4-phenanthrene; DIB: dibenzothiophene; C_1D: C_1-dibenzothiophene; C_2D: C_2-dibenzothiophene; C_3D: C_3-dibenzothiophene; FLA: fluoranthene; PYR: pyrene; CFP: methyl fluoranthene-pyrene; BAA: benz(a)anthracene; CHR: chrysene; C_1C: C_1-chrysene; C_2C: C_2-chrysene; C_3C: C_3-chrysene; C_4C: C_4-chrysene; BBF: benzo(b)fluoranthene; BKF: benzo(k)fluoranthene; BEF: benzo(e)pyrene; BAP: benzo(a)pyrene; PER: perylene; IDE: ideno(1,2,3-cd)pyrene; DBN: dibenzo(a,h)anthracene; BEQ: benzo(g,h,i)perylene

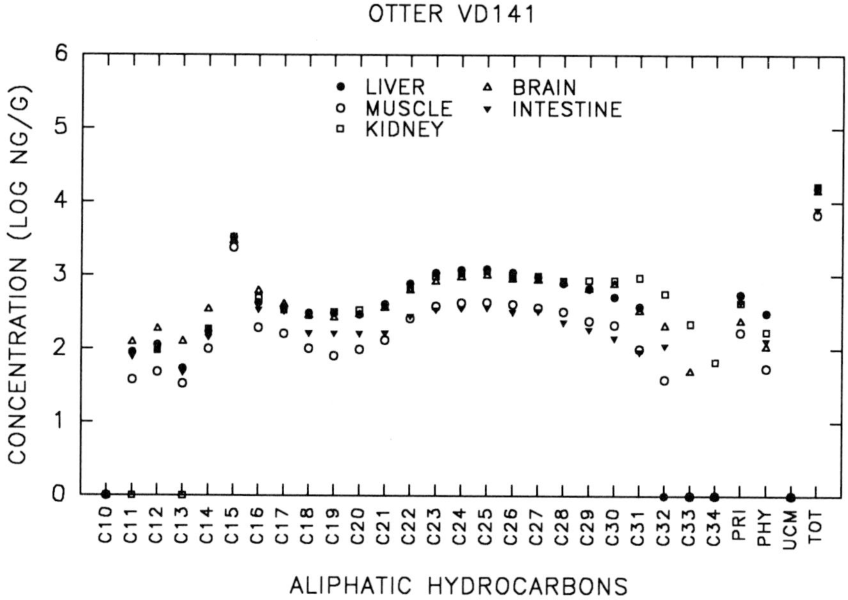
OTTER VD141
CONCENTRATION (LOG NG/G)
LIVER
MUSCLE
KIDNEY
BRAIN
INTESTINE
0
1
2
3
4
5
6
C10
C11
C12
C13
C14
C15
C16
C17
C18
C19
C20
C21
C22
C23
C24
C25
C26
C27
C28
C29
C30
C31
C32
C33
C34
PRI
PHY
UCM
TOT
ALIPHATIC HYDROCARBONS

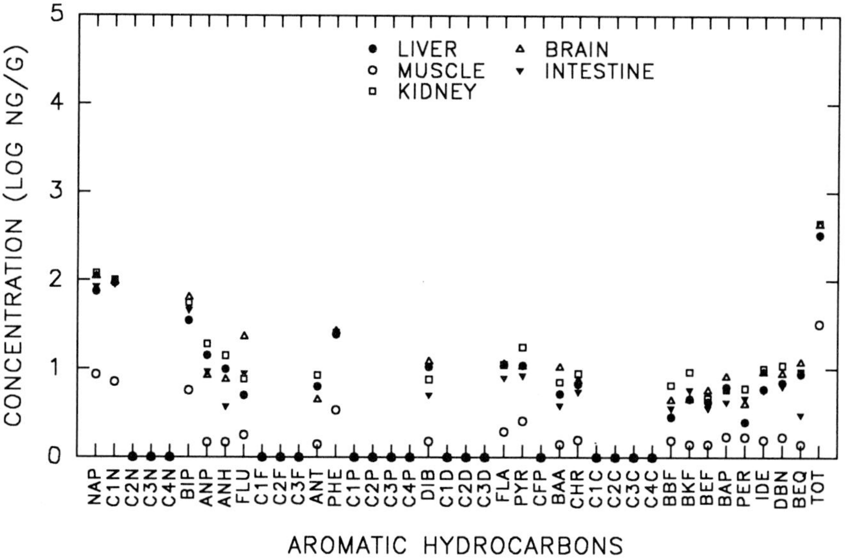
CONCENTRATION (LOG NG/G)
LIVER
MUSCLE
KIDNEY
BRAIN
INTESTINE
0
1
2
3
4
5
NAP
C1N
C2N
C3N
C4N
BIP
ANP
ANH
FLU
C1F
C2F
C3F
ANT
PHE
C1P
C2P
C3P
C4P
DIB
C1D
C2D
C3D
FLA
PYR
CFP
BAA
CHR
C1C
C2C
C3C
C4C
BBF
BKF
BEF
BAP
PER
IDE
DBN
BEQ
TOT
AROMATIC HYDROCARBONS

The individual hydrocarbon concentrations in each of the tissues sampled from sea otter VD141 are shown in Figure 18-1 as an example of this group. This pattern would be expected of animals that had died soon after being exposed to crude oil. The methylated derivatives of the parent aromatic compounds napthalene, fluorene, phenanthrene, dibenzothiophene, and chrysene were absent or present only in very low concentrations in these otters.

The second group of animals included sea otters VD056, VD065, and VD068. This group had a mean total aliphatic hydrocarbon concentration of 10,000 ng/g and a mean total aromatic hydrocarbon concentration of 2000 ng/g in the tissues. The hydrocarbon concentrations in the tissues of sea otter VD065 are shown in Figure 18-2. This pattern of hydrocarbon residues would be expected in animals which survived long enough following exposure to crude oil to move at least moderate concentrations of hydrocarbons into their tissues and to begin metabolizing them. Some of the methylated derivatives of napthalene, fluorene, phenanthrene, and dibenzothiophene were present in one or more of the tissues from each animal. The methylated derivatives of chrysene were absent from all tissues. There was a trend for aliphatic and aromatic hydrocarbon concentrations to be higher in fat than in other tissues.

The third group of animals included sea otters VD028, VD059, and VD074. The mean concentration of total aliphatic hydrocarbons was 11,000 ng/g and the mean total aromatic hydrocarbon concentration was 3000 ng/g. Concentrations of aliphatic hydrocarbons were especially high in intestinal tissues, and levels of aromatic hydrocarbons were highest in fat tissues. These concentrations of hydrocarbons would be expected in animals which were heavily oiled and lived long enough for hydrocarbons to move into tissues but died before sufficient metabolism occurred to reduce the concentrations of hydrocarbon residues. Otter VD074 (Fig. 18-3) had extremely high concentrations of aliphatic and aromatic hydrocarbons present in the intestinal sample. This tissue was the only sample that contained detectable concentrations of every aliphatic hydrocarbon analyzed. Concentrations of aliphatic hydrocarbons in the other tissues from this animal were not elevated relative to similar tissues from the other animals. The UCM fraction was found only in the intestine and was present at the highest concentration (941 µg/g) of any of the samples. Otter VD074 had the highest concentrations of total aromatic hydrocarbons of any of the sea otters. Concentrations of naphthalene and alkylated naphthalene, biphenyl, fluorene and alkylated fluorene, phenanthrene and alkylated phenanthrene, dibenzothiophene and alkylated dibenzothiophene, perylene, and C_1-fluoranthene-pyrene were generally higher in the intestine than in the other tissues from this otter. The intestinal sample from this animal had the only incidence of alkylated chrysenes. The concentration of total aromatic hydrocarbons was also very high in the intestine, and higher concentrations of alkylated

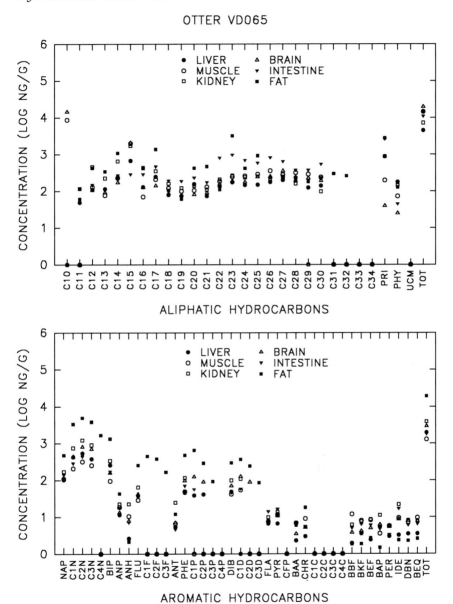

Figure 18-2. Aliphatic and aromatic hydrocarbons in tissues of sea otter VD065. See Figure 18-1 for abbreviations.

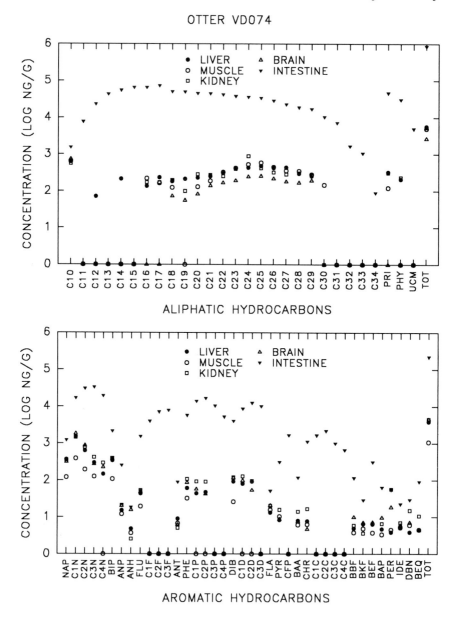

Figure 18-3. Aliphatic and aromatic hydrocarbons in tissues of sea otter VD074. See Figure 18-1 for abbreviations.

derivatives of naphthalene, phenanthrene, fluorene, dibenzothiophene, and chrysene than parent compounds were present.

Principal Components Analysis

Three principal components each were obtained for kidney, liver, and muscle samples from the oiled and unoiled sea otters. For the aliphatic hydrocarbons in the three tissues, the three principal components explained a total of 72-78% of the total variance. The hydrocarbons with the most influence in each of the three principal components were very similar among tissues. The first principal component for the three tissues explained from 32% to 38% of the total variance and the middle range of molecular weight n-alkanes (from about C_{19} to C_{29}) were the most influential hydrocarbons. The second principal component explained from 22% to 25% of the total variance and the upper molecular weight range of n-alkanes (from about C_{30} to C_{34}), C_{11}, C_{13}, and the UCM were the most influential hydrocarbons. The third principal component explained from 16% to 18% of the total variance and the lower molecular weight n-alkanes (C_{12} to C_{17}), pristane, and phytane were the most important hydrocarbons.

For the aromatic hydrocarbons, the three principal components for the three tissues explained from 63% to 74% of the total variance. The hydrocarbons with the most influence in each of the three principal components were somewhat similar between tissues. The first principal component for the three tissues explained from 33% to 37% of the total variance and naphthalene, C_1-naphthalene, C_2-naphthalene, C_3-naphthalene, C_4-naphthalene, phenanthrene, C_1-phenanthrene, C_2-phenanthrene, dibenzothiophene, C_1-dibenzothiophene, C_2-dibenzothiophene, fluorene, and biphenyl were the most influential hydrocarbons. The second principal component explained from 16% to 29% of the total variance and benzo(g,h,i)perylene, dibenzo(a,h)anthracene, ideno(1,2,3-cd)pyrene, chrysene, acenaphthene, and benzo(b)fluoranthene were the most influential hydrocarbons in common in the three tissues. In addition to these hydrocarbons, benzo(a)pyrene and perylene were important for kidney; benzo(a)pyrene and fluoranthene were important for liver; and pyrene, benz(a)anthracene, anthracene, acenaphthylene, benzo(k)fluoranthene, and benzo(e)pyrene were important for muscle. The third principal component explained from 9% to 12% of the total variance and the most important hydrocarbons varied between the three tissues. For kidney, the influential hydrocarbons were C_2-phenanthrene, C_2-dibenzothiophene, acenaphthylene, perylene, and anthracene; for liver, benzo(e)pyrene, ideno(1,2,3-cd)pyrene, perylene, benzo(k)fluoranthene, and anthracene were important; for muscle, C_2-phenanthrene, C_4-naphthalene, and benzo(a)pyrene were important. Plotting the first principal component against the second principal component for the aliphatic

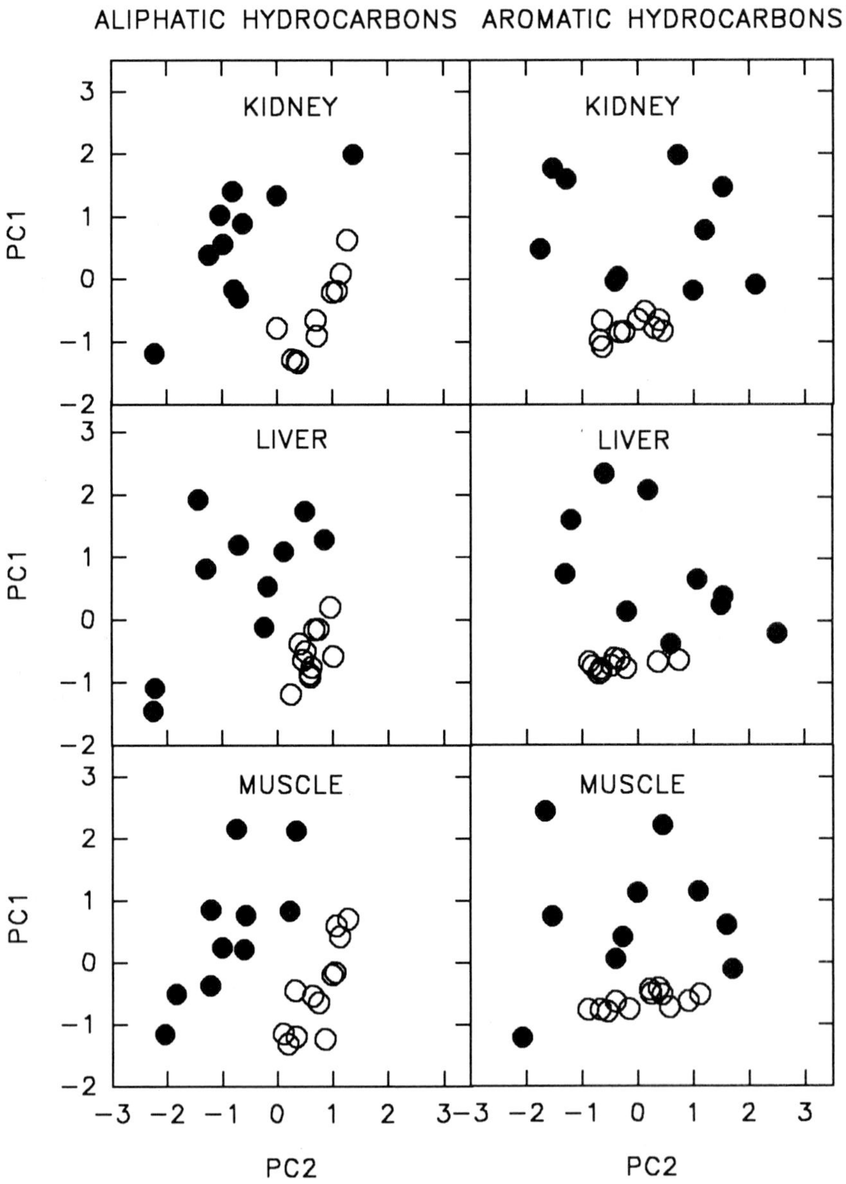

ALIPHATIC HYDROCARBONS   AROMATIC HYDROCARBONS

Figure 18-4. Principal components analysis of aliphatic and aromatic hydrocarbons present in kidney, liver, and muscle from sea otters oiled following the T/V *Exxon Valdez* oil spill (filled circles) and unoiled sea otters from southeastern Alaska (open circles). The hydrocarbon concentrations were analyzed as wet weights. PC1: first principal component; PC2: second principal component.

and aromatic hydrocarbons from each tissue clearly separated the oiled from the unoiled otters (Fig. 18-4).

Lesions

Several lesions were noted upon necropsy of the ten sea otters (Lipscomb et al. 1993; Chapter 16). Pulmonary, cranial mediastinal, or subcutaneous emphysema was present in all but two sea otters (VD015 and VD141). Gastrointestinal hemorrhages were present in every animal and gastric erosions were found in seven otters (VD056, VD059, VD065, VD068, VD074, VD141, and VD165). Oil was detected in the trachea and bronchi of six otters (VD028, VD056, VD059, VD065, VD074, and VD165), and in the gastrointestinal tracts of three otters (VD059, VD074, and VD165).

DISCUSSION

Little information has been published specific to the occurrence of petroleum hydrocarbons in mammals. Few marine mammals have been directly affected by oil spills in the marine environment, a fact that changed dramatically with the grounding of the T/V *Exxon Valdez*. Historically, when an oil spill has occurred in the marine environment, hydrocarbon studies have first concentrated on the oil itself, then on birds (because of their numbers and wide distribution in the marine environment), then on oil in sediments, invertebrates, fish, and finally mammals. This research has influenced the interpretation of hydrocarbon incidence and concentrations in marine mammals. Unlike most lower animals, marine mammals have the enzyme systems permitting the metabolism and excretion of systemic hydrocarbons (Engelhardt 1982; Addison and Brodie 1984; Addison et al. 1986). However, so little is known about the fate of hydrocarbons in mammals that the same interpretation of specific values of individual hydrocarbons, and many of the same sums and ratios of hydrocarbons used in studies of oiled sediments, are used for studies of oiled marine mammals (Hall and Coon 1988).

The presentation and discussion of hydrocarbon data which are quantitatively less than the calculated MDL for each hydrocarbon are controversial (Rhodes 1981; Berthouex 1993). The MDL is a statistical value estimated from repeated analysis of blank specimens. In the past, hydrocarbon concentrations which fall below the limit of detection have been presented as "trace," "not detected (ND)," "<MDL", zero, or some incremental number between zero and the MDL. However, other publication strategies, such as simply presenting the measured concentration regardless of its relationship to the MDL (our choice), presentation of both the

⚭   ⚭   ⚭

measured concentration and the MDL, or giving the measured concentration followed by a statistical estimate of its precision are considered superior (Berthouex 1993). These methods prevent the discarding of useful information which occurs with the former methods, all of which censor some of the data.

Aliphatic and aromatic hydrocarbons were detected in all tissues from each of the ten sea otters included in this study. Hydrocarbon concentrations were also measured in samples of kidney, liver, and muscle from 12 sea otters, not known to have been exposed to a crude oil spill, from southeastern Alaska (Ballachey and Mulcahy, unpublished data). The mean concentrations of the total aliphatic hydrocarbons measured in kidney, liver and muscle from sea otters from southeastern Alaska were 3800, 2800, and 3200 ng/g, respectively, compared to 7700, 6200, and 6100 ng/g in the same tissues from the oiled sea otters in this study. The mean concentrations of total aromatic hydrocarbons in kidney, liver, and muscle from sea otters from southeastern Alaska were 170, 160, and 170 ng/g, respectively, compared to 1500, 1100, and 600 ng/g in the same tissues from the oiled sea otters in this study. In kidney, liver, and muscle, respectively, total mean aliphatic hydrocarbon concentrations were 2.0, 2.2, and 1.9 times greater, and total mean aromatic hydrocarbon concentrations were 8.8, 6.9, and 3.5 times greater in the oiled sea otters compared to the unoiled sea otters from southeastern Alaska.

The pathology found in sea otters killed by the EVOS is described by Lipscomb et al. (1993, Chapter 16). Pulmonary, mediastinal, and subcutaneous emphysema, gastrointestinal hemorrhage, gastric ulceration, and the presence of oil in the gastrointestinal tract are characteristic in sea otters that died as a result of the EVOS. One or more of these lesions was present in each of the oiled sea otters in this study; no lesions were found in the unoiled sea otters from southeastern Alaska.

Based on our inspection of the samples in this study, some compounds were of greater importance than others in evaluating a tissue sample for the presence of oil due to exposure to crude oil. Both total aliphatic and total aromatic hydrocarbon concentrations were of interest and were elevated relative to the unoiled otters. However, total hydrocarbon concentrations may not be elevated in an animal which has lived long enough to begin metabolizing hydrocarbons. The UCM fraction, which was not present in every sample, proved useful because it should be both present and elevated in concentration in the presence of crude oil. Absence of the UCM fraction in tissues, however, was not a definitive indicator of the lack of exposure to oil. Of the aromatic hydrocarbons, the presence at elevated concentrations of the parent compounds, naphthalene, fluorene, phenanthrene, dibenzothiophene, chrysene, and especially the presence of their alkylated derivatives (except C_1-naphthalene, which was present in virtually all samples) at any concentration were the most useful compounds in determining exposure of an animal to oil.

The presence of observable oil, such as could be found in the gastrointestinal tract, should produce elevated concentrations of all hydrocarbons when that tissue is analyzed, presuming an oiled area of the tissue was actually taken for the sample. However, the necropsy reports indicated that, when present, oil existed as patches in the gastrointestinal tract, and sometimes required the use of ultraviolet light for visualization. It is likely that the levels of hydrocarbons found in an intestinal sample could vary greatly, depending on whether or not a patch of oiled tissue was included in the sample.

Examination of the samples from the oiled otters in this study, as well as many other samples from sea otters that died following the EVOS (Ballachey and Mulcahy, unpublished data), suggests an upper limit of 1 μg/g for the maximum concentration of individual hydrocarbons detected with the sampling and analytical methods used in this study. The effects of postmortem autolysis and freezing and thawing of tissues on concentrations of hydrocarbons are not known.

Examination of the patterns of aliphatic and aromatic hydrocarbons in the tissues, and consideration of gross pathology for each otter, make it possible to speculate on the interactions of each animal with the crude oil spilled by the T/V *Exxon Valdez*. All of the oiled otters, except VD015, had either pulmonary emphysema or gastric ulcerations, or both. All of the oiled otters, including VD015, had gastrointestinal hemorrhages present. Based on Lipscomb et al. (1993, Chapter 16), the ten oiled otters exhibited pathological lesions associated with crude oil contamination, in addition to the presence of heavy external oil contamination. Death from exposure to crude oil may have occurred due to hypothermia, inhalation, ingestion, or transdermal absorption of volatile hydrocarbons causing pulmonary or central nervous system dysfunction. Death due to the first two mechanisms would have occurred quickly without producing elevated hydrocarbon residues in the tissues of the animal. Survival for a few days should have resulted in higher concentrations of hydrocarbons in the tissues.

We propose that sea otter VD074 ingested crude oil, but died soon after ingestion, before the hydrocarbons could be extensively distributed to the other tissues. This animal had up to a one-hundred-fold elevation in concentrations across the range of aliphatic hydrocarbons in the intestine sample compared to other tissues. The UCM fraction was present only in the intestinal sample. High concentrations of aromatic hydrocarbons, including alkylated derivatives of parent compounds, were present in the intestine. Oil was visible in the gastrointestinal tract at necropsy and the animal was suffering pulmonary emphysema and gastric ulceration. Similar, but lower and less extensive, elevations in aliphatic hydrocarbon concentrations were seen in the intestines from sea otters VD028 and VD059.

Sea otter VD141 represented the hydrocarbon patterns expected from an animal with little exposure to crude oil, based on the very low concentrations and inci-

⋙　⋙　⋙

dences of both aliphatic hydrocarbons and aromatic hydrocarbons in the tissues, the near absence of alkylated derivatives from the tissues, and the absence of pulmonary emphysema and gastric ulceration. However, there were high concentrations of the UCM fraction in all tissues except brain. Sea otters VD015, VD018, and VD165 also represented animals with minimal exposure to oil, as judged by their hydrocarbon patterns. It is possible that these animals were exposed to oil but died from hypothermia before much hydrocarbon burden could be delivered to the internal organs.

The patterns of hydrocarbons seen in sea otter VD065 may have indicated exposure to hydrocarbons prior to the oil spill. The UCM fraction was absent from all tissues. Concentrations of total aliphatic and total aromatic hydrocarbons were much higher than in sea otter VD141 as were the concentrations of many aliphatic hydrocarbons and aromatic hydrocarbons, especially in fat tissue. Alkylated derivatives of several parent compounds (but not chrysene) were present, again, especially in fat. Fat is a storage tissue for hydrocarbons, and their presence in this tissue, at a concentration higher than the other tissues, may suggest a past exposure to oil for this animal.

Sea otter VD056 showed a similar pattern of higher concentrations of some aromatic hydrocarbons (including some alkylated derivatives); the highest concentration of total aromatic hydrocarbons in any of the organs was in fat. In this animal, concentrations of total aliphatic hydrocarbons and several of the individual aliphatic hydrocarbons were highest in the fat. However, the UCM fraction was present in high concentration in several tissues, but not fat. We speculate that this animal had some previous exposure to oil, reflected in the elevated concentrations of hydrocarbons in the fat, but that there was an additional recent exposure. Similar logic was applied to sea otters VD068 and VD056.

The storage capability of fat for hydrocarbons has been previously demonstrated in other animals, including the polar bear (Oritsland et al. 1981). Fat can serve as a depot for the chronic release of stored hydrocarbons or for a more rapid release, as might occur when fat is rapidly mobilized during periods of stress, starvation, or lactation. In the latter case, nursing females may serve as a source of hydrocarbons contaminating their nursing pups.

Exposure of fish to petroleum hydrocarbons is associated with altered susceptibility to infection with parasites (Khan and Kiceniuk 1983, 1988; Khan 1987, 1990, 1991; Kiceniuk et al. 1990). Sea otters are commonly parasitized by gastrointestinal parasites. All ten of the oiled otters in our study had trematodes in their gall bladders and all but one had intestinal acanthocephalin parasites. No studies similar to those done with fish on the interaction of parasites and exposure to petroleum hydrocarbons have been done with sea otters. Oiled sea otters brought to rehabilitation centers shed parasites in their feces, and the possible toxicity of ingested oil to gastrointestinal parasites was proposed (Wilson et al. 1990).

⊳⊳ ⊳⊳ ⊳⊳

Engelhardt et al. (1977) demonstrated a rapid uptake of hydrocarbons by ringed seals (*Phoca hispida*) exposed to crude oil by immersion and ingestion. Excretion rates are rapid, with biliary and renal routes thought to be most important. Following ingestion, radiolabeled hydrocarbon concentrations peak in blood after 2 days, with concentrations in plasma persisting for 5 days before declining rapidly (Engelhardt et al. 1977). Concentrations in liver and blubber decline to low concentrations over a 28-day period; concentrations in muscle persist longer. Liver is generally considered to be the principal organ for detoxification and excretion of lipophilic substances. Seals have a blubber layer that is lacking in sea otters, which rely on their pelage for insulation. Body fat functions as an energy storage organ in sea otters which may permit a more rapid accumulation and turnover of hydrocarbons than in seals. In several of the sea otters in this study, higher concentrations of a range of hydrocarbons were found in fat compared to other tissues.

Principal components analysis of the aliphatic and aromatic hydrocarbons present in kidney, liver, and muscle samples was useful in discriminating oiled from unoiled sea otters. Plots of the first two principal components showed that the concentrations of aliphatic and aromatic analytes from oiled sea otters were more variable for both the aliphatic and aromatic hydrocarbons than in the plots for the unoiled sea otters.

Although all of the sea otter carcasses were collected soon after the oil spill occurred, and the carcasses were in good condition and coated with moderate to heavy amounts of oil, no factual information is known about how long the otters lived following exposure. Some may have died from causes other than oiling and may have suffered postmortem oiling. These considerations, in the face of potential litigation, point out the importance of analyzing for hydrocarbon residues and performing necropsies even on animals that are obviously oiled. In view of the likelihood of future oil spills occurring in sea otter habitats, controlled studies are essential to define the relationship of the exposure of mammals to crude oil by different routes to the occurrence and concentration of specific hydrocarbons in tissues. Additional studies are required to examine the effect of metabolism on the patterns of hydrocarbons in mammalian tissues at various times after exposure.

CONCLUSIONS

All the sea otters examined in this study were early victims of the EVOS, all were moderately to heavily covered with oil, and all had gross lesions consistent with exposure to crude oil. Aliphatic and aromatic hydrocarbons were detected in all tissues examined. The patterns of hydrocarbon incidences and concentrations allowed us to divide the animals into three groups, perhaps related to the route of exposure and duration of life following exposure. Principal components analysis

∞ ∞ ∞

separated the hydrocarbon concentrations in kidney, liver, and muscle samples taken from the ten oiled sea otters from PWS from samples taken from unoiled sea otters from southeastern Alaska. Concentrations of aliphatic and aromatic hydrocarbons were two to eight times higher in tissues from the oiled sea otters compared to tissue from unoiled sea otters.

There was an upper limit of 1 µg/g of individual hydrocarbons in the tissues sampled. This low concentration in oiled sea otters probably resulted from the limited time for mobilization of hydrocarbons in sea otters that died quickly after exposure and metabolism by inducible enzyme systems of the original hydrocarbons into compounds that were not included as part of the laboratory analysis of the tissues.

Our study showed that measurements of the hydrocarbons found in crude oil did not give an optimum assessment of exposure of each animal to crude oil. Measurement of oil metabolites may provide a superior approach. Necropsies of affected carcasses by experienced veterinarians or veterinary pathologists should be considered an essential part of damage assessment following an oil spill.

ACKNOWLEDGMENTS

We thank B. Didrickson, Sitka, Alaska, for his help in supplying otter carcasses from southeastern Alaska. Necropsies were performed by veterinary pathologists from the U.S. Armed Forces Institute of Pathology, Walter Reed Army Medical Center, Washington, D.C. We thank E. Robinson-Wilson, Chief, Division of Environmental Contaminants, Alaska Region, U.S. Fish and Wildlife Service, Anchorage, Alaska, for organizing and coordinating the analysis of sea otter tissues.

REFERENCES

Addison, R. F., and P. F. Brodie. 1984. Characterization of ethoxyresorufin O-de-ethylase in grey seal *Halichoerus grypus*. Comparative Biochemistry and Physiology 79C:261–263.

Addison, R. F., P. F. Brodie, A. Edwards, and M. C. Sadler. 1986. Mixed function oxidase activity in the harbour seal (*Phoca vitulina*) from Sable Is., N.S. Comparative Biochemistry and Physiology 85C:121–124.

Bayha, K., and J. Kormendy (eds.). 1990. Sea otter symposium: Proceedings of a symposium to evaluate the response effort on behalf of sea otters after the T/V *Exxon Valdez* oil spill into Prince William Sound, Anchorage, Alaska, 17–19 April 1990. U.S. Fish and Wildlife Service Biological Report 90(12).

Berthouex, P. M. 1993. A study of the precision of lead measurements at concentrations near the method limit of detection. Water Environment Research 65:620–629.

Engelhardt, F. R. 1982. Hydrocarbon metabolism and cortisol balance in oil-exposed ringed seals, *Phoca hispida*. Comparative Biochemistry and Physiology 72C:133-136.

Engelhardt, F. R. 1983. Petroleum effects on marine mammals. Aquatic Toxicology 4:199–217.

≍   ≍   ≍

Engelhardt, F. R. 1985. Effects of petroleum on marine mammals. Pages 217–243, *in* F. R. Engelhardt (ed.), Petroleum effects in the Arctic environment. Elsevier Applied Science Publishers, London and New York.

Engelhardt, F. R., J. R. Geraci, and T. G. Smith. 1977. Uptake and clearance of petroleum hydrocarbons in the ringed seal, *Phoca hispida*. Journal of the Fisheries Research Board of Canada 34:1143–1147.

Galt, J. A., and D. L. Payton. 1990. Movement of oil spilled from the T.V. *Exxon Valdez*. Pages 4–17, *in* K. Bayha and J. Kormendy (eds.), Sea otter symposium: Proceedings of a symposium to evaluate the response effort on behalf of sea otters after the T/V *Exxon Valdez* oil spill into Prince William Sound, Anchorage, Alaska, 17–19 April 1990. U.S. Fish and Wildlife Service Biological Report 90(12).

Geraci, J. R., and T. G. Smith. 1977. Consequences of oil fouling on marine animals. Pages 399–410, *in* D. C. Malins (ed.), Effects of petroleum on arctic and subarctic marine environments and organisms. Academic Press, Inc., New York, New York.

Geraci, J. R., and D. J. St. Aubin. 1980. Offshore petroleum resource development and marine mammals: A review and research recommendations. Marine Fisheries Review 42:1–12.

Geraci, J. R., and D. J. St. Aubin (eds.). 1990. Sea mammals and oil: Confronting the risks. Academic Press, Inc., San Diego, California. 282 p.

Hall, R. J., and N. C. Coon. 1988. Interpreting residues of petroleum hydrocarbons in wildlife tissues. U.S. Fish and Wildlife Service Biological Report 88(15):1–7.

Jackson, T. J., T. L. Wade, T. J. McDonald, D. L. Wilkinson, and J. M. Brooks. In press. Polynuclear aromatic hydrocarbon contaminants in oysters from the Gulf of Mexico (1986–1990). Environmental Pollution.

Khan, R. A. 1987. Effects of chronic exposure to petroleum hydrocarbons on two species of marine fish infected with a hemoprotozoan, *Trypanosoma murmanensis*. Canadian Journal of Zoology 65:2703–2709.

Khan, R. A. 1990. Parasitism in marine fish after chronic exposure to petroleum hydrocarbons in the laboratory and to the *Exxon Valdez* oil spill. Bulletin of Environmental Contamination and Toxicology 44:759–763.

Khan, R. A. 1991. Influence of concurrent exposure to crude oil and infection with *Trypanosoma murmanensis* (Protozoa: Mastigophora) on mortality in winter flounder, *Pseudopleuronectes americanus*. Canadian Journal of Zoology 69:876–880.

Khan, R., and J. Kiceniuk. 1983. Effects of crude oil on the gastrointestinal parasites of two species of marine fish. Journal of Wildlife Diseases 19:253–258.

Khan, R. A., and J. Kiceniuk. 1988. Effect of petroleum aromatic hydrocarbons on monogeneids parasitizing Atlantic cod, *Gadus morhua* L. Bulletin of Environmental Contamination and Toxicology 41:94–100.

Kiceniuk, J. W., R. A. Khan, M. Dawe, and U. Williams. 1990. Examination of interaction of trypanosome infection and crude oil exposure on hematology of the longhorn sculpin (*Myoxocephalus octodecemspinosus*). Bulletin of Environmental Contamination and Toxicology 19:259–262.

Lipscomb, T. P., R. K. Harris, R. B. Moeller, J. M. Pletcher, R. J. Haebler, and B. E. Ballachey. 1993. Histopathologic lesions in sea otters exposed to crude oil. Veterinary Pathology 30:1–11.

McLeod, W. D., D. W. Brown, A. J. Friedman, D. G. Burrow, O. Mayes, R. W. Pearce, C. A. Wigren, and R. G. Bogar. 1985. Standard Analytical Procedures of the NOAA National Analytical Facility 1985-1986. Extractable Toxic Compounds. 2nd Edition. U.S. Department of Commerce, NOAA/NMFS. NOAA Technical Memorandum. NMFS F/NWC-92. 121 p.

Oritsland, N. A., F. R. Engelhardt, F. A. Juck, R. J. Hurst, and P. D. Watts. 1981. Effect of crude oil on polar bears. Environmental Studies No. 24. 166 p.

Rhodes, R. C. 1981. Much ado about next to nothing, or what to do with measurements below the detection limit. Pages 157–162, *in* Environmetrics 81: Selected papers, SIAM-SIMS Conference Series No. 8. Philadelphia, Pennsylvania.

Wade, T. L., E. L. Atlas, J. M. Brooks, M. C. Kennicutt, II, R. G. Fox, J. Sericano, B. Garcia-Romero, and D. DeFreitas. 1988. NOAA Gulf of Mexico status and trends program: Trace organic contaminant distribution in sediments and oysters. Estuaries 11:171–179.

Wade, T. L., T. J. Jackson, T. J. McDonald, W. L. Wilkinson, and J. M. Brooks. 1993. Oysters as biomonitors of the APEX Barge oil spill, Galveston Bay, Texas. Pages 313–317 *in* Proceedings, 1993 International Oil Spill Conference, 29 March–1 April 1993. Tampa, Florida.

Waldichuk, M. 1990. Sea otters and oil pollution. Marine Pollution Bulletin 21:10–15.

Williams, T. M., R. Wilson, P. Tuomi, and L. Hunter. 1990. Critical care and toxicological evaluation of sea otters exposed to crude oil. Pages 82–100, *in* T. M. Williams and R. W. Davis (eds.), Sea otter Rehabilitation Program: 1989 Exxon Valdez Oil Spill. International Wildlife Research.

Wilson, R. K., C. R. McCormick, T. D. Williams, and P. A. Tuomi. 1990. Clinical treatment and rehabilitation of sea otters. Pages 326–337, *in* K. Bayha and J. Kormendy (eds.), Sea otter symposium: Proceedings of a symposium to evaluate the response effort on behalf of sea otters after the T/V *Exxon Valdez* oil spill into Prince William Sound, Anchorage, Alaska, 17–19 April 1990. U.S. Fish and Wildlife Service Biological Report 90(12).

Chapter 19

Petroleum Hydrocarbons in Tissues of Harbor Seals from Prince William Sound and the Gulf of Alaska

Kathryn J. Frost, Carol-Ann Manen, and Terry L. Wade

INTRODUCTION

When the T/V *Exxon Valdez* ran aground in Prince William Sound (PWS) on 24 March 1989, approximately 11 million gallons of North Slope crude oil were released within a 5-hour period. Over the next 8 weeks, the oil impacted 1750 km of shoreline and extended more than 500 km south and west from where it was spilled. At the time of the spill and for most of the next 3 days, winds in PWS were 9–18 km/hr and the sea was calm. The oil slick floated in open water to the southwest of the grounded ship and spread over an area of approximately 300 km^2 (Morris and Loughlin, Chapter 1). During this time, the most volatile constituents, including benzenes, toluenes, and alkanes though C$_8$, evaporated and reached concentrations of approximately 9 ppm in the air over the slick (Wolfe et al. submitted). This period of calm was followed by 3 days with wind speeds of 35–45 km/hr and gusts of over 90 km/hr. The oil moved rapidly to the southwest and was driven ashore on beaches of islands in central PWS. For the next 3 weeks, oil was repeatedly deposited, refloated, and redeposited on shorelines of PWS as a result of local winds and tides. Traces of oil were still visible in the subsurface sediments of some of these beaches in 1992 (Wolfe et al. 1993).

The area impacted by the spill is inhabited by harbor seals (*Phoca vitulina richardsi*) throughout the year. Harbor seals contacted oil both in the water and on shore (Frost and Lowry 1993; Lowry et al., Chapter 12). In the early weeks of the spill, they swam through oil and breathed at the air/water interface where volatile hydrocarbon vapors were present. Possible exposure through inhalation has received little study, but this may be a significant route for exposure to toxic volatile hydrocarbon fractions (Geraci and St. Aubin 1987). Seals crawled over and rested on oiled rocks and algae throughout the spring and summer. Many seals were

331

heavily coated with oil from the time of the spill until their annual molt in August. Pups were born on haulout sites in May and June, when some of the sites still had oil on them. Some pups became oiled shortly after birth, and surface oiling may have persisted for many months, since pups do not molt during their first year of life. Many pups nursed on oiled mothers and may have ingested hydrocarbons from their mother's oiled pelage or through their mother's milk. Harbor seals feed on a variety of fishes and invertebrates (Pitcher and Calkins 1979), some of which were exposed to oil as a result of the spill (Varanasi et al. 1993a,b).

As part of the Natural Resources Damage Assessment (NRDA) program, a study was designed to investigate and quantify the effects of oil and the disturbance associated with cleanup from the *Exxon Valdez* oil spill (EVOS) on the distribution, abundance, and health of harbor seals. One of the objectives was to determine the levels of hydrocarbon contaminants in harbor seal tissues in order to characterize the magnitude and extent of exposure and to assist in the determination of the short-term and long-term effects resulting from this exposure.

METHODS

Harbor seals were collected from western PWS and the Gulf of Alaska (GOA) to conduct gross necropsies and to obtain samples for chemical and histopathological analyses. Seals were collected at and adjacent to sites impacted by the EVOS by shooting with a high-powered rifle. Each animal was necropsied and sampled as soon as possible after death by a qualified veterinary pathologist. Twenty seals were collected in PWS and the GOA in 1989 and six in PWS in 1990. Two seals were collected in August 1990 near Ketchikan, Alaska, a location more than 1000 km from areas impacted by the EVOS, to serve as reference specimens. These collections were supplemented with samples from seals that died in captivity at rehabilitation facilities, recently dead carcasses found during coastline searches, and animals killed by subsistence hunters.

For analyses, seals were assigned to one of five sample groups depending on the date and location of collection: PWS 1989 oiled, GOA 1989, PWS 1990 oiled, PWS 1989/1990 unoiled, and Ketchikan 1990. Seals in the PWS 1989 oiled group were collected or found in areas that were oiled by the EVOS; most had heavily oiled pelage, indicating that the seals had directly contacted oil. GOA 1989 seals were collected approximately 300 km downstream from where the spill occurred; two of six were externally oiled. Seals in the PWS 1990 oiled group showed no signs of external oiling but were collected in areas of PWS that had been oiled the year before. Information on date and location of collection, degree of oiling, sex, and sample group to which seals were assigned are given in Appendix 19-A.

The number and type of samples collected varied with the condition of the carcass at the time of sampling; as many tissues were sampled as the condition

>◇   >◇   >◇

permitted. Whenever possible, triplicate samples of tissues were dissected from the seals with chemically cleaned instruments and stored in chemically cleaned glass jars. Samples were cooled immediately and frozen as soon as possible. Unanalyzed samples and portions of samples remaining after analysis are archived at -60°C at the National Marine Mammal Laboratory, Seattle, Washington.

Bile samples were analyzed for the metabolites of petroleum-related aromatic compounds (ACs) by the Environmental Conservation Division (ECD), Northwest Fisheries Science Center, NMFS/NOAA in Seattle, Washington, using a high-performance liquid chromatographic (HPLC) technique with fluorescence detection (Krahn et al. 1986). Fluorescence responses were recorded at 260/380 nm and 290/335 nm excitation/emission wavelengths, which are indicative of phenanthrene-like and naphthalene-like compounds and give relative measures of AC metabolites that fluoresce at these wavelengths. The phenanthrene equivalent wavelength pair was used to estimate the total concentration of 3-ring aromatic compounds, and the naphthalene equivalent wavelength pair was used to estimate 2-ring aromatic compounds. Equivalents of fluorescent ACs have been shown to be highly correlated with the summed concentrations of the metabolites of phenanthrene and naphthalene/dibenzothiophene as determined by gas chromatography and mass spectroscopy (GC/MS) and are characteristic of Prudhoe Bay crude oil (Krahn et al. 1992).

Tissue samples were analyzed for a broad spectrum of aliphatic hydrocarbons (n-C_{10}–n-C_{34}, pristane, phytane, and unresolved complex mixture) and aromatic hydrocarbons (naphthalenes, dibenzothiophenes, phenanthrenes, fluorenes, and chrysenes) using capillary column gas chromatography with flame ionization (GC/FID) and mass spectrometry (GC/MS) detectors (Krahn et al. 1988; Brooks et al. 1990; Sloan et al. 1993). Analyses were done by the Geochemical and Environmental Research Group (GERG), Texas A&M University, College Station, Texas, or by ECD.

RESULTS

Sets of tissues, including liver, blubber, muscle, and brain, were analyzed for 27 seals collected from April 1989 to September 1990. Heart, kidney, lung, mammary gland, and milk were analyzed for some. Less complete sets of tissues, including mostly bile, liver, and/or blubber, were analyzed from two fetuses from the collected seals and for 15 additional seals (Frost and Lowry 1993; Varanasi et al. 1993a,b).

Analyses of petroleum-related metabolites in bile from 36 harbor seals collected at various times and locations indicated a wide range in the concentrations of fluorescent ACs at phenanthrene (PHN) and naphthalene (NPH) wavelengths

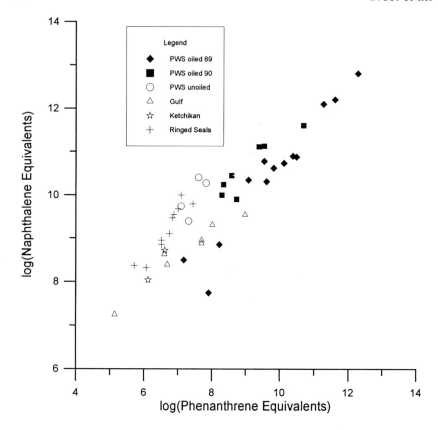

Figure 19-1. Graph showing log-transformed concentrations of fluorescent ACs (ng/g of naphthalene and phenanthrene equivalents) in the bile of harbor seals collected in Prince William Sound, the Gulf of Alaska, and near Ketchikan. Values for 10 ringed seals collected near Barrow are also shown.

(Fig. 19-1, Table 19-1, Appendix 19-B). Harbor seals from oiled areas of PWS in 1989 and 1990 had significantly higher concentrations of fluorescent ACs than conspecifics from unoiled areas of PWS, the GOA, and Ketchikan (one-way ANOVA of log-transformed values, P<0.001). The highest concentrations were measured in three seals collected in oiled areas of PWS in 1989: a heavily oiled pregnant female collected in April in Herring Bay (TS-HS-1), a heavily oiled pup of an oiled female collected in June in Bay of Isles (TS-HS-8), and a heavily oiled subadult female collected in Herring Bay in May (AF-HS-1). Fluorescent ACs in these seals ranged from 79,000 to 215,000 ng/g phenanthrene equivalents and from 180,000 to 365,000 ng/g naphthalene equivalents.

There were no significant differences in the concentrations of fluorescent ACs in seals from oiled areas of PWS in 1989 and 1990 (P>0.16 for PHN equivalents;

Table 19-1. Mean concentrations of fluorescent aromatic compounds (ACs), measured at 260/380nm (phenanthrene equivalents) and 290/335nm (naphthalene equivalents), and average ratios of naphthalene (NPH) to phenanthrene (PHN) equivalents in the bile of harbor seals. Ranges are given in parentheses. All samples were analyzed by the ECD/NMFS using an HPLC screening method (Krahn et al. 1986). Individual data for all seals are shown in Appendix 19-B.

| Area/sample | Sample size | Equivalents of fluorescent ACs ng/g wet weight | | NPH/PHN |
		Phenanthrene equivalents	Naphthalene equivalents	
PWS 1989 oiled area[a]	13	43,192 (1300-215,000)	81,708 (2300-365,000)	2.2 (0.9-3.8)
PWS 1990 oiled area[a]	7	12,821 (4000-44,000)	49,929 (20,000-110,000)	5.0 (2.5-6.7)
PWS 1989/1990 unoiled area	4	1800 (1200-2500)	22,750 (12,000-33,000)	12.5 (8.0-16.5)
GOA 1989	7	2433 (170-8000)	7329 (1400-14,000)	4.8 (1.7-8.2)
Ketchikan 1990	2	598 (455-740)	4600 (3100-6100)	7.5 (6.8-8.2)

[a] Sample groups are as specified in Appendix 19-1, except as follows: the PWS 1989 oiled sample group does not include two pups found dead in PWS (LL-HS-1, MH-HS-3); the PWS 1990 oiled group does not include the fetus of TS-HS-23 (TS-HS-23F).

$P > 0.81$ for NPH equivalents), although the mean concentrations were considerably lower in 1990. There was considerable variability among seals. Biliary concentrations of fluorescent ACs in a pregnant female collected in April 1990 were among the highest measured during our study (TS-HS-23; 44,000 ng/g PHN equivalents and 110,000 ng/g NPH equivalents). Comparisons of seals from the GOA, which could have been exposed to oil, to seals from unoiled areas of PWS and Ketchikan indicated no significant differences in the concentrations of fluorescent ACs ($P > 0.79$ for PHN equivalents; $P > 0.21$ for NPH equivalents).

Fluorescent ACs in bile were markedly different in the mother and the pup/fetus for the three pairs for which we have data. In 1989, concentrations were much higher in the two pups when compared to their heavily oiled mothers. In contrast, the fetus collected in the spring of 1990 had much lower concentrations than did its mother.

There was a marked difference among sample groups in the ratio of naphthalene to phenanthrene equivalents (Table 19-1). Seals (not including fetuses) from oiled areas of PWS had a mean ratio of naphthalene to phenanthrene equivalents of 2.2

Table 19-2. Concentrations of polynuclear aromatic hydrocarbons in harbor seals from Prince William Sound (PWS), the Gulf of Alaska (GOA), and Ketchikan, 1989-1990. Values are given in parts per billion (= ng/g). Sample sizes are given in parentheses[a]. The analytical laboratory is shown in parentheses after each tissue type (N = ECD/NMFS, TX = GERG). (--- = no sample was analyzed; nd = not detected)

Tissue	PWS 1989 oiled seals		GOA 1989 most unoiled		PWS 1990 oiled areas		PWS 1990 unoiled areas		Ketchikan unoiled	
	mean	range	mean	range	mean	range	mean	range	mean	range
Liver (TX, N)	6	nd-45 (19)	3	nd-5 (7)	2	nd-15 (7)	4	1-6 (7)	nd	nd (2)
Blubber (TX,N)	191	nd-800 (17)	5	nd-23 (7)	46	21-101 (7)	5	2-10 (5)	nd	nd (2)
Muscle (N)	2	nd-10 (10)	nd	nd (7)	nd	nd-2 (7)	nd	nd (3)	nd	nd (2)
Brain (TX)	1	nd-11 (11)	23	nd-39 (5)	10	nd-37 (7)	---	---	nd	nd (2)
Kidney (TX,N)	nd	nd (4)	nd	nd (6)	---	---	---	---	nd	nd (2)
Heart (TX)	nd	nd (3)	---	---	---	---	---	---	nd	nd (2)
Lung (TX)	nd	nd (3)	---	---	---	---	---	---	nd	nd (2)
Blood (TX)	48	nd-73 (4)	---	---	19	19 (1)	---	---	16	11-22 (2)

[a] Sample size is the number of seals in the sample. If replicates were run for a particular tissue, the average of the values for the replicates was used as the value for that seal.

in April–June 1989 compared to 5.0 in April 1990. Seals collected in the GOA in 1989 had a mean naphthalene to phenanthrene ratio of 4.8. The ratios for seals from unoiled parts of PWS and Ketchikan were 12.5 and 7.5, respectively.

Concentrations of polycyclic aromatic hydrocarbons (PAHs) were near or below detection limits for muscle, liver, and brain in all seals regardless of location, year, or oiling status (Table 19-2, Appendix 19-C). Very few kidney, heart, and lung samples were analyzed, but concentrations of PAHs in all of them were also near or below detection limits.

The highest PAH concentrations were found in the blubber of seals in the PWS 1989 oiled sample group. PAH concentrations of over 100 ppb were found in the blubber of 8 of the 17 seals found dead or collected in oiled parts of PWS during April-July 1989 (two of eight pups and six of nine nonpups). The highest PAH concentrations were found in a heavily oiled adult female from Herring Bay (TS-HS-1), a heavily oiled pup found dead on Applegate Rocks (MH-HS-5), a heavily oiled subadult female collected in Herring Bay (AF-HS-1), and a heavily oiled mother–pup pair (TS-HS-7 and TS-HS-8) collected in the Bay of Isles. Concentrations of blubber PAHs were also slightly higher for seals from the oiled parts of PWS in April 1990 when compared to seals from unoiled areas during the same time period, animals collected in the GOA in June–July 1989, or the two seals collected near Ketchikan in August 1990. The average ratio of high-molecular-weight aromatic compounds [4- to 7-ring aromatic compounds, e.g., chrysenes and benzo(a)pyrene] to low-molecular-weight aromatic compounds (2- to 3-ring aromatic compounds, naphthalenes, and phenanthrenes) for oiled seals collected in PWS in 1989 was 0.02, indicating that most PAHs were low-molecular-weight compounds. By 1990, this ratio increased to 0.25, indicating either the loss of the low-molecular-weight compounds from the exposure source, the metabolism of these compounds by seals, or a combination of the two.

A one-way ANOVA conducted on log-transformed values for PAH concentrations in blubber indicated significant differences between seals from oiled and unoiled areas (PWS oiled 1989 + 1990 versus GOA 1989 + PWS unoiled + Ketchikan; $P<0.001$) and between oiled PWS 1989 versus oiled PWS 1990 ($P<0.01$). The difference in PAH concentrations between GOA seals and seals from unoiled areas of PWS was not significant ($P>0.10$). Sample groups used in these statistical analyses did not include blubber samples from premature pups or pups that were only a few days old.

Mammary tissue and/or milk were analyzed from eight adult females and two pups collected in PWS and the GOA in 1989 (Table 19-3). Average PAH concentrations in mammary tissue (16 ppb, range nd-87 ppb) and mother's milk (3 ppb, range nd-11 ppb) were similar. The PAH values for mammary tissue and milk in the mother and milk in the pup were similar for one heavily oiled mother-pup pair (TS-HS-3 and 4). For the other (TS-HS-7 and 8), PAHs were not detected in

>⋄ ⋙ ⋄

Table 19-3. Concentrations of polynuclear aromatic hydrocarbons (PAH) in blubber, mammary tissue, and milk samples (parts per billion) (= ng/g). Equivalents of fluorescent aromatic compounds in bile (phenanthrene equivalents, ng/g wet weight) are also shown. Milk and mammary samples were analyzed by GERG; bile samples were analyzed by ECD/NMFS; and blubber samples were analyzed by both ECD/NMFS and GERG. (— = no sample was analyzed; nd = not detected). If replicates were run, the average values for the replicates are shown.

		PAHs				
			Adult females		Pups	
Specimen number	Phenanthrene equivalents	Blubber	Mammary	Milk	Milk	Comments
TS-HS-3	2700	48	nd	nd		
TS-HS-4	25,000	26			nd	Pup of TS-3
TS-HS-5	32,500	90	nd	11		
TS-HS-7	36,000	501	44	—		
TS-HS-8	215,000	436			1142	Pup of TS-7
TS-HS-10	18,500	151	nd	nd		
TS-HS-11	15,000	106	nd	—		
TS-HS-13	2200	nd	87	—		
TS-HS-16	800	1	nd	nd		
TS-HS-17	2200	2	nd	—		

milk from mammary tissue of the mother, while milk from the pup's stomach had the highest PAH value (1142 ppb) in any tissue from any seal that we analyzed. Concentrations of fluorescent ACs in the bile of TS-HS-8 were also the highest for any seal examined in this study.

The aliphatic hydrocarbon phytane occurred at high concentrations (>1000 ppb) in the brains of 7 of 11 oiled seals collected in PWS in 1989 (Table 19-4, Appendix 19-D). One seal (TS-HS-5) had a brain phytane concentration of 7839 ppb. The pristane/phytane ratio in these seals averaged 0.02. In the brains of all other seals, the concentration of phytane was very low or absent: not detected to 29 ng/g in four seals collected in the GOA, not detected in the brains of seven seals collected in PWS in November 1989–April 1990, and not detected in the Ketchikan seals (Table 19-4). Pristane/phytane ratios were 4.2–8.2, or could not be calculated because phytane was not detected.

The phytane concentrations in the other 73 tissue samples ranged from not detectable to 334 ppb. The highest concentration occurred in the milk of pup TS-HS-8. Phytane exceeded 100 ppb in tissues of five oiled seals from PWS 1989, and included mammary (*n*=2), milk (*n*=2), blubber (*n*=1), ovary (*n*=1), and liver (*n*=1) samples.

The unresolved complex mixture (UCM) is sometimes used as an indicator of the presence of petroleum hydrocarbons in sediments. UCM was low in all tissues except the brain, exceeding 10 ppm in only 7 of 73 nonbrain samples. The mean UCM in the brains of oiled PWS seals in 1989 was 264 ppm, compared to 826 ppm in GOA 1989 seals and 778 ppm in PWS 1990 seals from oiled areas. The UCM was 0–6 ppm in the brains of two Ketchikan seals.

DISCUSSION

Crude oil is a complex mixture of hydrocarbons ranging widely in structure, molecular weight, and toxicity. In general, the aromatic compounds (ring structures such as benzene, naphthalene, and phenanthrene) are more toxic than the aliphatic compounds (straight and branched chain structures such as phytane and C_{10} alkanes). Low-molecular-weight aromatic compounds (naphthalenes and phenanthrenes) are more acutely toxic than the high-molecular-weight aromatic compounds [chrysenes and benzo(a)pyrene], which have been identified as carcinogens and mutagens (Cerniglia and Heitkamp 1989). Every oil spill is unique as a result of a combination of weather, location and timing of the spill, and the type of oil spilled. Once crude oil is released into the environment, composition changes rapidly with the dissolution and evaporation of the low-molecular-weight compounds. Effects of the oil depend on the composition of the oil, and the route and length of exposure; all of these parameters are highly variable and cannot be controlled in the sampling design of a study such as this.

⋈   ⋈   ⋈

Table 19-4. Concentrations of pristane and phytane in tissues from harbor seals. Values are sample averages given in parts per billion (=ng/g). Sample sizes are shown in parentheses. All samples were analyzed by GERG. (— = no sample was analyzed).

Sample		Brain	Blubber	Blood	Kidney	Liver	Lung	Mammary
PWS 1989	Pristane	2	100,773	185	5117	7402	407	96,324
(TS-HS-1-11)	Phytane	2176	32	0	9	18	0	98
		(11)	(4)	(4)	(4)	(4)	(3)	(5)
GOA 1989	Pristane	181	—	—	—	—	—	56,607
(TS-HS-12-18)	Phytane	18	—	—	—	—	—	0
		(5)						(3)
PWS 1990	Pristane	135	—	291	—	—	—	—
(TS-HS-19-25)	Phytane	0	—	0	—	—	—	—
		(7)		(1)				
Ketchikan	Pristane	35	41,287	26	211	64	2024	—
(TS-HS-26-27)	Phytane	0	25	0	0	0	0	—
		(2)	(2)	(2)	(2)	(1)	(2)	

Petroleum hydrocarbons may be taken into the body of seals through dermal contact and absorption, ingestion, and inhalation (Engelhardt et al. 1977; Klassen 1986; Engelhardt 1987). Mammals are able to metabolize hydrocarbons through mixed-function oxidases in the liver that convert the hydrocarbons to metabolites that are excreted in urine and bile (Addison et al. 1986). The mechanisms used by seals for detoxification and excretion were described by Engelhardt et al. (1977). As they noted, at some hydrocarbon concentration it is likely that the detoxifying and excretory mechanisms would cease to function, but the concentration at which this might occur is not known.

The presence of petroleum-related AC metabolites in bile is a useful indicator of relatively recent exposure to oil (Krahn et al. 1992). Furthermore, HPLC chromatograms and GC/MS analyses of bile can be used to suggest the source of contamination (Krahn et al. 1993). The HPLC chromatograms of bile from PWS seals following the EVOS were similar to those of fish experimentally exposed to Prudhoe Bay crude oil. The GS/MS analysis of hydrolyzed seal bile confirmed that petroleum-related ACs, including dibenzothiophenol marker compounds characteristic of Prudhoe Bay crude oil, were present in high proportions (Krahn et al. 1993).

The concentrations of AC metabolites in bile were greatly elevated in harbor seals from oiled areas of PWS: the mean concentration of PHN equivalents for oiled seals from PWS was over 70 times greater than that for Ketchikan seals and approximately 20 times greater than that for seals from unoiled PWS areas or the GOA. The highest phenanthrene equivalent concentrations in oiled PWS seals were over 1000 times greater than for unexposed seals. Mean concentrations of fluorescent ACs in bile from 10 ringed seals (*Phoca hispida*) collected 3000 km away near Point Barrow, Alaska, (882 ng/g PHN equivalents, range 300–1700; 11,510 ng/g NPH equivalents, range 4100–22,000; Becker, unpublished data for samples analyzed by ECD) were 1/50th of the concentrations found in seals from oiled areas of PWS in 1989 and were similar to concentrations in reference seals from Ketchikan. The low concentrations of fluorescent ACs in seals from the GOA, which was in the path of the spill, and their similarity to levels recorded for seals from remote unoiled areas suggests that either the GOA seals that we sampled had little exposure to oil, or that most of the aromatic fraction of the oil had evaporated by the time it reached the GOA.

The significance of the marked differences among sample groups in the ratio of naphthalene to phenanthrene equivalents in bile is unknown, but it is clear that the ratios were lowest for the seals that were heavily and recently exposed to oil and highest for those collected in unoiled areas. Ringed seals from Barrow, like harbor seals from unoiled areas, also had very high ratios of naphthalene to phenanthrene equivalents in bile (12.7, range 9.5–18.3).

⋊⋉ ⋊⋉ ⋊⋉

One year after the EVOS, petroleum-related metabolites were still substantially elevated in the bile of seals from areas of PWS that had been oiled, indicating either that seals continued to encounter oil in the environment or that they were metabolizing stored fat reserves that had elevated levels of hydrocarbons. This is not unexpected since shoreline surveys in spring 1990 documented the presence of oil on and in many originally oiled beaches (Maki 1991). Fish collected in PWS during spring 1990 also showed elevated levels of AC metabolites in bile, indicating recent exposure to crude oil (Krahn et al. 1992).

Deleterious effects of the systemic accumulation of petroleum-related hydrocarbons are difficult to determine. Because all vertebrates rapidly and efficiently metabolize PAHs, analyses of tissues for these compounds usually result in values at or below the detection limits for most analytical methods. Consequently, the very low concentrations of PAHs in most seal tissues, despite the elevated concentrations of AC metabolites in bile, were not surprising. The elevated PAH concentrations in the blubber of seals collected in oiled areas of PWS in 1989 and 1990, and in milk from the stomach of a pup, are indicative of exposure to hydrocarbon concentrations so high that the seals could not completely metabolize them, and that PAHs were instead accumulated in lipid-rich blubber or milk. Hydrocarbons are lipophilic and may concentrate in such lipid-rich tissues. The PAHs stored in blubber may have been a mechanism for chronic exposure at a later date.

The lipid-rich milk produced by seals provides a possible mechanism for transfer of stored hydrocarbon contaminants from the mother to the pup (Addison et al. 1986). PAHs were detectable in samples of milk and mammary tissue from oiled seals, but in most cases the concentrations were not particularly high. However, the highest concentration of aromatic hydrocarbons in any tissue that we examined was in milk from the stomach of a pup. Both the pup and its mother were heavily oiled. Both had high PAH concentrations in their blubber, and bile from the pup had the highest concentration of PAH metabolites of any seal tested. The mother had obviously assimilated hydrocarbons, some of which were probably present in her milk and transferred to, metabolized by, and stored in her pup. However, because PAH concentrations in mammary tissue from the mother were less than those in the milk from the pup's stomach, and since PAH concentrations were similar in mammary tissue and milk in the four other females that were analyzed, it is likely that the pup ingested oil from the mother's fur during suckling (Lowry et al., Chapter 12). Concentrations of biliary AC metabolites were higher in the pup than in the mother for both this pup and another that had also nursed on an oiled female. This observation further suggests that pups were exposed to a source of hydrocarbons that their mothers were not, which was probably oiled fur that they contacted during suckling.

Phytane is a petrogenic aliphatic hydrocarbon characteristic of petroleum which is relatively resistant to metabolism and is usually found in low concentrations

✖ ✖ ✖

(<1 ppb) in unoiled sediments. Pristane is a biogenic hydrocarbon that occurs commonly in biota and in recent sediments as a degradation product of chlorophyll. A very low ratio of pristane to phytane is a commonly accepted indicator of the presence of petroleum hydrocarbons and a measure of the extent of petroleum hydrocarbon input (National Academy of Sciences 1985). The pristane/phytane ratios of 0.01–0.04 that were found in the brains of seals collected in PWS following the EVOS would, in sediments, be considered indicative of heavy contamination by petroleum hydrocarbons (National Academy of Sciences 1985). This is in contrast to ratios of 4.2–8.2 in GOA 1989 seals, which are similar to usual background ratios in sediments. The pristane/phytane ratio was 1.4 in oil spilled during the EVOS and 0.3–1.2 in residue of that oil found on beaches (Kvenvolden et al. 1993). The absence of phytane in some seals that had other indications of exposure to oil (nerve damage, high PAHs in blubber, high PAH metabolites in bile) is unexplained. Without more information on how and when seals were exposed, it is not possible to relate variability in phytane to exposure history or to make any statement as to the cause or effect of the presence of this compound in brain samples.

Following the EVOS, the rapid release of crude oil in calm weather allowed the more volatile components of the oil to reach very high concentrations over the slick. The fact that oil fumes were causing eye irritation and headaches in some researchers 8 days after the grounding, together with the oiling patterns seen on seals at this time (Lowry et al., Chapter 12), suggest that harbor seals were surfacing through the slick and inhaling these vapors. Acute high-level exposure to volatile petroleum-related hydrocarbons, at concentrations such as those calculated by Wolfe et al. (submitted) for the EVOS, can cause narcosis and death in mammals. Disorientation, euphoria, confusion, unconsciousness, paralysis, convulsion, and death from respiratory or cardiovascular failure are typically observed in mammals upon acute high-level exposure to these compounds (Browning 1965; Engelhardt 1987). In humans, short-term exposure to low-molecular-weight aliphatics usually causes symptoms for several weeks, followed by recovery (Pleasure and Schotland 1984). Because breathing in marine mammals is voluntary, respiratory effects on harbor seals may have been accentuated.

It is likely that the primary impact on harbor seals of exposure to crude oil following the EVOS was due to inhalation of volatile, short-chain aromatic hydrocarbons. Seals may have been exposed to levels of volatile hydrocarbons sufficient to cause respiratory or cardiac arrest or to interfere with normal breathing patterns. Seven of the eight nonpup seals collected in PWS following the EVOS had brain lesions, including intramyelinic edema, neuronal swelling and necrosis, and axonal degeneration (Spraker et al., Chapter 17), which are consistent with central nervous system damage caused by highly volatile C_5–C_8 hydrocarbons (Cornish 1980). Both females of the two heavily oiled mother–pup pairs that we collected had significant brain lesions, but no lesions were found in the brain of either pup. Since most of the aromatic hydrocarbons would probably have dissipated by late May or early June when pups were born, pups should have had a much lower level of exposure to those compounds. If inhalation was a primary route of exposure, this could explain why lesions were found in some seals that were collected near areas affected by the spill but that showed little or no evidence of external oiling.

We examined the principal indicators of exposure to oil measured during this study to determine whether any combination occurred consistently in oiled seals (Table 19-5). The incidence of each of the indicators was greater in oiled than in unoiled sample groups, but the relationships among the different indicators were not entirely consistent. Six of the seven seals that had high PAHs in the blubber also showed elevated hydrocarbon metabolites in the bile. Six of the seals that had high levels of phytane in the brain were examined histologically, and four of them showed signs of nerve damage. Of 11 seals with brain damage, 8 showed evidence of a high level of hydrocarbon exposure based on metabolites in the bile, but blubber PAH levels were elevated in only 4.

The lack of complete correlation among indicators can be explained by the physiology of seals and the timing of sample collections. Initial exposure to volatile compounds in the oil was probably intense but relatively short lived. Exposure may have been sufficient to cause nerve damage even though metabolic processes were able to break down the compounds and excrete them in the bile. Several weeks after acute exposure, the brain damage would remain, but bile metabolite levels would have dropped to much lower levels. Accumulation of PAHs in the blubber would have occurred only when exposure exceeded the ability of liver enzyme systems to metabolize hydrocarbons. The one seal that was collected relatively soon after the spill showed all indicators of exposure to and assimilation of oil. All other seals were collected three or more months after the spill, long enough for nerve damage to persist but for other indicators of exposure to be more variable. Clearly, in the future it would be important to collect sufficient samples as soon as possible after a spill and at regular intervals thereafter.

Other studies of pinnipeds have produced equivocal and sometimes contradictory results regarding effects of exposure to oil (St. Aubin 1990). Mortality of seals has been attributed to the effects of oil released during the *Torrey Canyon* (Spooner 1967), *Arrow* (Anonymous 1970), and *Kurdistan* (Parsons et al. 1980) spills but the cause of death, while not clearly established, was thought to be the result of external oiling and not systemic accumulation. Most laboratory studies have involved short-term exposure to relatively small doses of oil. Geraci and Smith (1976) conducted a study in which three ringed seals were put in a tank, the surface of which was covered with a 1-cm-thick layer of Norman Wells crude oil. The seals immediately showed signs of extreme distress, and they died after exposures of 21, 60, and 71 minutes. Although the results of this experiment may have been

Table 19-5. Summary of indications of exposure to and damage caused by oil in harbor seals collected in Prince William Sound, the Gulf of Alaska, and near Ketchikan, 1989-1990. Data about nerve damage and interleukin levels in blood are from Spraker et al. (Chapter 17) and Frost and Lowry (1993). (--- = no sample was analyzed)

Specimen number	External oiling	Nerve damage in brain	Fluorescent ACs in bile >20,000	PAHs in blubber >100 ppb [a]	Phytane in brain >1000 ppb	Interleukin in blood
TS-HS-1	yes	yes	yes	yes	yes	yes
TS-HS-2	yes	yes	yes	no	yes	---
TS-HS-3	yes	yes	no	no	yes	yes
TS-HS-4	yes	---	yes	no	yes	no
TS-HS-5	yes	yes	yes	no	yes	yes
TS-HS-6	yes	no	no	no	yes	no
TS-HS-7	yes	yes	yes	yes	no	no
TS-HS-8	yes	no	yes	yes	yes	yes
TS-HS-9	yes	no	no	yes	no	yes
TS-HS-10	yes	yes	yes	yes	no	yes
TS-HS-11	yes	yes	yes	yes	no	no
TS-HS-12	no	---	no	no	---	no
TS-HS-13	no	no	no	no	no	---
TS-HS-14	yes	yes	yes	no	no	no
TS-HS-15	no	no	no	no	---	no
TS-HS-16	no	yes	no	no	no	no
TS-HS-17	yes	yes	no	no	no	no
TS-HS-18	no	no	yes	no	no	yes
TS-HS-19	no	yes	yes	no	no	yes
TS-HS-20	no	no	yes	no	no	yes
TS-HS-21	no	no	yes	no	no	no
TS-HS-22	no	no	yes	no	no	no

(Table continues)

Table 19-5. Continued

Specimen number	External oiling	Nerve damage in brain	Fluorescent ACs in bile > 20,000	PAHs in blubber > 100 ppb[a]	Phytane in brain > 1000 ppb	Interleukin in blood
TS-HS-23	no	no	yes	no	no	yes
TS-HS-24	no	no	yes	no	no	no
TS-HS-25	no	no	yes	yes	no	yes
TS-HS-26	no	no	no	---	no	no
TS-HS-27	no	no	no	---	no	no

[a] Based on values from analyses done by ECD/NMFS. Blubber samples from TS-HS-26 and TS-HS-27 were not analyzed by ECD/NMFS and results therefore were not used in this table.

influenced by the captive setting, it nonetheless indicates that contact with oil such as occurred as a result of the EVOS may contribute to the death of seals. Furthermore, the EVOS, unlike laboratory experiments where seals were exposed to oil and then returned to clean water, in many cases resulted in long-term exposure to heavy concentrations of oil.

CONCLUSIONS

As a consequence of the EVOS, harbor seals in PWS and in parts of the GOA were exposed to petroleum-related hydrocarbons. Harbor seal tissue samples collected for chemical analyses following the spill and presently archived at the National Marine Mammal Laboratory in Seattle form the most extensive collection of tissues and data from phocid seals exposed to crude oil that is currently available. Over 1300 samples representing 15 tissues from 44 seals are archived and constitute an unparalleled resource for the study of marine contaminants in this species.

Some tissues from seals found dead or collected in oiled areas of PWS in 1989 contained elevated concentrations of petroleum-related hydrocarbons (i.e., phytane in brain samples and PAHs in blubber samples) when compared to tissues from seals collected in the GOA or out of the path of the spilled oil. Concentrations of AC metabolites in the bile of oiled seals collected in oiled areas of PWS in 1989 were significantly higher than concentrations in the bile of seals collected in the GOA and near Ketchikan. One year after the EVOS, none of the tissues from seals collected in the spill area showed significantly elevated concentrations of petroleum-related hydrocarbons. However, average concentrations of AC metabolites in bile were still significantly higher than those observed for seals from the GOA or unoiled PWS areas. These data support the hypothesis that harbor seals in PWS were exposed to high concentrations of petroleum-related hydrocarbons in the spring and summer of 1989. The level of this exposure in PWS declined in 1990 but was still greater than exposure in the GOA or outside of the spill path.

The implications of the results of hydrocarbon analyses for the health of the seals are unknown. Levels of hydrocarbons in seal tissues were low. However, since seals metabolize hydrocarbons very efficiently, the levels remaining in tissues when they were sampled underestimate the actual degree of exposure. Essentially, no information is available on the likely effects of hydrocarbons on seals for anything other than short-term experimental exposure. It is important to note that chemical analyses did not measure the most volatile and acutely toxic C_5–C_8 hydrocarbons, which have been documented to cause mortality in other mammals and which were the most likely cause of the nerve damage that we observed in oiled seals (Spraker et al., Chapter 17).

The poor correlation between different indicators of exposure to and metabolism of petroleum hydrocarbons, particularly in seals collected several months or more after the spill, complicates interpretation of the data from the EVOS. Although the prevalence of each of the indicators was greater in the oiled group than in other seals, no one indicator adequately indicated exposure in all seals. While this was not unexpected because of the different exposure histories of each seal, it makes it very difficult to recommend a simple sampling protocol or a single parameter to use as a measure of exposure in the event of a future spill. Based on the data from this study, we recommend that a suite of indicators be measured, including fluorescent AC metabolites in bile, PAHs in blubber (and milk when available), and phytane in brain. Because of the very efficient metabolism of hydrocarbons by seals, it is probably not very informative to analyze tissues other than blubber and milk for PAHs. It is essential that chemical analyses be accompanied by careful necropsy and histopathology of affected seals, with particular emphasis on the brain.

ACKNOWLEDGMENTS

We thank the crews of the ADF&G research vessel *Resolution* and the NOAA vessel *1273* for their assistance and support in field studies, and all the people who assisted in collecting and handling samples, especially J. Lewis, M. Bates, C. Armisted, R. Haebler, K. Pitcher, A. Franzmann, and E. Sinclair. Thanks to laboratory personnel at GERG and ECD for their careful analysis of samples and to J. Ver Hoef for analysis of the data. M. Krahn was especially helpful in patiently answering a myriad of questions about the analysis of bile. Special thanks to L. Lowry who assisted with all aspects of this study. This study was conducted in cooperation with the National Marine Fisheries Service, National Marine Mammal Laboratory, as part of the Natural Resource Damage Assessment Study, funded by the *Exxon Valdez* Oil Spill Trustee Council. Helpful comments on the draft manuscript were provided by T. Loughlin, P. Becker, and two anonymous reviewers.

REFERENCES

Addison, R. F., P. F. Brodie, A. Edwards, and M. C. Sadler. 1986. Mixed function oxidase activity in the harbour seal (*Phoca vitulina*) from Sable Is., N.S. Comparative Biochemical Physiology 85C(1):121–142.

Anonymous. 1970. Report of the Task Force—Operation Oil (clean-up of the *Arrow* oil spill in Chedabucto Bay). Ottawa: Canadian Ministry of Transportation. 164 p.

Brooks, J. M., M. C. Kennicutt II, T. L. Wade, A. D. Hart, G. J. Denoux, and T. McDonald. 1990. Hydrocarbon distribution around a shallow water multiwell platform. Environmental Science and Technology 24:1079–1085.

Browning, E. 1965. Toxicity and metabolism of industrial solvents. Elsevier Applied Science Publishers, New York, New York. 65 p.

Cerniglia, C. E., and M. A. Heitkamp. 1989. Microbial degradation of polycyclic aromatic hydrocarbons (PAH) in the aquatic environment. Pages 41–69, *in* U. Varanasi (ed.), Metabolism of polycyclic aromatic hydrocarbons in the aquatic environment. CRC Press, Inc., Boca Raton, Florida.

Cornish, H. H. 1980. Solvents and vapors. Pages 468–496, *in* J. Doull, C. D. Klaassen, and M. O. Amdur (eds.), Casarett and Doull's toxicology. MacMillan Publishing Company, New York, New York.

Engelhardt, F. R. 1987. Assessment of the vulnerability of marine mammals to oil pollution. Pages 101–115, *in* J. Kiuper and W. J. Van Den Brink (eds.), Fate and effects of oil in marine ecosystems. Martinus Nijhoff Publishing, Boston, Massachusetts.

Engelhardt, F. R., J. R. Geraci, and T. J. Smith. 1977. Uptake and clearance of petroleum hydrocarbons in the ringed seal, *Phoca hispida*. Journal of the Fisheries Research Board of Canada 34:1143–1147.

Frost, K. J., and L. F. Lowry. 1993. Assessment of injury to harbor seals in Prince William Sound, Alaska, and adjacent areas following the *Exxon Valdez* oil spill. Final Report, Marine Mammal Study Number 5, State-Federal Natural Resource Damage Assessment. 95 p.

Geraci, J. R., and T. J. Smith. 1976. Direct and indirect effects of oil on the ringed seals (*Phoca hispida*) of the Beaufort Sea. Journal of the Fisheries Research Board of Canada 33:1976–1984.

Geraci, J. R., and D. J. St. Aubin. 1987. Effects of offshore oil and gas development on marine mammals and turtles. Pages 587–617, *in* D. F. Boesch and N. N. Rabalais (eds.), Long-term environmental effects of offshore oil and gas developments. Elsevier Applied Science, New York, New York.

Klassen, C. D. 1986. Distribution, excretion and absorption of toxicants. Pages 11–32, *in* C. D. Klassen, M. O. Amdur, and J. Doull (eds.) Toxicology: the basic science of poisons. Macmillan Publishing Co., New York, New York.

Krahn, M. M., L. K. Moore, and W. D. MacLeod, Jr. 1986. Standard analytical procedures of the NOAA National Analytical Facility, 1986: metabolites of aromatic compounds in fish bile. U.S. Department of Commerce, NOAA Technical Memorandum, NMFS F/NWC 102. 25 p.

Krahn, M. M., L. K. Moore, R. G. Bogar, C. A. Wigren, S.-L. Chan, and D. W. Brown. 1988. High-performance liquid chromatographic method for isolating organic contaminants from tissue and sediment extracts. Journal of Chromatography 437:161–175.

Krahn, M. M., D. G. Burrows, G. M. Ylitalo, D. W. Brown, C. A. Wigren, T. K. Collier, S.-L. Chan, and U. Varanasi. 1992. Mass spectrometric analysis for aromatic compounds in bile of fish sampled after the *Exxon Valdez* oil spill. Environmental Science and Technology 26:116–126.

Krahn, M. M., G. M. Ylitalo, J. Buzitis, S.-L. Chan, and U. Varanasi. 1993. Review: Rapid high-performance liquid chromatographic methods that screen for aromatic compounds in environmental samples. Journal of Chromatography 642:15–32.

Kvenvolden, K. A., F. D. Hostettler, J. B. Rapp, and P. R. Carlson. 1993. Hydrocarbons in oil residues on beaches of islands of Prince William Sound, Alaska. Marine Pollution Bulletin 26:24–29.

Maki, A. W. 1991. The *Exxon Valdez* oil spill: Initial environmental impact assessment. Environmental Science and Technology 25:24–29.

National Academy of Sciences. 1985. Oil in the sea: Inputs, fates and effects. National Academy Press, Washington, D.C. 601 p.

Parsons, J., J. Spry, and T. Austin. 1980. Preliminary observations on the effect of Bunker C fuel oil on seals on the Scotian shelf. Pages 193–202, *in* J. H. Vandermeulen (ed.), Scientific studies during the *Kurdistan* tanker incident: Proceedings of a workshop. Report series BI-R-80-3. Bedford Institute of Oceanography, Dartmouth, Nova Scotia, Canada.

Pitcher, K. W., and D. G. Calkins. 1979. Biology of the harbor seal, *Phoca vitulina richardsi*, in the Gulf of Alaska. U.S. Department of Commerce, Environmental Assessment of the Alaskan Continental Shelf, Final Reports of Principal Investigators 19(1983):231–310.

✄ ✄ ✄

Pleasure, D. E., and D. L. Schotland. 1984. Acquired neuropathies. Pages 484–498, *in* L. P. Rowland (ed.), Merritt's textbook of neurology. 7th edition. Lea & Febiger, Philadelphia, Pennsylvania.

Sloan, C. A., N. G. Adams, R. W. Pearce, D. W. Brown, and S.-L. Chan. 1993. Northwest Fisheries Science Center organic analytical procedures. Pages 53–97, *in* G. G. Lauenstein and A. Y. Cantillo (eds.), Sampling and analytical methods of the National Status and Trends Program, National Benthic Surveillance and Mussel Watch Projects 1984–1992, Vol. IV, Comprehensive descriptions of trace organic analytical methods. U.S. Department of Commerce, NOAA Technical Memorandum NOS ORCA 71.

Spooner, M. F. 1967. Biological effects of the *Torrey Canyon* disaster. Journal of the Devonshire Trust Nature Conservancy Supplement 1967:12–19.

St. Aubin, D. J. 1990. Physiologic and toxic effects of oil on pinnipeds. Pages 103–127, *in* J. R. Geraci and D. J. St. Aubin (eds.), Sea mammals and oil: Confronting the risks. Academic Press, San Diego, California.

Varanasi, U., D. W. Brown, T. Hom, D. G. Burrows, C. A. Sloan, L. J. Field, J. E. Stein, K. L. Tilbury, B. B. McCain, and S.-L. Chan. 1993a. Volume I: Survey of Alaskan subsistence fish, marine mammal and invertebrate samples collected 1989–91 for exposure to oil spilled from the *Exxon Valdez*. U.S. Department of Commerce, NOAA Technical Memorandum NMFS-NWFSC-12. 110 p.

Varanasi, U., D. W. Brown, T. Hom, D. G. Burrows, C. A. Sloan, L. J. Field, J. E. Stein, K. L. Tilbury, B. B. McCain, and S.-L. Chan. 1993b. Volume II: Supplemental information concerning a survey of Alaskan subsistence fish, marine mammal and invertebrate samples collected 1989–91 for exposure to oil spilled from the *Exxon Valdez*. U.S. Department of Commerce, NOAA Technical Memorandum NMFS-NWFSC-13. 173 p.

Wolfe, D. A., M. J. Hameedi, J. A. Galt, G. Watabayasi, J. W. Short, C. E. O'Clair, S. Rice, J. Michel, J. R. Payne, J. F. Braddock, S. Hanna, and D. M. Sale. 1993. Fate of the oil spilled from the T/V *Exxon Valdez* in the Prince William Sound, Alaska. Pages 6–9, *in Exxon Valdez* oil spill symposium, 2–5 February 1993, Anchorage, Alaska. (Available, Oil Spill Public Information Center, 645 G Street, Anchorage, AK 99501.)

Wolfe, D. A., M. J. Hameedi, J. A. Galt, G. Watabayasi, J. W. Short, C. E. O'Clair, S. Rice, J. Michel, J. R. Payne, J. F. Braddock, S. Hanna, and D. M. Sale. Submitted. Fate of the oil spilled from the T/V *Exxon Valdez* in the Prince William Sound, Alaska. Environmental Science and Technology.

Appendix 19-A. Information on harbor seals from which tissues were collected and analyzed for the presence of hydrocarbon contaminants. Sample groups used for data analysis were: PWS 1989 oiled = AF-HS-1-2, LL-HS-1, MH-HS-5, 6, 10-15, TS-HS-1-11; GOA 1989 = TS-HS-12-18; PWS 1990 = TS-HS-19-25; PWS unoiled = MH-HS-4, HBSL-1-3, JT-2-5; Ketchikan 1990 = TS-HS-26-27. Data for HBSL-1-3 and JT-2-5 were provided by Paul Becker of NOAA/NMFS. All specimen numbers marked with an asterisk represent seals collected by this project. Others were found dead, killed by subsistence hunters, or died at rehabilitation facilities.

Specimen number	Date found/ collected	Location	Degree of oiling	Comments
AF-HS-1*	16 May 1989	Herring Bay, PWS	very heavy	subadult female
AF-HS-2	16 May 1989	Herring Bay, PWS	unoiled	pup
LL-HS-1	15 May 1989	Herring Bay, PWS	light	pup
MH-HS-3	19 April 1989	Green Island, PWS	unoiled	premature pup
MH-HS-4	20 April 1989	Tatitlek Narrows, PWS	unoiled	subsistence kill, juvenile
MH-HS-5	21 April 1989	Applegate Rocks, PWS	heavy	premature pup, scavenged
MH-HS-6	1 May 1989	Herring Bay, PWS	heavy	captured alive and died, adult
MH-HS-7	28 April 1989	Windy Bay, GOA	heavy	predated or scavenged, adult
MH-HS-10	30 May 1989	Herring Bay, PWS	heavy	pup
MH-HS-12	2 May 1989	Herring Bay, PWS	moderate	in lanugo when caught, rehabilitated pup, died May 31 in captivity
MH-HS-13	3 May 1989	PWS	heavy	rehabilitated pup, died May 31 in captivity
MH-HS-15	9 July 1989	Herring Bay, PWS	heavy	pup
TS-HS-1*	29 April 1989	Herring Bay, PWS	very heavy	adult female, pregnant
TS-HS-1F				fetus of TS-HS-1
TS-HS-2*	16 June 1989	Bay of Isles, PWS	very heavy	adult male
TS-HS-3*	16 June 1989	Seal Island, PWS	heavy	adult female
TS-HS-4*	16 June 1989	Seal Island, PWS	heavy	pup of TS-HS-3
TS-HS-5*	17 June 1989	Bay of Isles, PWS	very heavy	adult female

(Appendix continues)

Appendix 19-A. Continued

Specimen number	Date found/ collected	Location	Degree of oiling	Comments
TS-HS-6*	17 June 1989	Applegate Rocks, PWS	light	pup
TS-HS-7*	17 June 1989	Bay of Isles, PWS	very heavy	adult female
TS-HS-8*	17 June 1989	Bay of Isles, PWS	very heavy	pup of TS-HS-7
TS-HS-9*	18 June 1989	Herring Bay, PWS	very heavy	adult male
TS-HS-10*	18 June 1989	Herring Bay, PWS	very heavy	adult female
TS-HS-11*	18 June 1989	Herring Bay, PWS	very heavy	adult female
TS-HS-12*	25 June 1989	Perenosa Bay, GOA	unoiled	adult female
TS-HS-13*	25 June 1989	Perenosa Bay, GOA	unoiled	female
TS-HS-14*	29 June 1989	W Amatuli Island, GOA	moderate	adult male
TS-HS-15*	30 June 1989	Ushagat Island, GOA	unoiled	adult male
TS-HS-16*	30 June 1989	Ushagat Island, GOA	unoiled	adult female
TS-HS-17*	6 July 1989	Perl Island, GOA	light	adult female
TS-HS-18*	26 October 1989	Big Fort Island, GOA	unoiled	subadult male
TS-HS-19*	1 November 1989	Agnes Island, PWS	unoiled	adult male
TS-HS-20*	11 April 1990	Herring Bay, PWS	unoiled	subadult male
TS-HS-21*	12 April 1990	Herring Bay, PWS	unoiled	subadult male
TS-HS-22*	12 April 1990	Herring Bay, PWS	unoiled	adult male
TS-HS-23*	12 April 1990	Eleanor Island, PWS	unoiled	adult female, pregnant
TS-HS-23F				fetus of TS-HS-23
TS-HS-24*	12 April 1990	Herring Bay, PWS	unoiled	adult male
TS-HS-25*	13 April 1990	Bay of Isles, PWS	unoiled	adult male
TS-HS-26*	15 August 1990	Ketchikan	unoiled	adult female
TS-HS-27*	16 August 1990	Ketchikan	unoiled	adult male
HBSL-1	10 March 1990	New Year Island, PWS	unoiled	subadult male
HBSL-2	22 April 1990	Galena Bay, PWS	unoiled	adult female

(Appendix continues)

Appendix 19-A. Continued

Specimen number	Date found/ collected	Location	Degree of oiling	Comments
HBSL-3	22 April 1990	Galena Bay, PWS	unoiled	adult male
JT-2	15 March 1990	Little Green Island, PWS	unoiled	adult male
JT-3	15 March 1990	Little Green Island, PWS	unoiled	adult male
JT-4	15 March 1990	Little Green Island, PWS	unoiled	adult male
JT-5	15 March 1990	Little Green Island, PWS	unoiled	adult male

353

Appendix 19-B. Results of HPLC fluorometric analysis of bile from harbor seals collected in Prince William Sound and the Gulf of Alaska, 1989-1990. Equivalents of fluorescent aromatic compounds (ACs) were measured at 260/380nm (phenanthrene wave lengths) and 290/335nm (naphthalene wave lengths). All samples were analyzed by ECD/NMFS. Data for HBSL-1-3 were provided by Paul Becker, NOAA/NMFS.

Specimen number	Equivalents of fluorescent ACs, ng/g wet weight		Comments
	Phenanthrene equivalents	Naphthalene equivalents	
AF-HS-1	79,000	180,000	subadult female, heavily oiled
LL-HS-1	2000	13,000	dead pup, lightly oiled
MH-HS-3	4000	51,000	dead pup, unoiled
MH-HS-4	2000	33,000	subadult male, unoiled
MH-HS-6	14,000	48,000	adult female, heavily oiled
TS-HS-1	110,000	200,000	pregnant female, heavily oiled
TS-HS-2	8800	31,000	adult male, heavily oiled
TS-HS-3	2700	2300	adult female, heavily oiled
TS-HS-4	25,000	46,000	pup of TS-3, heavily oiled
TS-HS-5	32,500	54,000	adult female, heavily oiled
TS-HS-6	3700	7000	pup, lightly oiled
TS-HS-7	36,000	53,000	adult female, heavily oiled
TS-HS-8	215,000	365,000	pup of TS-7, heavily oiled
TS-HS-9	1300	4900	adult male, heavily oiled
TS-HS-10	18,500	41,000	adult female, heavily oiled
TS-HS-11	15,000	30,000	adult female, heavily oiled
TS-HS-12	730	5600	adult female, unoiled
TS-HS-13	2200	7200	subadult female, unoiled
TS-HS-14	8000	14,000	adult male, moderately oiled
TS-HS-15	3000	11,000	adult male, unoiled
TS-HS-16	800	4400	adult female, unoiled
TS-HS-17	2200	7700	adult female, lightly oiled
TS-HS-18	170	1400	adult male, unoiled
TS-HS-19	6200	20,000	adult male, unoiled
TS-HS-20	14,000	68,000	juvenile male, unoiled
TS-HS-21	4000	22,000	juvenile male, unoiled
TS-HS-22	4200	28,000	adult male, unoiled
TS-HS-23	44,000	110,000	adult female, unoiled
TS-HS-23F	1800	3300	fetus of TS-HS-23
TS-HS-24	5350	34,500	adult male, unoiled
TS-HS-25	12,000	67,000	adult male, unoiled
TS-HS-26	455	3100	adult female, unoiled
TS-HS-27	740	6100	adult male, unoiled
HBSL-1	1500	12,000	subadult male, unoiled
HBSL-2	1200	17,000	adult female, unoiled
HBSL-3	2500	29,000	adult male, unoiled

Appendix 19-C. Results of GC/MS analysis of tissue samples from harbor seals collected in Prince William Sound, the Gulf of Alaska, and near Ketchikan, 1989-1990. Values are expressed in parts per billion (=ng/g) for low molecular weight aromatic compounds (LAC) and high molecular weight aromatic compounds (HAC). All brain tissues were analyzed by GERG. For other samples the lab is indicated after the specimen number (TX=GERG; N=ECD/NMFS). Data for HBSL-1, 2, 3 and JT-2, 3, 4, 5 were provided by Paul Becker, NOAA/NMFS. (—=sample was collected but not analyzed; a blank=no sample was taken; nd=the compound was not detected)

Specimen number	Liver LAC	Liver HAC	Blubber LAC	Blubber HAC	Muscle LAC	Muscle HAC	Brain LAC	Brain HAC	Kidney LAC	Kidney HAC	Heart LAC	Heart HAC	Lung LAC	Lung HAC
AF-HS-1-TX	nd	nd	339	nd				—		—				—
AF-HS-2-TX	nd	nd	—	—				—		—				—
LL-HS-1-TX	nd	24	—	—	—	—		—		—				—
MH-HS-5-TX	—	—	377	nd				—		—				—
MH-HS-10-TX	nd	nd	47	nd	—	—		—		—				—
MH-HS-12-TX	nd	18	43	nd				—		—				—
MH-HS-13-TX	11	17	nd	nd	—	—		—		—				—
MH-HS-15-TX	—	—	nd	<1				—		—				—
TS-HS-1-N	2	nd	800	—	—	—	nd	—		—				—
TS-HS-1-TX	87	nd	—					—		—		—		—
TS-HS-1F-N	45	5			—			—		—				—
TS-HS-2-N	nd	nd	77	2	4	nd	nd	nd	—	—	—	—		—
TS-HS-3-N	nd	nd	21	2	4	nd	nd	nd	—	—	—	—		—
TS-HS-3-TX	—	—	73	nd	—	—		—	nd	nd	nd	nd	nd	nd
TS-HS-4-N	nd	nd	26	nd	10	<1	nd	nd	—	—	—	—		—
TS-HS-5-N	nd	<1	85	1	nd	nd	nd	—	—	—	—	—		—
TS-HS-5-TX	nd	nd	94	nd	—	—	nd		nd	nd	nd	nd	nd	nd
TS-HS-6-N1	nd	nd	18	<1	nd	nd	11	nd	—	—	—	—		nd

(Appendix continues)

Appendix 19-C. Continued

Specimen number	Liver LAC	Liver HAC	Blubber LAC	Blubber HAC	Muscle LAC	Muscle HAC	Brain LAC	Brain HAC	Kidney LAC	Kidney HAC	Heart LAC	Heart HAC	Lung LAC	Lung HAC
TS-HS-6-N2	nd	nd	19	1	nd	nd	—	—	—	—	—	—	—	—
TS-HS-7-N1	2	nd	420	1	4	<1	nd	nd	—	—	—	—	—	—
TS-HS-7-N2	nd	nd	520	4	5	nd	—	—	—	—	nd	nd	—	—
TS-HS-7-TX	nd	nd	559	nd	—	—	—	—	nd	nd	nd	nd	—	—
TS-HS-8-N	<1	nd	210	nd	1	nd	nd	nd	—	—	—	—	—	—
TS-HS-8-TX	nd	nd	662	nd	—	—	—	—	nd	nd	—	—	nd	nd
TS-HS-9-N	nd	nd	170	7	<1	<1	nd	nd	—	—	—	—	—	—
TS-HS-10-N	nd	nd	150	1	nd	nd	nd	nd	—	—	—	—	—	—
TS-HS-11-N	nd	nd	98	9	nd	nd	nd	nd	—	—	—	—	—	—
TS-HS-12-N	4	<1	4	nd	nd	nd	—	—	nd	nd	—	—	—	—
TS-HS-13-N	4	<1	nd	nd	<1	nd	28	nd	nd	nd	—	—	—	—
TS-HS-14-N	nd	nd	1	2	nd	nd	32	nd	nd	nd	—	—	—	—
TS-HS-15-N	nd	nd	1	nd	nd	nd	—	—	nd	nd	—	—	—	—
TS-HS-16-N	3	2	2	nd	nd	nd	15	nd	nd	nd	—	—	—	—
TS-HS-17-N	5	<1	21	2	nd	nd	39	nd	—	—	—	—	—	—
TS-HS-18-N	nd	nd	21	3	nd	nd	37	nd	—	—	—	—	—	—
TS-HS-19-N	nd	nd	19	2	nd	<1	nd	nd	—	—	—	—	—	—
TS-HS-20-N	nd	nd	19	2	nd	nd	nd	nd	—	—	—	—	—	—
TS-HS-21-N	nd	nd	26	7	nd	<1	34	nd	—	—	—	—	—	—
TS-HS-22-N	15	nd	28	2	nd	nd	nd	nd	—	—	—	—	—	—
TS-HS-23-N	nd	nd	20	4	nd	nd	17	nd	—	—	—	—	—	—
TS-HS-23F-N	nd	nd			6	1			—	—	—	—	—	—

(Appendix continues)

Appendix 19-C. Continued

Specimen number	Liver LAC	Liver HAC	Blubber LAC	Blubber HAC	Muscle LAC	Muscle HAC	Brain LAC	Brain HAC	Kidney LAC	Kidney HAC	Heart LAC	Heart HAC	Lung LAC	Lung HAC
TS-HS-24-N	nd	nd	51	39	nd	2	nd	nd	--	--	--	--	--	--
TS-HS-25-N	nd	nd	86	15	nd	<1	14	3	--	--	--	--	--	--
TS-HS-26-TX	nd	nd	nd	nd	nd	nd	nd	nd	nd	nd	nd	nd	nd	nd
TS-HS-27-TX	nd	nd	nd	nd	nd	nd	nd	nd	nd	nd	nd	nd	nd	nd
HBSL-1-N	3	nd	nd	2	nd	nd			nd	<1				
HBSL-2-N	4	nd	5	5	nd	nd			nd	nd				
HBSL-3-N	1	nd	4	<1	nd	nd			<1	nd				
JT-2-N	1	<1	4	nd	--	--			nd	nd				
JT-3-N	5	1	nd	3	--	--			1	nd				
JT-4-N	6	<1	--	--	--	--			nd	<1				
JT-5-N	6	<1	--	--	--	--			<1	<1				

357

Appendix 19-D. Concentrations of pristane phytane, unresolved complex mixture (UCM) and total alkanes in the brains of harbor seals. Values are given in parts per billion (pristane and phytane) or parts per million (UCM). All samples were analyzed by GERG.

Sample	Pristane	Phytane	Pris:Phy	UCM	Total Alkanes
TS-HS-1	37	4735	<0.1	246	25,110
TS-HS-2	50	1325	<0.4	0	11,048
TS-HS-3	48	3849	<0.1	327	17,612
TS-HS-4	33	3669	<0.1	228	20,550
TS-HS-5	119	7839	<0.1	403	25,170
TS-HS-6	32	1294	<0.1	80	12,605
TS-HS-7	0	0		398	6937
TS-HS-8	34	1228	<0.1	0	13,500
TS-HS-9	49	0		248	16,376
TS-HS-10	76	0		417	25,599
TS-HS-11	98	0		560	60,569
TS-HS-13	238	29	8.2	515	57,586
TS-HS-14	93	22	4.2	659	44,716
TS-HS-16	93	22	4.3	407	46,052
TS-HS-17				1721	46,678
TS-HS-18				1911	42,251
TS-HS-19	50	0		1624	70,164
TS-HS-20	61	0		1229	46,086
TS-HS-21	128	0		1146	57,788
TS-HS-22	39	0		897	39.553
TS-HS-23	117	0		0	32,411
TS-HS-23F	175	0		64	4441
TS-HS-24	93	0		1085	47,281
TS-HS-25	456	0		311	25,528
TS-HS-26	0	0		0	286
TS-HS-27	69	0		6	693

⋈ ⋈ ⋈

Chapter 20

Tissue Hydrocarbon Levels and the Number of Cetaceans Found Dead after the Spill

Thomas R. Loughlin

INTRODUCTION

Nineteen species of cetaceans occupy marine habitats in the Gulf of Alaska (Calkins 1986). Of these, fin (*Balaenoptera physalus*), sei (*Balaenoptera borealis*), minke (*Balaenoptera acutorostrata*), humpback (*Megaptera novaeangliae*), gray (*Eschrichtius robustus*), and killer whales (*Orcinus orca*), and Dall's (*Phocoenoides dalli*) and harbor porpoise (*Phocoena phocoena*) are often encountered in waters affected by the *Exxon Valdez* oil spill (EVOS).

Baleen whales are most vulnerable to the effects of an oil spill due to their generally low numbers, peculiar feeding mode (such as at the surface or on the bottom of the ocean nearshore), and dependence on selected, localized habitats for feeding and reproduction (Würsig 1990). Nearly the entire population of eastern Pacific gray whales migrate along Alaska's coast each March through May en route to summer feeding grounds in the Bering and Chukchi Seas (Braham 1984). Some gray whales apparently feed in Alaskan waters during their northward migration (Braham 1984), and others are known to remain in the Gulf of Alaska throughout the summer (Nerini 1984). They rarely enter Prince William Sound (PWS).

Gray whale carcasses are routinely seen along the north coast of the Alaska Peninsula each spring, a presumed result of natural mortality. By mid-May 1989, four carcasses were found at Tugidak Island, south of Kodiak Island. These findings increased concern about the possible impact of the spilled oil on gray whales and other cetaceans and prompted this study to: (1) locate and count stranded cetaceans from Kayak Island through Bristol Bay and if possible conduct necropsies on each carcass and collect tissue samples for chemical analysis; (2) determine if examined carcasses recently came in contact with oil, in particular *Exxon Valdez* crude oil; and (3) determine if stranded cetaceans died as a result of crude oil contamination.

METHODS

Aerial surveys were used to search for stranded cetaceans on 13 June and from 23 to 30 June 1989. Surveys were flown from Kayak Island north to PWS, along all coastlines and island groups west to Cape Sarichef, and east to King Salmon in Bristol Bay. Unscheduled surveys occurred from March to October and concentrated in PWS. Surveys were conducted in single- or twin-engine float planes at approximately 200 m altitude and at about 80 knots airspeed. As a follow-up to the 1989 survey, and because dead gray whales were seen there in 1989, Tugidak and Sitkinak Islands were surveyed by helicopter during 14–16 July 1990 and from 28 August to 6 September 1990.

Once a stranded cetacean was located, and was in an acceptable condition (not putrefied) for necropsy and tissue collection, a crew was flown by helicopter to the stranding site. Animals were considered inappropriate for necropsy if they had signs of being scavenged, showed visible signs of decomposition, or were known to have been on the beach longer than 2 weeks. Where feasible, each whale was examined by a certified veterinary pathologist for gross evidence of oil-related or other causes of death. Multiple tissue samples were collected according to a prescribed standard protocol (Appendix I) developed for the EVOS and measurements were obtained according to accepted cetacean necropsy procedures (Committee on Marine Mammals 1961; Fay et al. 1979; Aguilar 1985). This project was funded only in 1989, and tissue samples were collected only from carcasses examined then.

Tissue samples were analyzed for a broad spectrum of aliphatic and aromatic hydrocarbons by the Geochemical and Environmental Research Group (GERG), Texas A&M University, College Station, Texas, using capillary column gas chromatography with detection by flame ionization and mass spectrometry (GC/MS). Conclusions as to the presence or absence of petroleum in the sample were based on a preponderance of evidence such as the presence of the aromatic compounds naphthalene, dibenzothiophene, phenanthrene, fluorenes, and chrysenes as both unsubstituted compounds and the alkylated homologues; the presence of an unresolved complex mixture (UCM); and a pristane/phytane ratio <5 (Boehm et al. 1987; Wade et al. 1988; Varanasi et al. 1990; Krahn et al. 1992; Wade et al. 1993).

A full suite of tissue samples was collected as prescribed in the protocol (cetaceans have no gall bladder, thus no bile was collected). However, because marine mammals extensively metabolize aromatic compounds in their livers and the metabolites are excreted, the metabolites were measured to establish exposure to Prudhoe Bay crude oil (Krahn et al. 1992). Liver and blubber samples were analyzed as the most likely tissues to indicate the presence of petroleum hydrocarbons.

⋉⋊ ⋉⋊ ⋉⋊

Histological examinations were performed by a veterinary pathologist (Dr. Terry Spraker, Colorado State Diagnostic Laboratory, Colorado State University, Fort Collins, Colorado). Tissues were imbedded in paraffin, sectioned at 5–6 μ, and stained with hematoxylin and eosin.

All samples were stored at the National Marine Mammal Laboratory, Seattle, in a secure freezer or secure storage facility (formaldehyde samples) with chain of custody forms attached (Appendix I). A record was kept listing the location and status of each sample. Samples and data were archived by tissue sample number which included the initials of the collector and the species.

RESULTS

Over 9600 km of coastline were surveyed to locate 37 dead cetaceans, of which 26 were gray whales, 5 were harbor porpoise, 3 were unidentified, 2 were minke whales, and 1 was a fin whale (Fig. 20-1). Nearly all of the specimens were moderately to very decomposed, which roughly corresponds to several weeks to a few months postmortem. Double-counting may have occurred in three cases as animals were moved by currents from one beach to the next (Table 20-1). Only seven animals were in satisfactory condition (minimal tissue degradation) for tissue collection and analysis: three gray whales, one minke whale, and three harbor porpoise (Table 20-2).

Tugidak and Sitkinak Islands had the greatest number of gray whale carcasses (10; 27%), probably because of the circulating current systems near the confluence of Shelikof Strait and the Alaska Stream (Reed and Schumacher 1986). During July 1990, 9 gray whales were stranded at Sitkinak Island and 14 were stranded at Tugidak Island; in August 3 more were seen at Tugidak Island. A total of 36 gray whale carcasses were counted at these two islands in 2 years.

Details of the necropsies provide no indication of cause of death for the sampled cetaceans. For the minke whale stranded in Turnagin Arm (Cook Inlet) on 10 October 1989, the veterinarian at the site noted that there were no signs of obvious infection, toxicological effects, or physical damage that may have caused the whale's death. The whale was alive when it stranded. Similarly, a pregnant harbor porpoise stranded alive near Portage Creek on Turnagin Arm. It was determined that this porpoise and fetus could not survive on their own and they were humanely euthanized. The mother had dry and deeply cracked skin on the dorsal surface, torn flesh hanging from the oral cavity, breathing was shallow and irregular, and fluke movements were weak and spasmatic. There were no external signs of oil contamination. No cause for the stranding was determined at necropsy. The fetus was unremarkable in size and appearance.

✂ ✂ ✂

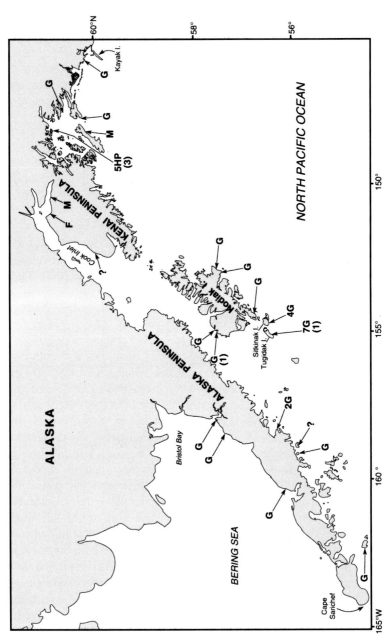

Figure 20-1. Map depicting the location of 37 cetaceans found dead on beaches from Kayak Island to Bristol Bay from March to October 1989. Tissue samples were collected from seven animals and are indicated by (). F=fin whale; G=gray whale; HP=harbor porpoise; M=minke whale; ?=unidentified cetacean.

Table 20-1. Chronological list of cetacean carcasses documented during aerial surveys March to October 1989, from Kayak Island to King Salmon, Alaska. Animals from which samples were obtained are noted with an *.

Location	Lat.	Long.	Exam date	Sex	Condition[a]
Gray whale					
Hinchinbrook	60.15	146.45	31 March	?	floating
Tugidak, lagoon	56.30	154.31	13 May	M	md
Tugidak, lagoon	56.30	154.32	13 May	?	md
Tugidak, NW	56.24	154.46	17 May	F	md
Tugidak, NW	56.26	154.45	18 May	F*	md
Sanak	54.30	162.51	24 May	?	fresh
Copper River	60.10	144.40	26 May	F	md
Kodiak	57.30	152.10	3 June	?	ad
Kodiak, Karluk	57.35	154.30	3 June	M*	md
Sitkinak	56.30	154.10	10 June	M	md
Sitkinak	56.29	154.15	10 June	F?	md
Tugidak, SE	56.23	154.46	12 June	M	md
Kodiak	57.25	152.25	12 June	M*	md
Tugidak	56.23	154.46	24 June	?	md
Kodiak	56.46	153.52	24 June	?	ad
Port Moller	56.10	160.26	25 June	?	ad
Ugashik Bay	57.36	157.40	25 June	?	md
Cinder River	57.20	158.12	25 June	?	md
Chignik	56.20	158.30	5 July	?	fresh
Chignik, lagoon	56.20	158.30	17 July	M	fresh
Port Gravina	60.40	146.30	19 July	?	floating
Sitkinak[b]	56.30	154.29	31 July	?	md
Tugidak[b]	56.23	154.45	31 July	?	md
Lake Beach	56.18	158.17	23 August	?	fresh
Chiachi Island[c]	55.31	159.08	23 August	?	ad
Shelikof Strait	57.40	154.33	30 October	?	ad
Minke whales					
Montague Island	60.10	147.15	7 July	?	md
Turnagin	60.50	148.59	10 October	F*	fresh
Fin whale					
Cook Inlet	61.24	149.59	25 July	M	fresh
Harbor porpoise					
Kodiak area	location not known		3 July	M	calf, md
Kodiak area	location not known		3 July	F	juv., md
Turnagin Arm	50.08	148.58	6 June	F*	fresh
Turnagin Arm	fetus of above female			?*	fresh

(Table continues)

Table 20-1. Continued

Location	Lat.	Long.	Date	Sex	Condition[a]
Harbor porpoise					
Valdez	location not known			F*	fresh
Undetermined species					
Happy Valley	59.56	151.44	6 June	?	ad
Unga Island	55.14	160.32	27 June	?	ad
Mitrofania Is.	55.51	158.51	28 June	?	ad

[a] md = moderate decomposition; ad = advanced decomposition.
[b] Could be a repeat of third animal above, with southwest movement.
[c] Could be a repeat of Mitrofania animal under unknown whales with southwest movement.

The skin on the flipper of gray whale KP-GW-1 at Tugidak Island examined in May 1989 and scrapings from the beach nearby showed oil contamination, implying the presence of petroleum in the area where the whale had beached (Table 20-2). Moreover, scrapings from the oral surface of the mandible also showed oil contamination but analysis of the stomach contents and liver did not. It is probable that oil on the mandible may have entered postmortem, suggesting that this animal died before the oil spill.

Chemical analyses of blubber from AF-GW-1 showed a PAH level of 467 ng/g wet weight. These PAHs contained a suite of hydrocarbons broadly consistent with a petrogenic source, which may have been either a crude oil or diesel oil (Fig. 20-2). Reported values for all other tissues were below detection limits or did not indicate exposure to petrogenic hydrocarbons.

Histological examination of tissues were unremarkable and provided no information on the cause of death for any of the cetaceans examined.

DISCUSSION

The 26 gray whale carcasses observed in the study area in 1989 exceeded numbers in earlier years when only 6 were reported between Kayak Island and Cape Sarichef during 1975–1987 (Zimmerman 1989). The reason for the greater number observed in 1989 is unexplained, but may be attributed to the timing of the search effort which coincided with the northern migration of gray whales and to the increased activities associated with the oil spill. The spill occurred during the northern migration of gray whales as they moved along the eastern Pacific coast

Table 20-2. Results of analysis of cetacean tissue samples collected during 1989. Values are expressed as parts per billion (ng/g wet weight). LAC[*], HAC[*], and PAH[*] values were obtained by summing all values within the fraction. CPI[*] was calculated as per Boehm et al. (1987).

Species	Coll. number	Tissue	Pristane	Phytane	UCM[a]	PAH	CPI
Gray whale	MH-GW-1	liver	880	0[b]	160	116	0
Gray whale	KP-GW-1	liver	121	0	7	161	0
Gray whale	AF-GW-1	blubber	6505	0	63	467	2
Minke whale	RM-MW-1	blubber	8025	0	24	202	0
Minke whale	RM-MW-1	liver	33	0	0	105	0
Harbor porpoise	MH-HP-1	blubber	174,783	45	50	149	8
Harbor porpoise	MH-HP-1	liver	11,614	0	36	172	0
Harbor porpoise	MH-HP-2	blubber	234	0	4	446	0
Harbor porpoise	MH-HP-2	liver	19	0	0	93	0
Harbor porpoise	MH-HP-1	liver	185	0	1	90	0

[a] Abbreviations: LAC = light aromatic compounds; HAC = heavy aromatic compounds; PAH = polycyclic aromatic hydrocarbons; CPI = carbon preference index; UCM = unresolved complex mixture
[b] A value of zero indicates the measured level was not detected.

from Mexico to the Bering Sea and adjacent waters. They were seen swimming through oil slicks near Montague Island and other portions of the Gulf of Alaska during April and June.

The stranding of 26 gray whales in 1989, although a high number, is not unusual compared to other areas in the whales' range. Fay et al. (1978) observed 31 gray whale carcasses on St. Lawrence Island, 17 of which apparently washed ashore during the ice-free season (June–November) in 1976. The other 14 carcasses were more decomposed and likely had been dead longer. No cause of death was determined for any of these whales. In Puget Sound and the outer Washington coast, 21 gray whale carcasses were found from 1989 to 1991 (B. Norberg, National Marine Fisheries Service, Northwest Region, Seattle, Washington, personal communication). Sixteen of these were analyzed for PAHs, chlorinated hydrocarbons, and toxic metals by Varanasi et al. (in press). Their analyses showed that the concentrations of potentially toxic chemicals in gray whale tissues were low when compared with the concentrations in tissues of marine mammals feeding on higher trophic level species, such as fish.

Brownell (1971) summarized information on the possible effect of the 1969 Santa Barbara Channel oil spill on cetaceans and concluded that reports that gray whales and other cetaceans died from the spill were incorrect. The spill occurred after the majority of the California gray whale herd had passed on their southern migration; however, the entire northern migration passed through the area while it was still contaminated. Analyses of tissue samples from three gray whale carcasses found after the spill were unable to detect the presence of crude oil. Brownell (1971) concluded that the report of dead cetaceans on the California coast following the spill was likely a result of natural mortality.

Geraci (1990) summarized available information on the physiological and toxic effects of oil on cetaceans including a summary table (his Table 6-1) reporting cetaceans associated with oil. In spite of numerous observations of cetaceans in spills, no deleterious effects were recorded with certainty (Geraci 1990).

Griffiths et al. (1987) concluded that migrating whales probably would ingest little oil, although actively feeding whales, particularly surface filter-feeders, could take in larger quantities, but the effect of ingested oil was equivocal. They also speculated that an oil spill in the Barents Sea would result in few cetaceans killed after oil spillage, while the discomfort caused to certain individuals might be significant.

I saw gray whales swimming in oil during March 1989 near Montague Island, as did other researchers (Harvey and Dahlheim, Chapter 15). I did not record quantitative behavioral notes, but my subjective appraisal did not detect any alteration in swimming speed, direction, and breathing behavior by whales in or near oil on the water's surface.

∞   ∞   ∞

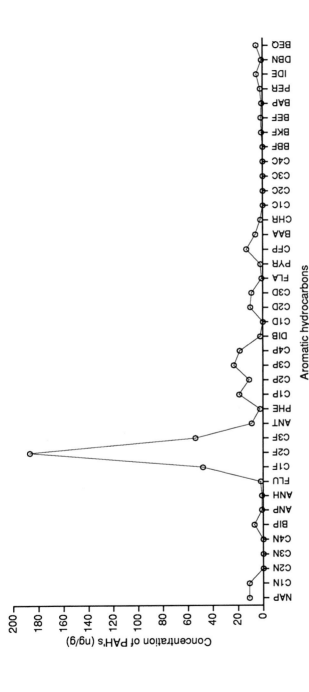

Figure 20-2. Polycyclic aromatic hydrocarbons in blubber of gray whale AF-GW-1. Abbreviations: NAP=naphthalene, C_1N=C_1-naphthalene, C_2N=C_2-naphthalene, C_3N=C_3-naphthalene, C_4N=C_4-naphthalene, BIP=biphenyl, ANP=acenaphthylene, ANH=acenaphthene, FLU=fluorene, C_1F=C_1-fluorene, C_2F=C_2-fluorene, C_3F=C_3-fluorene, ANT=anthracene, PHE=phenanthrene, C_1P=C_1-phenanthrene, C_2P=C_2-phenanthrene, C_3P=C_3-phenanthrene, C_4P=C_4-phenanthrene, DIB=dibenzothiophene, C_1D=C_1-dibenzothiophene, C_2D=C_2-dibenzothiophene, C_3D=C_3-dibenzothiophene, FLA=fluoranthene, PYR=pyrene, CFP=methyl fluoranthene-pyrene, BAA=benz(a)anthracene, CHR=chrysene, C_1C=C_1-chrysene, C_2C=C_2-chrysene, C_3C=C_3-chrysene, C_4C=C_4-chrysene, BBF=benzo(b)fluoranthene, BKF=benzo(k)fluoranthene, BEF=benzo(e)pyrene, BAP=benzo(a)pyrene, PER=perylene, IDE=ideno(1,2,3-cd)pyrene, DBN=dibenzo(a,h)anthracene, BEQ=benzo(g,h,i)perylene.

Both mysticete and odontocete whales are able to detect oil, yet they seem not to avoid it (Geraci 1990). Geraci speculated that the stimulus from oil may not be noxious enough to elicit a negative (avoidance) response, or perhaps the cetacean disregards oil when engaged in more engrossing activities.

Although gray whale carcasses were found during and after EVOS in the Gulf of Alaska, the cause of their death could not be directly linked to the spill. One whale did have hydrocarbon residues in its blubber (AF-GW-1), suggesting exposure to a petrogenic source, which may have been either crude oil or diesel oil, but the PAH concentrations were too low to discriminate further. Gray whales migrate along the eastern Pacific coast where many natural oil seeps occur and ingestion of hydrocarbons through prey or in the water column may have been the source for detected residues.

CONCLUSIONS

Thirty-seven cetaceans were found dead on Alaskan beaches from Kayak Island (eastern Gulf of Alaska) to King Salmon (Bristol Bay) during March through October 1989. The carcasses included 26 gray whales, 5 harbor porpoise, 3 unidentifiable whales, 2 minke whales, and 1 fin whale. Tissue samples were obtained from 7 of the 37 carcasses for hydrocarbon analysis and histological examination. Results of these analyses and on-site necropsies could not determine the cause of death for any of the seven. The large number of stranded gray whales (n=26) was attributed to the timing of the search effort coinciding with the northern migration of gray whales, augmented by increased survey effort in the study area associated with the oil spill.

ACKNOWLEDGMENTS

I am grateful to R. Ferrero and D. Rugh for their assistance on the 1989 aerial surveys and to B. Kelly, J. Sease, and L. Rea for the 1990 Tugidak Island surveys. U. Varanasi provided information on gray whale KP-GW-1; T. Spraker conducted the histological analysis and T. Wade the chemical analysis. The manuscript was improved by comments from H. Braham, D. Brown, G. Duker, J. Lee, C. Manen, J. Short, and U. Varanasi. This study was funded by the *Exxon Valdez* Oil Spill Trustee Council.

⋈ ⋈ ⋈

REFERENCES

Aguilar, A. 1985. Compartmentation and reliability of sampling procedures in organochloride pollution surveys of cetaceans. Residue Reviews 95:91–114.

Boehm, P. D., M. S. Steinhauer, D. R. Green, B. Fowler, B. Humphrey, D. L. Fiest, and W. J. Cretney. 1987. Comparative fate of chemically dispersed and beached crude oil in subtidal sediments of the Arctic nearshore. Arctic 40 (Suppl. 1):133–148.

Braham, H. W. 1984. Distribution and migration of gray whales in Alaska. Pages 249–266, *in*, M. L. Jones, S. L. Swartz, and S. Leatherwood (eds.), The gray whale. Academic Press, San Diego, California.

Brownell, R. L., Jr. 1971. Whales, dolphins and oil pollution. Pages 255–279, *in* Biological and oceanographical survey of the Santa Barbara Channel Oil spill, 1969–1970. Allen Hancock Foundation, University of Southern California, Los Angeles, 1971, Vol. 1 (Sea Grant Publication No. 2).

Calkins, D. G. 1986. Marine mammals. Pages 527–558, *in* D. W. Hood and S. T. Zimmerman (eds.), The Gulf of Alaska, physical environment and biological resources. U.S. Government Printing Office, Washington, D.C.

Committee on Marine Mammals. 1961. Standardized methods for measuring and recording data on the smaller cetaceans. Journal of Mammalogy 42:471–476.

Fay, F. H., R. A. Dieterich, L. M. Shults, and B. P. Kelly. 1978. Morbidity of marine mammals. Pages 39–79, *in* Environmental assessment of the Alaskan continental shelf, annual report of principal investigators for the year ending March 1978, Volume I.

Fay, F. H., L. M. Shults, and R. A. Dieterich. 1979. A field manual of procedures for postmortem examination of Alaskan marine mammals. Institute of Marine Sciences Report No. 79-1, University of Alaska, Fairbanks. 51 p.

Geraci, J. R. 1990. Physiologic and toxic effects on cetaceans. Pages 167–197, *in* J. R. Geraci and D. J. St. Aubin (eds.), Sea mammals and oil: Confronting the risks. Academic Press, San Diego, California.

Griffiths, D. J., N. A. Øritsland, and T. Øritsland. 1987. Marine mammals and petroleum activities in Norwegian waters. Fisken og Havet, Serial B, no. 1. 179 p.

Krahn, M. M., G. M. Ylitalo, D. G. Burrows, J. Buzitis, S.-L. Chan, and U. Varanasi. 1992. Methods for determining crude oil contamination in sediments and biota after the *Exxon Valdez* oil spill. Pages 60–62, *in Exxon Valdez* oil spill symposium, 2–5 February 1993, Anchorage, Alaska. (Available, Oil Spill Information Center, 645 G Street, Anchorage, Alaska 99501.)

Nerini, M. 1984. A review of gray whale feeding ecology. Pages 423–450, *in* M. L. Jones, S. L. Swartz, and S. Leatherwood (eds.), The gray whale. Academic Press, San Diego, California.

Reed, R. K., and J. D. Schumacher. 1986. Physical oceanography. Pages 57–75, *in* D. W. Hood and S. T. Zimmerman (eds.), The Gulf of Alaska, physical environment and biological resources. U.S. Government Printing Office, Washington, D.C.

Varanasi, U., S.-L. Chan, W. D. MacLeod, J. E. Stein, D. W. Brown, D. G. Burrows, K. L. Tilbury, J. T. Landahl, C. A. Wigren, T. Hom, and S. M. Pierce. 1990. Survey of subsistence fish and shellfish for exposure to oil spilled from the *Exxon Valdez* first year: 1989. U.S. Department of Commerce, NOAA Technical Memorandum NMFS F/NWC-191. 151 p.

Varanasi, U., J. E. Stein, K. L. Tilbury, J. P. Meador, C. A. Wigren, R. C. Clark, and S-L. Chan. In press. Chemical contaminants in gray whales (*Eschrichtius robustus*) stranded along the west coast of North America. Science of the Total Environment.

◇ ◇ ◇

Wade, T. L., E. L. Atlas, J. M. Brooks, M. C. Kennicutt II, R. G. Fox, J. Sericano, B. Garcia, and D. DeFreitas. 1988. NOAA Gulf of Mexico status and trends program: Trace organic contaminant distribution in sediments and oysters. Estuaries 11:171-179.

Wade, T. L., T. J. Jackson, T. M. McDonald, D. L. Wilkenson, and J. M. Brooks. 1993. Oysters as biomonitors of the APEX Barge oil spill, Galveston Bay, Texas. *In* Proceedings, 1993 International oil spill conference, 29 March-1 April 1993, Tampa, Florida.

Würsig, B. 1990. Cetaceans and oil: Ecologic perspectives. Pages 129–165, *in* J. R. Geraci and D. J. St. Aubin (eds.), Sea mammals and oil: Confronting the risks. Academic Press, San Diego, California.

Zimmerman, S. T. 1989. A history of marine mammal stranding networks in Alaska, with notes on the distribution of the most commonly stranded cetacean species, 1975–1987. Pages 43–53, *in* J. Reynolds and D. Odell (eds.), Marine mammal strandings in the United States: Proceedings of the second marine mammal workshop, 3–5 December 1987, Miami, Florida. NOAA Technical Report NMFS 98.

Chapter 21

Summary and Conclusions

David J. St. Aubin and Joseph R. Geraci

The grounding of the *Exxon Valdez* in Prince William Sound (PWS) unleashed a flurry of activity to contain and clean up the crude oil, rescue and rehabilitate affected animals, and determine the impact of the spill on the environment. Marine mammals drew special attention, with $18 million invested in rehabilitation and $5 million for damage assessment and research during 1989–1991. Driven largely by the need to answer questions of legal liability, the priority of much of this effort was to quantify the losses. Seven chapters in this book deal with this issue. The other chapters report on the findings of behavioral, pathologic, and toxicologic studies undertaken within a forensic framework, providing new information on how oil can disrupt the life of a marine mammal.

An early priority was to relieve the distress and suffering brought about by oil contamination to the area's wildlife. The rescue program was mounted within 24 hours, and over the next 6 months continued to draw heavily on time, funding, and resources. Implementing the impact studies took longer. Equipment had to be assembled, foul weather overcome, and various private, state, and federal agencies coordinated, all in a crisis environment in which leadership was yet to be defined. In the first week alone, with no plan in effect, valuable opportunities to collect information were lost. Surveys during the first few hours or days of the spill could have resolved the hotly debated issue of whether seven missing killer whales had succumbed to oil or had even been in the vicinity at the time of the spill (Matkin et al., Chapter 8; Dahlheim and Matkin, Chapter 9). Impact studies began quickly thereafter and over the next 3 years yielded considerable data. The reader nevertheless will sense the frustration of researchers attempting to satisfy scientific inquiry in a sometimes chaotic environment with competing priorities and limited resources.

As in any investigation of this magnitude, the many approaches used by investigators did not always produce corroborative results. This was particularly evident in the attempt to establish a final death toll. Body counts yielded one tally—1011 sea otters, 19 harbor seals, 12 Steller sea lions, and 37 cetaceans. These numbers can of course be misleading because not all victims were recovered, and not all those recovered were victims of the spill. Population surveys and models were redesigned to provide a broader perspective, but even with solid baseline data and broad coverage, the models were not sensitive enough to project anything less than a substantial loss. Add, in PWS, the obstacle of dealing with two populations of marine mammals already in decline—harbor seals and Steller sea lions—and the estimates become even less certain.

Frost et al. (Chapter 6), using direct counts of animals on haulout sites, determined that the number of molting harbor seals decreased more on oiled sites than on unoiled ones in 1989. The following year, counts on oiled areas had increased above previous levels, whereas those on unoiled areas showed a continued decline. The number of seals in 1991 was at or above the 1989 levels, but a decrease was again apparent in 1992. What then was the impact on the seal population? Investigators calculated a spill-related loss of 135 seals from seven haulout sites, and extrapolated the death toll in PWS to be at least 302 animals. This, coupled with a 25% decrease in pup production in 1989, amounts to a significant loss to a population already in a tenuous state.

Calkins et al. (Chapter 7) had the complicated task of assessing the effects of the oil spill on Steller sea lions. In the years prior to the spill, pup production had decreased, premature births increased, and the population was in decline. The team had to determine whether these trends were being augmented by the oil spill. Within the wide confidence limits of their estimates, no effect could be demonstrated, and the trends continued at prespill rates. Although some sea lions were oiled, the population appeared to have escaped from suffering any significant impacts, primarily because its habitat lies mostly outside of PWS, where the oil was most concentrated.

Surveys of sea otters revealed the inherent difficulty in using population census techniques to detect an impact. Helicopter surveys in some areas indicated a decrease in the number of sea otters, but the difference was within the confidence limits of the method, and hence not considered significant (DeGange et al. 1993). Boat-based surveys of PWS indicated a 35% decrease in the sea otter population in oiled areas and a 13% increase in unoiled areas (Burn, Chapter 4). One estimate placed the losses at 2800 individuals (Garrott et al. 1993); another calculation using the same data suggested that sea otters were at least as abundant following the spill as they had been 4 or 5 years earlier (Garshelis, in Ballachey et al., Chapter 3) when the PWS population was thought to number 10,000 animals (EVOS Trustees 1992). Still another approach used the actual number of dead animals retrieved, corrected by a carcass recovery factor of 20%, for a total loss estimate of over 4000 sea otters (Doroff and DeGange 1993). Whatever the mortality, preliminary studies in 1993 suggest that the sea otter population is recovering.

Little attempt was made to assess cetacean populations on a broad scale. Von Ziegesar et al. (Chapter 10) logged 9600 miles of transects in 260 days to count humpback whales in the area. They could find no change from the expected number of animals, and there seemed to have been no residual impact on humpback whales judging by the increase in the reproductive rate the following year. Killer whale surveys took a different approach by using the association patterns of known individuals that frequent the study area (Matkin et al., Chapter 8). Animals missing from pods were presumed to have died as a consequence of the spill. If so, the loss of 14 whales from one pod represents the largest known cetacean mortality due to oil. The association, however, remains circumstantial: no animals were seen in distress, the missing whales had last been sighted 6 months before the spill, and no carcasses were found for pathologic or toxicologic studies.

How marine mammals behave around an oil spill largely determines how severely they might be affected. The issue has been widely explored (Geraci and St. Aubin 1990), but many questions still remain. What was learned from this event? At one time or another, sea otters, harbor seals (Lowry et al., Chapter 12), sea lions (Calkins et al., Chapter 7), porpoises, dolphins, and killer and gray whales (Harvey and Dahlheim, Chapter 15; Loughlin, Chapter 20) were observed swimming in oil-covered waters. Oiled seals showed no tendency to seek out clean beaches (Lowry et al., Chapter 12), and even chose to pup on contaminated sites 2 months or more after the spill. Quantitative studies to determine avoidance behavior under the prevailing conditions did not command, or perhaps even warrant, high priority. Experience shows that such studies are difficult to interpret in any event, even under fairly controlled circumstances (Geraci 1990). Monnett et al. (1990) found that individually tracked sea otters tended to remain in oiled habitats. Yet distributional studies by Burn (Chapter 4) suggest that sea otters were leaving oiled areas in favor of clean ones. Despite strong indications that some marine mammals can detect oil (Geraci and St. Aubin 1990), the observations here, as in the past, give no assurance that marine mammals will avoid an oil spill.

A new understanding of the toxic effects of oil was gleaned from pathological findings, and also from evaluating animals brought to rehabilitation centers (Lipscomb et al., Chapter 16; Spraker et al., Chapter 17; Williams et al., Chapter 13). The consequences of inhaling petroleum vapors were particularly evident during the early phase of the spill, underscoring the harmful character of fresh oil. Sea otters developed pulmonary emphysema and harbor seals exhibited neurologic lesions. Much of the damage to liver, kidney, gastrointestinal, and hematopoietic systems was attributed to starvation and shock secondary to hydrocarbon exposure. The direct effects of oil on these tissues should not be discounted, in view of the physiological differences between marine mammals and the terrestrial animals on which the interpretation was based. For example, blood circulation in marine mammals during diving bypasses the liver and its detoxifying activity, and it may deliver toxic substances directly to the brain or other sensitive tissues (Geraci et al. 1989). Any study on marine mammals needs to consider these and other special-ized adaptations for hypoxia, thermoregulation, and fasting that will influence the animals' response to a toxic substance.

Seals exposed to oil in PWS behaved abnormally. They were lethargic, could be easily approached, and remained hauled out in the face of activities that would normally cause them to enter the water (Lowry et al., Chapter 12). There have been other accounts of oiled phocid seals behaving the same way (Anonymous 1971; Sergeant 1991). Here for the first time, this clinical effect could be traced to degenerative lesions in the brain (Spraker et al., Chapter 17). The correlation must have been satisfying to those drawing together findings from disparate studies—testimony to the power of collaboration.

Data were obtained on the distribution of aliphatic and aromatic hydrocarbons in tissues of harbor seals (Frost et al., Chapter 19) and sea otters (Mulcahy and Ballachey, Chapter 18). The small number of animals tested made it difficult to correlate hydrocarbon levels with specific pathologic findings. Nevertheless, the observed patterns did provide clues to the source of the hydrocarbons, the route, and, in some cases, the duration of exposure. These findings may serve as a foundation for broader analyses in future events, and may someday help to establish the degree of exposure in situations where other corroborating evidence is equivo-cal. For instance, Loughlin (Chapter 20), using the same techniques that were used in the sea otter and harbor seal studies, found unremarkable levels of hydrocarbons in gray whale carcasses that had shown some evidence of surface contamination. His conclusion that the oiling was incidental or occurred after death is strengthened by comparisons with findings in sea otters and harbor seals.

The investigation also leaves us with an enigma. Did the dorsal fins of two killer whales collapse as a result of the spill, and if so, by what mechanism? Observers (Matkin et al., Chapter 8) invoked stress and a tenuous link with the same phenomenon in some captive killer whales. It would be simpler to consider changes in the mechanical forces that maintain the fin erect rather than to suggest the complex web of hormonal and neurological events that can somehow lead to permanent disfiguration. Yet the observation is undeniable and the association with the spill appears convincing. To forge the link, studies will first have to determine the nature of the biomechanical support of the fin and the forces that influence its posture.

Can we expect to learn more from future oil spills, given the experience gained from the *Exxon Valdez* incident? The observations and conclusions described in these chapters should better prepare others to mobilize and channel resources for research. Sampling protocols developed from this event will go a long way toward improving the efficiency of any investigative response, particularly during the critical early phases. Survey strategies can be mapped out in advance to determine

⋙ ⋙ ⋙

effects at the population level, though the approach may not be sensitive to anything less than large-scale changes. Even under the best circumstances, researchers may have to compete with clean-up and containment crews for limited resources such as boats, aircraft, and other equipment. Contingency plans should be realistic in expectations for logistic support. Despite all the preparation, there will always be problems in assembling an expert scientific team on such short notice and securing their time for long-term studies. Funding may be less of a concern in subsequent events. Scientists working early in the *Exxon Valdez* oil spill were uncertain of the source and amount of funds that would be made available. The Oil Pollution Act of 1990 now provides some assurance that the scientific community will be able to respond more effectively. Those responsible for apportioning funds can now appreciate how costly such studies can become and rank them to get the best information in the most efficient way.

It is apparent from the studies presented in this volume that good baseline data yield better answers sooner—a reminder of the value of ongoing studies into fundamental biological questions. The EVOS investigation also showed that some questions can be answered in rehabilitation centers, which will invariably form a major part of future responses. Researchers, while working with care-givers at these centers, can use information on recovery, residual effects, and mechanisms of toxicity to supplement field data for a more comprehensive understanding of how oil affects wild marine mammal populations.

REFERENCES

Anonymous, 1971. Arrow oil spill. Smithsonian Institution, Center for Short-Lived Phenomenon Annual Report, 1970 Event No. 15-70:134–136.

DeGange, A. R., D. C. Douglas, D. H. Monson, and C. Robbins. 1993. Surveys of sea otters in the Gulf of Alaska in response to the *Exxon Valdez* oil spill. NRDA Report, Marine Mammal Study No. 6. U.S. Fish and Wildlife Service, Anchorage, Alaska.

Doroff, A., and A. R. DeGange. 1993. Experiments to determine drift patterns and rates of recovery of sea otter carcasses following the *Exxon Valdez* oil spill. NRDA Report, Marine Mammal Study No. 6. U.S. Fish and Wildlife Service, Anchorage, Alaska.

EVOS Trustees. 1992. *Exxon Valdez* Oil Spill Restoration Framework. Anchorage, Alaska. 52 pp.

Garrott, R. A., L. L. Eberhardt, and D. M. Burn. 1993. Mortality of sea otters in Prince William Sound following the *Exxon Valdez* oil spill. Marine Mammal Science 9:343–359.

Geraci, J. R. 1990. Physiologic and toxic effects of oil on cetaceans. Pages 167-197 *in* J. R. Geraci and D. J. St. Aubin (eds.), Sea mammals and oil: Confronting the risks. Academic Press, San Diego, California.

Geraci, J. R., and D. J. St. Aubin (eds.). 1990. Sea mammals and oil: Confronting the risks. Academic Press, San Diego, California. 282 pp.

Geraci, J. R., D. J. St. Aubin, D. M. Anderson, R. J. Timperi, G. A. Early, J. H. Prescott, and C. A. Mayo. 1989. Humpback whales (*Megaptera novaeangliae*) poisoned by dinoflagellate toxin. Canadian Journal of Fisheries and Aquatic Science 46 (11):1895–1898.

✕ ✕ ✕

Monnett, C., L. M. Rotterman, C. Stack, and D. Monson. 1990. Postrelease monitoring of radio-instru-
    mented sea otters in Prince William Sound. Pages 400–420, *in* K. Bayha and J. Kormendy (eds.),
    Sea otter symposium: Proceedings of a symposium to evaluate the response effort on behalf of sea
    otters after the T/V *Exxon Valdez* oil spill into Prince William Sound, Anchorage, Alaska, 17–19
    April 1990. U.S. Fish and Wildlife Service, Biological Report 90(12).
Sergeant, D. E. 1991. Harp seals, man, and ice. Canadian Special Publication of Fisheries and Aquatic
    Sciences 114. 153 pp.

Appendix I

Sample Collection, Storage, and Documentation

Thomas R. Loughlin and Elizabeth H. Sinclair

SAMPLE COLLECTION PROTOCOL

Soon after the *Exxon Valdez* oil spill (EVOS) federal and state government scientists developed a marine and terrestrial mammal tissue and organ sampling protocol. The protocol was distributed to all individuals and organizations expected to collect samples. The Trustees (see Morris and Loughlin, Chapter 1) also organized an Analytical Chemistry Group and a Histopathology Technical Group for Oil Spill Assessment Studies. The two groups provided additional technical advice on sample collection, preservation, and processing of samples for toxicological and histopathological analysis to all Natural Resource Damage Assessment (NRDA) scientists. The following is a summary of these sampling protocols and of advice provided by interested experts.

Individuals collecting marine mammal samples must recognize that the tissues autolyze rapidly following death due to their high initial body temperature and the insulating properties of fur, skin, and blubber. To be of analytical value, marine mammal tissue samples should be collected as early as possible after death and must be preserved and processed under strict guidelines.

Histological Analysis

Prepare a solution of buffered formalin in a 19-liter (5 gallon) bucket as follows:

76 g of monobasic sodium phosphate,
123 g of dibasic sodium phosphate,
1900 cc of 37% formaldehyde, and
16,900 cc tap water.

If the sodium phosphate salts are not available, make the solution with nine parts sea water and one part formaldehyde.

Collect the appropriate tissue or organ samples (see below) using clean cutting tools (sterile, disposable surgical blades for each animal; knives, scissors, forceps rinsed in methyl chloride). The samples should be at least 2X2X1 cm. Place the sample in a large container (i.e., interlocking-sealed plastic bag, jar, or bucket) then add formalin and labels. All tissues from the same animal can go into the same container, but make sure that there is sufficient formalin to totally immerse the samples, perhaps 10:1. After 6–8 hours, change the solution with fresh formalin, then again after 24 hours for the next few days. Use standard specimen labels that will not disintegrate in solution, such as plastic or waterproof field notebook paper. Record data with a permanent marking pen (rapidograph) or pencil. Information on the label must include the sample number, species, sex, date sampled, and location. Extra information should include age, time, and location of death. Include condition of the carcass and note whether it was oiled. Prevent contamination of the sample with oil, tar balls, and the like. If an organ or tissue appears damaged or irregular, take samples of both the unhealthy tissue and normal tissue.

Preferred tissues for histological analysis include: skin, brain, pituitary, liver, lung, kidney, thyroid, adrenal, bone marrow, stomach, blubber, spleen, muscle (body and heart), esophagus and tonsil (if present), small and large intestine with attached pancreas, gonads (include epididymis, testes, prostate, uterus, ovaries).

Toxicological Analysis

Samples taken under this protocol must be collected with care since the slightest amount of contamination may result in erroneous results. EXTREME CARE MUST BE TAKEN TO AVOID HYDROCARBON CONTAMINATION. THESE SAMPLES MUST NOT CONTACT ANY PLASTIC OR PETROLEUM-DERIVED PRODUCTS!

Samples collected for toxicology should be placed in glass jars thoroughly washed in distilled water then rinsed with reagent-grade methylene chloride and air dried, or in Teflon-lined glass jars. The lids of the jars should be lined with Teflon sheeting. Labels should be attached outside the jars. Methylene chloride is toxic and should be handled under a hood or outside in a well-ventilated area. Do not breath the fumes! If methylene chloride is not available, rinse with another organic solvent (acetone, ethanol). Clean the knives, scissors, and forceps in the same manner. Rinse with ethanol after each sample and with methylene chloride after each animal. If rubber or surgical gloves are used while sampling, be sure that all tissue samples are handled with solvent-cleaned forceps and that the samples do not contact the gloves. Gloves without talc are preferred. When possible, remove the sample from the center of the organ, avoiding possible contaminating material. Cool the samples immediately and freeze when possible (-20°C or less if possible).

≫⊂ ≫⊂ ≫⊂

Marine mammals have rapid hydrocarbon degrading capabilities, making it rare that oil can be detected in tissues, and even rarer that oil can be linked to a source. High-priority tissues are bile [for measurement of hydrocarbon metabolites; place bile in a small (4 ml) amber glass vial], stomach and lungs, where prey and inspired oil may not have been degraded, and other tissues such as blubber and liver, where tissue loading may have exceeded metabolism in an acute death. Next highest priority tissues include the brain and kidneys. Take other samples as time and supplies permit, similar to those taken for histology. If prey are in the stomach, take samples but clearly label them as prey.

Sample Identification

Tags and identification numbers for all samples must be adequately documented with the sample information described above plus your name and a unique field number for each sample. The unique field number should correspond to a detailed set of field notes for each animal. For example, if a dead seal is found, assign it a field number that would consist of the collector's initials, consecutive seal number, and date. Using this code, the first harbor seal that Tom Loughlin found and sampled would be TL-HS-1-24/3/89. This would be the first harbor seal sampled by Tom Loughlin on 24 March 1989. His second sea otter would be TL-SO-2-24/3/89. The most important consideration is that each sample be clearly identified by species and the collector's unique number, and that careful notes are kept which clearly match the collector's number with respective, detailed collection and necropsy records.

Collect tissues from a consistent, easily described location within the organ or body part of each specimen. When consistency is not possible due to specimen condition, make sure to record excision location in field notes. Although not specifically recommended within the protocols, document prey items found in the stomach to determine if changes in feeding have occurred. Collect the appropriate samples needed for aging, and measure animals using accepted procedures (e.g., Committee on Marine Mammals 1961; Scheffer 1967).

Last, make sure that CHAIN OF CUSTODY forms accompany all collected specimens and samples (Fig. AI-1). Chain of custody forms help trace the path of each sampled tissue from collection to analysis and document where and when individuals handled the sample and what action they performed. This is very important in terms of the evidentiary nature of the samples and possible litigation, and provides a formal tie of the samples to the principal investigator. These forms are independent of the normal labels placed with the samples.

⋙ ⋙ ⋙

N.M.F.S.	Prince William Sound Oil Assessment					Chain of Custody Form		

VAL-89-   Serial #

Page ——— of ———

NOTE: Use ballpoint pen, waterproof ink (eg Rapidograph)
or fine-tip waterproof marker

Alaska Fisheries Science Center
National Marine Mammal Laboratory
7600 Sand Point Way N.E., Bin C15700
Seattle, Washington 98115-0070
(206) 526-4045

Date Collected	Sample # (collector's)	Assigned # (leave blank)	Type (tissue, water, sediment, etc)	Location Collected	Latitude	Longitude	Remarks

Chain of Custody

Samples collected by ——— print name ——— agency ——— signature ——— date

Transferred to ——— print name —— of —— agency —— at —— place ——— signature ——— date

Transferred to ——— print name —— of —— agency —— at —— place ——— signature ——— date

Transferred to ——— print name —— of —— agency —— at —— place ——— signature ——— date

Transferred to ——— print name —— of —— agency —— at —— place ——— signature ——— date

Transferred to ——— print name —— of —— agency —— at —— place ——— signature ——— date

Figure AI-1. A chain of custody form used to accompany samples collected during *Exxon Valdez* oil spill studies.

STORAGE AND DISTRIBUTION OF SAMPLES

Short-Term Storage (Postnecropsy)

Histological samples should be retained in buffered formalin before analysis. Toxicological samples should be kept frozen (at -20°C or less) until analysis. Samples have been held for over 5 years continuously frozen at -20°C or less with no loss of data quality for high-molecular-weight aromatic hydrocarbons. All samples must be marked with tamper-proof tape and maintained in locked storage facilities.

Handling and Shipment to Long-Term Storage

The field sampler is responsible for the care and custody of samples until they are transferred. When samples are transferred from one individual's custody to another's, the individuals relinquishing and receiving will sign and date the chain of custody record. Shipping containers must be custody sealed for shipment such that access to the package is obtained only by breaking the seal. Whenever samples are split, a separate chain of custody record must be prepared for those samples and marked to indicate for whom the samples are being split. If samples are sent by common carrier, copies of the bills of lading or air bills must be retained as part of the permanent record. Frozen samples must remain in that state throughout the shipping process.

Long-Term Storage

Ideally, tissues can be shipped directly from the field to the analytical laboratory. After the EVOS, the volume of material collected was so great that long-term storage facilities were required to maintain tissues for up to 3 years before analysis.

Storage facilities should house specimen material and records in locked freezers (at -20°C or less) and cabinets accessible only by designated personnel. Tissues should be security taped and updated records maintained of when, why, and by whom the tape is broken and reapplied. Tissues should be checked periodically for condition by designated personnel. Freezer units should have a secondary power supply, temperature-change alarms, and posted emergency contact numbers in case of power failure.

Marine mammal tissues and all original collection notes and necropsy records should be stored at one designated institution. It is a good safeguard to designate a second institution for storage of duplicate marine mammal tissue sets and duplicates of accompanying collection and necropsy records.

All individual tissues within both sets must be accompanied by a chain of custody form. It is critical that an updated inventory of specimens be maintained by each storage facility. Therefore, any activity involving transfer of a tissue (e.g.,

✕◇   ✕◇   ✕◇

shipment to the analytical laboratory) requires notification of the second storage facility. Since the detail involved in maintaining updated specimen activity records at one or more storage facilities is extensive, we recommend the following:

1. Establish individual hard-copy files for each specimen, organized alphabetically by specimen number. For example, file AB-GW-1 contains all original records of collection, necropsy, and chain of custody records for the first gray whale collected by Al Baker. The file also contains original correspondence regarding this animal or its tissues, and original records of tissue-transfer arrangements and receipts. Any reports of analytical results are also maintained in the file.

2. Make duplicates of all hard-copy records contained in each specimen file and store in a secondary locked cabinet under general categories such as Cetacea: Gray Whales. This secondary file serves not only as backup in case of damage to the originals, but it can also be made available to parties requesting the originals.

3. Maintain either a computer or hard-copy notebook of tissue exchange, including location of material and stage of analysis. The notebook should also include the updated status of pertinent information on the specimen, such as missing forms or missing necropsy information. It is the responsibility of the long-term storage institution to find missing information on individual specimens.

4. Maintain computerized summary tables of information such as the number of cetaceans and pinnipeds found dead, collected, and necropsied during and after the spill, as well as preliminary results of analysis. It is best to organize these tables so that they can be categorized by species. This information was frequently requested of storage institutions in various forms by the media, attorneys, and public interest groups after the EVOS. Computerizing the summary information that can be distributed (considering litigation) saves considerable time and effort.

Besides maintenance of specimen records within independent long-term storage facilities, a protocol was established for maintenance of an inventory of all categories of samples collected for analytical chemistry. This inventory served as the Trustee Management Team's basis for developing a plan and budget for sample analysis. The information required for purposes of the inventory was similar to that recorded on chain of custody forms. Each long-term storage institution submitted the required inventory to the Natural Resources Damage Assessment team which then incorporated it into the main inventory. Tissues were then ranked for analysis based on the quality of records, sample condition, and time of collection relative to date of the spill.

REFERENCES

Committee on Marine Mammals. 1961. Standardized methods for measuring and recording data on smaller cetaceans. Journal of Mammalogy 42:471–476.

Scheffer, V. B. 1967. Standard measurements of seals. Journal of Mammalogy 48:459–462.

✂︎ ✂︎ ✂︎

Appendix II

Oil Tanker Accidents

Over 25 million barrels (3.3 million metric tons) of crude oil have been spilled since 1960 as a result of accidents involving tankers. Some of the largest spills, such as the 260,000 tons lost by the *ABT Summer* in the Atlantic Ocean have had little attention when compared to those taking place closer to shore (i.e., *Exxon Valdez* and *Braer*). A table identifying 51 tanker accidents and the quantity of oil lost (in metric tons) follows. Some accidents may have occurred that are not listed. The table is not intended to be inclusive of all accidents but rather to provide a perspective of the relative magnitude of the *Exxon Valdez* spill compared to others in the world during the last 30 years. The summary was prepared by William Folsom, National Marine Fisheries Service, Office of International Affairs, Silver Spring, Maryland. For conversion, there are about 7.4 barrels/metric ton of oil and about 42 gallons/barrel. The *Exxon Valdez* spill was officially listed by the Alaska Department of Conservation as 268,194 barrels (11,264,148 gallons; 36,242 metric tons). Exxon placed the amount spilled at 240,000 barrels.

Table II-1. Accidents involving oil tankers, including the name of the vessel, the year and location of the accident, and the quantity of oil lost (metric tons) 1960-93.

Name	Year	Location	Oil Lost (Metric Tons)
Sinclair Petrolore	1960	Brazil	57,000
Torrey Canyon	1967	United Kingdom	124,000
Mandoil II	1968	U.S.A.	41,000
World Glory	1968	South Africa	48,000
Odyssey	1968	Canada	132,000
Ennerdale	1970	Seychelles	41,500
Pacific Glory	1970	France	3500
Chryssi	1970	Bermuda	32,000
Wafra	1971	South Africa	62,000
Texaco Denmark	1971	Belgium	107,000

(Table continues)

Table II-1. Continued

Name	Year	Location	Oil Lost (Metric Tons)
Golden Drake	1972	Bermuda	32,000
Trader	1971	Greece	36,000
Sea Star	1972	Gulf of Oman	123,000
Napier	1973	Chile	36,000
Metula	1974	Argentina	45,000
Vuyo Maru #10	1974	Japan	42,000
British Ambassador	1975	Japan	46,000
Jakob Maersk	1975	Portugal	41,000
Corintos	1975	U.S.A.	36,000
Epic Colocotronis	1975	Caribbean Sea	58,000
Saint-Peter	1976	Ecuador	34,000
Urquiola	1976	Spain	91,000
Bohlen	1976	France	9800
Argo Merchant	1976	U.S.A.	28,000
Hawaiian Patriot	1977	U.S.A.	101,000
Amoco Cadiz	1978	France	223,000
Andros Patria	1978	Spain	50,000
Betelguese	1979	Ireland	27,000
Gino	1979	France	32,000
Atlantic Empress	1979	Brazil	257,000
Ioannis Angelicoussis	1979	Angola	32,000
Burmah Agate	1979	Gulf of Mexico	41,000
Independenta	1979	Turkey	95,000
Irenes Serenade	1980	Greece	100,000
Tanio	1980	France	6000
Juan Antonio Lavalleja	1980	Algeria	38,000
Assimi	1983	Oman	50,000
Castillo de Beliver	1983	South Africa	239,000
Pericles GC	1983	Persian Gulf	46,000
Nova	1985	Persian Gulf	68,000
Exxon Valdez	**1989**	**U.S.A.**	**36,000**
Khark V	1989	Morocco	76,000
Aragon	1989	Madeira, Portugal	24,000
Mega Borg	1990	Italy	13,000
Haven	1991	Italy	144,000
ABT Summer	1991	Atlantic Ocean	260,000
Kirki	1991	Australia	17,000
Katina P.	1992	Mozambique	66,000
Aegean Sea	1992	Spain	80,000
Braer	1993	United Kingdom	84,000
Maersk Navigator	1993	Indonesia	1000[a]

[a] No estimates on the total spill were available when this table was prepared in April 1993. The ship's oil-carrying capacity was 255,000 tons of crude oil.

Source: W. Folsom, NMFS, summarized from *La Marin* 1993.

Subject Index

Boldface page entries indicate figures; *italic* page entries indicate tables.

A

acanthocephalan, *Corynosoma* spp. 289, 326
acidosis 267
adrenal 290, 307
adrenal cortices 267
Aialik Bay 111
Air National Guard, Alaska 26, 28
alanine aminotransferase (ALT) 232
Alaska coastal current 4, **5**
Alaska Peninsula, currents 4
    gray whale carcasses on 359, **363**
    oil on 9, 16, 23, 47, 82, 193, 313
    sea otters on 30, 48, 49
    Steller sea lions on 119
Aleutian Islands 2, 4, 118, 173
Aleutian storm track 3
algae, *Fucus gardneri* 220
alphanumeric code, *see* humpback
    whales, killer whales
Alyeska Pipeline Service Company 6, 8, 18
Alyeska Oil Spill Contingency Plan 37
analysis of variance (ANOVA)
    see Data analysis
Analytical Chemistry Group 377
Anchorage 2, 37, 40, 41, 43
anemia, in harbor seal 233, 286
    in sea otter 267, 273
anorexia 267, 268, 273

Applegate Rocks 29, 32, 109, 220, 223, 337
arcsine transformation
    see Data analysis
Arlington, Oregon 10
Armed Forces Institute of Pathology
    (AFIP) 265
aromatic compounds (ACs)
    see Hydrocarbons
aspartate aminotransferase (AST) 232
azotemia 267, 273

B

Baikal seal, *Phoca sibirica* 238
Bainbridge Island 84
baleen 257
    see also Mysticete
barnacles, *Balanus* spp. 201
Barren Islands 4, 16, 124, 220
bathymetry
    contours 64
    data layers 78
Bay of Isles 222
    harbor seal samples from 228, 334, 337
    observations of harbor seals at 210, 217, 220
    oiled harbor seals on 29
behavior,
    see Harbor seal
    see Sea otter
    see Cetaceans
benthic feeding 64, 193
benzene 17, 272, 331, 339

in sea otter prey 203-205, **204**
fluorescent aromatic contaminants
    (FACs) 124, 130, 137, 333
high-molecular-weight aromatic
    compounds (HACs) 125
    in gray whales 360, 364-367
    in harbor seals 331-358
    in sea otters 314, 316-328
    in sea otter prey 195, 203-206, 314
low-molecular-weight aromatic
    compounds (LACs) 125
polycyclic aromatic hydrocarbons
    (PAHs) 132, 337-339, 342-348,
    364-366
unresolved complex mixture (UCM)
    125, 133, 197, 316, 333, 360
hyperkalemia, in sea otter 267
hypoglycemia, in sea otter 267
hypothermia, in
    harbor seal 232
    sea otter 266, 277, 313, 325

I

Icy Bay 180
Iktua Bay 84
Inipol 35
innkeepers, *Echiurus alaskensis* 201
Interagency Shoreline Cleanup
    Committee 9, 32
International Bird Rescue Research
    Center 7
intersection model,
    see Data analysis
interstitial pulmonary emphysema, *see*
    Emphysema
intertidal zone 16, 205
intramyelinic edema, in harbor seal 286,
    294, 303, 304, 309, 343

J

Jakolof Bay 38
jejunum 314
jingles, *Pododesmus* spp. 201

Johnstone Point 149
Junction Island 220

K

Kachemak Bay 4, 124, 130
Kayak Island 359, 364, 368
Kenai Fiords National Park 2
Kenai Peninsula 2, 9, 16
    cleaning of 9
    marine mammals on 30, 48-50, 81-94,
        119, 124, 220, 298
    oil on 18, 37, 313
kidney, as sample 51, 378
    analysis of in harbor seals 232, 289,
        292, 303, 333, 373
    in sea otters 51, 267, 314, 321-327
    in Steller sea lions 132
killer whale, *Orcinus orca*
    alphanumeric code 142, 145
    directional effects of masking 250-253
        sound propagation 250
        vessel noise and pure tones 250
        vessel noise and whale sounds 248-250
    dorsal fin as ID 144
        collapse off 156
    echolocation 243
    hearing 243, **249**, 250
    in oil, *see* Cetacean
    interactions with fisheries 167-168
    masking of noise 243-255
    maternal assemblage 141-142, 158
    mortality estimates *151, 153-154,* 166
    movements 147, 165
    pods 141
    recovery of AB pod *169*
    reproductive rate 146
    saddlepatch as ID 144
    sound pressure 244
    status in Prince William Sound 141-160,
        143
    strandings of 155, 168
    survey biases 164-165
    vital rates
        resident pods 141, 144-147, 150-
            155, *153*

><> ><> ><>

∞ ∞ ∞

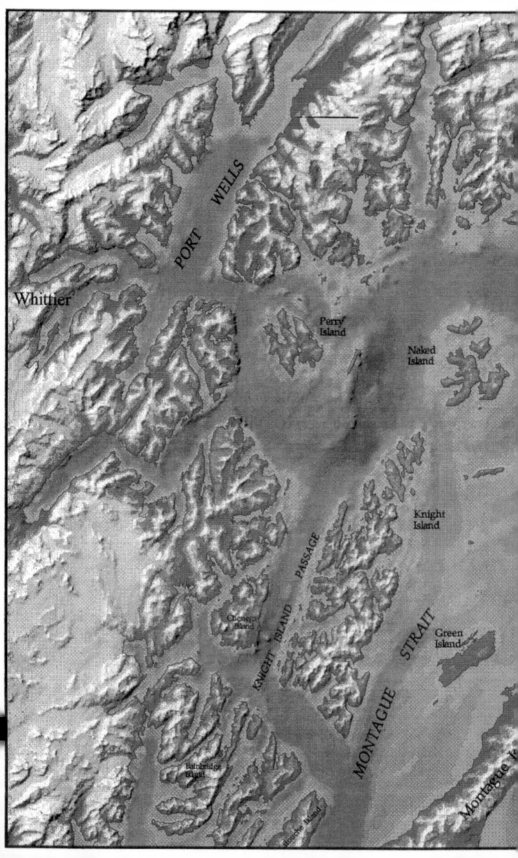

Whittier

PORT WELLS

Perry Island

Naked Island

Knight Island

KNIGHT ISLAND PASSAGE

Chenega Island

MONTAGUE STRAIT

Green Island

Bainbridge Island

Montague I